Stochastic Models for Social Processes

3rd Edition

Probability and Mathematical Statistics (Continued)

WILLIAMS • Diffusions, Markov Processes, and Martingales, Volume I: Foundations

ZACKS • Theory of Statistical Inference

Applied Probability and Statistics

ANDERSON, AUQUIER, HAUCK, OAKES, VANDAELE, and WEISBERG • Statistical Methods for Comparative Studies

ARTHANARI and DODGE • Mathematical Programming in Statistics

BAILEY • The Elements of Stochastic Processes with Applications to the Natural Sciences

BAILEY • Mathematics, Statistics and Systems for Health

BARNETT • Interpreting Multivariate Data

BARNETT and LEWIS • Outliers in Statistical Data

BARTHOLOMEW • Stochastic Models for Social Processes, *Third Edition*

BARTHOLOMEW and FORBES • Statistical Techniques for Manpower Planning

BECK and ARNOLD • Parameter Estimation in Engineering and Science

BELSLEY, KUH, and WELSCH • Regression Diagnostics: Identifying Influential Data and Sources of Collinearity

BENNETT and FRANKLIN • Statistical Analysis in Chemistry and the Chemical Industry

BHAT • Elements of Applied Stochastic Processes

BLOOMFIELD • Fourier Analysis of Time Series: An Introduction

BOX • R. A. Fisher, The Life of a Scientist

BOX and DRAPER • Evolutionary Operation: A Statistical Method for Process Improvement

BOX, HUNTER, and HUNTER • Statistics for Experimenters: An Introduction to Design, Data Analysis, and Model Building

BROWN and HOLLANDER • Statistics: A Biomedical Introduction

BROWNLEE • Statistical Theory and Methodology in Science and Engineering, *Second Edition*

BURY • Statistical Models in Applied Science

CHAMBERS • Computational Methods for Data Analysis

CHATTERJEE and PRICE • Regression Analysis by Example

CHERNOFF and MOSES • Elementary Decision Theory

CHOW • Analysis and Control of Dynamic Economic Systems

CHOW • Econometric Analysis by Control Methods

CLELLAND, BROWN, and deCANI • Basic Statistics with Business Applications, *Second Edition*

COCHRAN • Sampling Techniques, *Third Edition*

COCHRAN and COX • Experimental Designs, *Second Edition*

CONOVER • Practical Nonparametric Statistics, *Second Edition*

CORNELL • Experiments with Mixtures: Designs, Models and The Analysis of Mixture Data

COX • Planning of Experiments

DANIEL • Biostatistics: A Foundation for Analysis in the Health Sciences, *Second Edition*

DANIEL • Applications of Statistics to Industrial Experimentation

DANIEL and WOOD • Fitting Equations to Data: Computer Analysis of Multifactor Data, *Second Edition*

DAVID • Order Statistics, *Second Edition*

DEMING • Sample Design in Business Research

DODGE and ROMIG • Sampling Inspection Tables, *Second Edition*

DRAPER and SMITH • Applied Regression Analysis, *Second Edition*

DUNN • Basic Statistics: A Primer for the Biomedical Sciences, *Second Edition*

DUNN and CLARK • Applied Statistics: Analysis of Variance and Regression

ELANDT-JOHNSON • Probability Models and Statistical Methods in Genetics

ELANDT-JOHNSON and JOHNSON • Survival Models and Data Analysis

continued on back

Stochastic Models for Social Processes

3rd Edition

D. J. BARTHOLOMEW

Professor of Statistics
London School of Economics and
Political Science

1807 1982

JOHN WILEY & SONS

Chichester · New York · Brisbane · Toronto

First edition 1967
Reprinted August 1970

Second edition 1973
Reprinted February 1978

British Library Cataloguing in Publication Data:

Bartholomew, D. J.
    Stochastic models for social processes.—3rd ed.—
    (Wiley series in probability and mathematical
    statistics)
    1. Social sciences—Mathematical models
    2. Stochastic models
    I. Title
    300.07'2   H61

ISBN 0 471 28040 2

Photo Typeset by
Macmillan India Ltd., Bangalore
Printed at Page Bros, Norwich

Acknowledgements

The passage of time makes it increasingly difficult to acknowledge adequately all the help one has received. Even a simple list is bound to betray lapses of memory. Nevertheless, I am happy to thank David Balmer, Gerald Davies, Burton Singer, Tony Shorrocks, Gustav Feichtinger, and Henry Daniels who have given help and encouragement in the preparation of this edition.

Permission to quote from their publications was given by D. Van Nostrand Company Inc. in respect of Kemeny and Snell's *Finite Markov Chains*, and by the New York State School of Industrial and Labour Relations, Cornell University, in respect of Blumen, Kogan and McCarthy's *The Industrial Mobility of Labour as a Probability Process*.

The preparation of the typescript, bibliography, and indexes has been widely shared. At the L.S.E. I have had the assistance of Susan Pratt, Siân Turner, Margaret Walter, and Ruth Pickering. At home my wife and daughters have joined in as the need arose.

It is a particular pleasure to acknowledge, somewhat belatedly, the help and efficiency of the publisher and especially of Jamie Cameron who has been associated with this book in its successive editions for almost as long as I have.

Contents

Preface

Although its purpose remains unchanged, the form and content of this edition are substantially different from its predecessors. The second edition was essentially an expanded version of the first with a good deal of new material but covering the same ground. To have followed the same course with the third edition would not only have made the book too long but would have given an unbalanced view of contemporary work in the field. The aim behind the revision has been threefold:

(a) To give a broader and more balanced account of the subject as it now exists.
(b) To make the book more suitable for use as a graduate text.
(c) To give a full bibliography.

Users of earlier editions will find much that is familiar. Earlier numerical examples and derivations have been left intact wherever possible. To help students, most of the historical material, guides to mathematical sources, and comments on the literature have been collected together at the end of chapters, in Complements sections. It was my original intention to include exercises. These, inevitably, would have been largely manipulative and would have thus tended to emphasize the deductive aspects of modelling. Many years' experience of teaching this material to graduate students has convinced me that, especially if they come from a mathematical background, the main difficulty is on the model-building side. The delicate art of constructing and using models to gain insight into the processes of the real world can, perhaps, only be fully learnt by long experience. Nevertheless, it is clear that a formal mathematical education can often be a disadvantage to those entering applied work and there seemed to be real danger in treating the subject in a way which reinforced the handicaps under which students labour. The student must be encouraged to entertain a variety of hypotheses, to look at the problem from many angles, to balance realism against tractability, and so on. The perceptive teacher will find many ways of encouraging this attitude by reference to the text and the Complements. There are also many gaps and loose ends in the argument which would provide practice in the routines of model-solving.

The relationship with the second edition is as follows:

Chapter 1 has been rewritten.

Chapter 2 is an expanded version of the old Chapter 2 with much new material on measuring social mobility. Most of the former Section 2.3 on occupational mobility is now found in the new Chapter 4.

Chapter 3 is substantially the same as the old Chapter 3, but with a new approach to the study of limiting behaviour.

Chapter 4 is a new chapter on closed continuous time Markov models incorporating some material from the old Chapters 2 and 5.

Chapters 5 is new, but includes much material on open continuous time Markov models from the old Chapter 5.

Chapter 6 is a rewritten chapter on control theory for Markov models. Most of the old material on attainability has been dropped and replaced by new work on stochastic control.

Chapter 7 contains a condensed version of the old Chapter 6 on durations together with new material on size distributions.

Chapter 8 is based on the old Chapters 7 and 8. Much of the exact theory of the second edition has been omitted and greater emphasis placed on simple models for which a fairly complete analysis is possible. New material on vacancy chains has been added.

Chapter 9 is the same as in the second edition apart from a new introduction and some minor changes.

Chapter 10 contains a shortened account of all the models previously treated, together with new work on recurrent and competing epidemics.

Since there are now several specialist treatments available, less attention has been paid to applications in manpower planning. This has allowed a broader spectrum of applications to be covered.

London

D. J. BARTHOLOMEW

CHAPTER 1

Stochastic Modelling

1.1 SOCIAL PHYSICS

In his *Letters on the Theory of Probabilities as Applied to the Moral and Political Sciences*, written in the early nineteenth century, Adolphe Quetelet distinguished statistics from what he called social physics. In his view, statistics was concerned with presenting a more or less faithful picture of the social body at a particular instant. In contrast, the study of those laws of society which existed independently of time and the 'caprices of men' were to be regarded as a separate science of social physics. In his *Sociale Physique*, Quetelet set out some of those laws and their empirical basis. Physics and statistics have made enormous advances in the 150 years which have passed since then and few statisticians today would wish to take so limited a view of their subject. Nevertheless, Quetelet was right in recognizing the analogy between physical and social systems and in seeking to bring probability theory to bear on the latter. His apparent belief that society exhibits laws as fixed as those which govern the heavenly bodies may now sound naïve and over-optimistic—Kendall (1961) gives a more realistic assessment. But the basic insight remains valid. Unpredictability at the individual level combined with large numbers produces the regularities displayed by both the gas laws and the life table. The recognition that there are regularities and patterns in the aggregate behaviour of human societies was the first step along the road to the stochastic modelling of social processes.

The analogy with physics is, of course, incomplete and potentially misleading. Even large human populations are small compared with the number of air molecules in, say, a bicycle pump. This means that, in social systems, the randomness will not be entirely smoothed out and the pattern may be only dimly discernible. Furthermore, people are not entirely unpredictable in their behaviour; they respond to changes in their environment and by interaction with one another. There is also a marked lack of homogeneity in most human populations which may result in different patterns of behaviour in different sub-populations. Few of even the best established regularities in society can be regarded as existing independently of time, though they may well persist for long enough to be worthy of note. Progress is limited not only by the nature of the subject-matter but by the means available to study the phenomena. Unlike the physicist the social scientist can rarely experiment in order to identify the role of each variable. Even more than in physics, observation of the system may affect what is observed. The social scientist is usually confined to passive observation of

1

the flux of social systems and what can be learnt is correspondingly limited. If a model can be constructed which reproduces the main qualitative features of the real system, the social scientist may be well satisfied. If predictions made with the model are borne out by experience there are grounds for even greater satisfaction. The underlying aim is to render the complex patterns exhibited by real social systems intelligible by showing how far they are explicable in terms of a few simple postulates about individual behaviour.

Time and chance are the two features of social processes which point to the theory of stochastic processes as an appropriate tool for their study. Such systems evolve in time under the changing and partly unpredictable behaviour of their members. This is formalized in the theory by defining a set of states to which the system may belong and probabilistic laws which govern movement between those states. The analysis of the formal process then allows us to attach probabilities to possible future states and to discover what kinds of behaviour are possible. Processes of this kind occur widely in nature and have provided a powerful stimulus for the development of the theory. It is only relatively recently that the social sciences have become a major field of application.

Quetelet's initiative had to await the development of the modern theory of stochastic processes before its potential could be realized. In psychology, an early example in stochastic learning theory was provided by Bush and Mosteller (1955). Coleman's (1964b) influential work on mathematical sociology showed how relatively simple stochastic models could be used to illuminate substantive problems. In economics Steindhl (1965) used the theory to show how some of the simple but persistent patterns exhibited by the growth of firms could be explained as consequences of stochastic mechanisms of growth. When the first edition of this book appeared in 1967 it was already clear that there was great potential for harnessing the rapidly growing theory of stochastic processes to the study of social phenomena. At that time, however, significant developments had taken place in only a few areas and the treatment was uneven in coverage and depth. The greater size and wider scope of this edition is some measure of the growth and maturity which the subject has acquired in little more than a decade.

1.2 MODELS AND MODELLING

In common with most branches of applied mathematics, we lay the foundations for the theoretical study by constructing mathematical models. The non-mathematical reader will be aware of the use made of actual physical models by engineers. For example, facts about the behaviour of an aircraft are deduced from the behaviour of the model under simulated flight conditions. The accuracy of these deductions will depend on the success with which the model embodies the features of the actual aircraft. The aircraft and the model differ in many respects. They may be of different size and made of different materials; many of the detailed fittings not directly concerned with flight behaviour will be omitted from the model. The important thing about the model is not that it should be exactly like the real thing but that it should behave in a wind tunnel like the aircraft in flight. The basic requirement is thus that the aircraft and the model should be *isomorphous* in all *relevant* respects.

A mathematical model is used in an entirely analogous way. In the social systems which we shall discuss the constituent parts are interrelated. When one characteristic is changed there will be consequential changes in other parts of the system. Provided that the changes in question can be quantified, these interrelationships can be described, in principle at least, by mathematical equations. A set of equations which purports to describe the behaviour of the system is a mathematical model. Put in another way, such a model is a set of assumptions about the relationships between the parts of the system. Its adequacy is judged by the success with which it can predict the effects of changes in the social system which it describes and by whether or not it can account for changes which have occurred in the past. The model is an abstraction of the real world in which the relevant relations between the real elements are replaced by similar relations between mathematical entities.

Mathematical models may be deterministic or stochastic. If the effect of any change in the system can be predicted with certainty the system is said to be deterministic. In practice, especially in the social sciences, this is not the case. Either because the system is not fully specified or because of the unpredictable character of much human behaviour there is usually an element of uncertainty in any prediction. This uncertainty can be accommodated if we introduce probability distributions into the model in place of mathematical variables. More precisely, this means that the equations of the model will have to include random variables. Such a model is described as stochastic.

The necessity of using stochastic models in this book can be seen by considering some of the phenomena which we shall discuss. In the case of social mobility, for instance, no one can predict with certainty whether or not a son will follow in his father's footsteps. Similarly there is, in general, no certain means of telling when a man will decide to change his job or whether a student will pass his examination. In the same way the diffusion of news in a social group depends, to a large extent, on chance encounters between members of the group. It is the inherent uncertainty brought about by the freedom of choice available to the individual which compels us to formulate our models in stochastic terms. As we shall see later, there are often advantages in using deterministic methods to obtain approximations to stochastic models, but this does not affect the basic nature of the process. There has often been debate among sociologists about the rival merits of deterministic and stochastic methods. It should be clear from the above discussion that any model describing human behaviour should be formulated in stochastic terms. When it comes to the solution of the models it may be advisable to use a deterministic approximation. The greater simplicity of the deterministic version of a model may also make it easier to grasp the nature of the phenomena in question. Nevertheless, these are tactical questions which do not affect the basic principle which we have laid down.

The range of stochastic models covered in this book is somewhat narrower than the definition given above suggests. We shall be concerned here with what may be called 'explanatory' models as opposed to 'black-box' models. This division is somewhat arbitrary and it would be difficult to define either kind precisely. It is nevertheless worth making the distinction for practical purposes. A black-box model is concerned with the relationship between the output of a system and the input. It is stochastic if the output cannot be exactly predicted from the input. A model in this case usually takes the form of a regression equation (or equations) linking output variables to input variables. The form of the regression equation is purely empirical, being chosen from among a simple class of functions (linear, say) because it fits the output most closely. It makes no pretence of explaining how the input is turned into the output. An explanatory model, on the other hand, aims to describe the mechanism by which the input to the system becomes the output. A black-box model may be sufficient if our object is to predict the output or to control it by manipulating the input. If we wish to understand how the system works we shall need a model which attempts to explain what goes on inside the 'black box'.

To illustrate the point, consider recruitment and wastage in a firm. Over a long period we would probably observe a relationship between the series of recruitment and wastage figures. An estimate of this relationship would be a black-box model, and it could be used to predict future recruitment from past records of wastage and recruitment. An explanatory model would attempt to describe the movement of individuals within the firm and, in particular, the dependence of loss and recruitment on such things as grade and length of service. It will be clear from the next section that our purposes in this book are better served by explanatory models. (We are aware that what is an explanation at one level may be a mystery when viewed at a deeper level. The only claim made here is that our models aim to take us below the surface.) A comprehensive account of what we have called black-box modelling is contained in Box and Jenkins (1971).

Stochastic models of social phenomena have been constructed in the past with different objects in view. For our purposes it will be useful to distinguish four main functions of models. The first is to give *insight* into and *understanding* of the phenomenon in question. This is the activity characteristic of the pure scientist. The investigation begins with the collection of data on the process and the formulation of a model which embodies the observed features of the system. This we shall describe as the *model-building* stage. The next step is to use the model to make predictions about the system which can be tested by observation. This activity will require the use of mathematical reasoning to make deductions from the model and will be referred to as *model-solving*. The final step is to compare the deductions with the real world and to modify the original model if it proves to be inadequate; this is the procedure of *model-testing*. An investigation of this kind is not complete until the operation of the system has been accurately and comprehensively described in mathematical terms and when the solution of the equations which arise has been obtained.

The second and third objectives in model-building fall within the province of the applied social scientist. Of these the use of models for *prediction* is widely recognized. The social planner wants to know what is likely to happen if specific policies are implemented. The manager wishes to know, in advance, the consequences of various recruitment or promotion policies for the staffing of his firm. A model which adequately describes the behaviour of the system will be capable of providing answers to such questions. It is important to emphasize that no prediction is unconditional; we are always forecasting what will happen *if* certain conditions are met.

Closely related to the problem of prediction is the question of the *design* of social systems and their mode of operation. Some of the models for hierarchical organizations described in Chapters 3, 5, and 8 were first constructed to establish principles for the design of recruitment and promotion policies. The management structure of a firm has to be such that it provides the correct number of people with the requisite skills and experience at each level to carry out the functions of the organization efficiently. Whether or not this object is achieved depends on the manpower policies of the firm. Hence it is necessary to evaluate competing policies in terms of their success in attaining the stated objectives.

The art of model-building is to know where and when to simplify. The object, as Coleman (1964b) has expressed it, is to condense as much as possible of reality into a simple model. Not only does this give us understanding of the system but it provides a convenient base-line with which to compare the behaviour of actual systems. These remarks made above should not be interpreted as justifying unlimited simplification. They apply, of course, to the model-building stage of the process, but this cannot be considered in isolation from the model-solving aspects. In the past the temptation to trim the model to a form having a tractable mathematical solution has been strong. The situation has been radically altered by the wide availability of high-speed computers, but the habits of thought of the old era linger on. It is no longer necessary for a solution to be obtainable in closed form in terms of simple functions in order to yield useful information about the process.

A further contribution which stochastic modelling can make to social research is in the field of measurement. This statement may appear paradoxical in view of our earlier remark that lack of suitable measures has itself hindered the development of stochastic models. Both statements are true, but at different levels of sophistication, as the following examples will show. In several applied fields research workers have become aware that the obvious but crude measures they were using were inadequate. For example, the success of a treatment for cancer is often assessed by calculating the proportion of patients who survive for a given period, say five years. However, this measure is influenced by irrelevant factors such as 'natural' mortality which should, ideally, be eliminated from the measure. Fix and Neyman (1951) showed how this could be done by constructing a stochastic model of the post-treatment period and estimating one of its parameters (see Chapter 5). Another example arose in the study of labour turnover. A widely used method of measuring turnover is to express the number of people leaving a firm per unit time as a percentage of the average labour force during the same period. Large values of the measure have often been taken as indicative of low morale. The measure is inadequate because it ignores the strong dependence of propensity to leave on length of service. This dependence is such that a firm with a large number of new recruits will have a misleadingly high figure of turnover. By constructing the renewal theory model of Chapter 8, it was possible to demonstrate the limitations of the usual index and to determine the conditions under which it could be meaningfully used. The results of this study emphasized the advantages of measures of turnover based on such things as the mean or median length of service.

One of the most important practical benefits of having available a range of stochastic models is that they greatly facilitate the statistical analysis of social data. This is particularly useful when, as often happens, the data are fragmentary. A model enables us to write down the likelihood function and so estimate the parameters in which we are interested. Statistical methods appropriate for stochastic models in the social sciences are relatively undeveloped. Methods for Markov chains have been available for some time and a first attempt on a limited front to cover some of the ground is given in Bartholomew (1977a). A detailed and elementary account of methods used in manpower planning is provided by Bartholomew and Forbes (1979). Treatments of other aspects are scattered throughout the literature, but a full-length discussion in parallel with the present book would be a timely contribution to the advance of the subject.

1.3 SOME BASIC ISSUES

We have already warned that the analogy with the natural sciences must not be pressed too far and we have also indicated some of the differences. Here we shall discuss, in more detail, some of the basic issues which arise in applying stochastic theory to social phenomena. Some of these are of a purely practical kind, but others are more serious and raise questions about the possibility or propriety of what is being attempted.

At the practical level there has been a shortage of social scientists with the degree of mathematical knowledge necessary to enter the field and contribute effectively to research. Kruskal (1970) and Rosenbaum (1971) are symptomatic of the concern and indicate steps that were being taken to overcome the problem a decade ago. However, until mathematics is seen to be as important for the social scientist as the engineer or chemist, the rate of progress will be limited. Experience shows that it is much harder for a graduate social scientist to acquire a mathematical and statistical training than vice versa. In the following section we have given some guidance to help the non-mathematical reader gain some benefit from the book, but it is no more than first aid.

On a more fundamental level it is sometimes objected that with so many of the basic measurement problems unsolved it is premature, if not foolhardy, to attempt anything so sophisticated as stochastic modelling. In physics there are relatively few basic scales of measurement—time, mass, electric charge, and suchlike—and instruments exist which enable them to be measured with great precision. In the social sciences, by contrast, there are few natural scales of measurement and little agreement on the meaningfulness of many scales that are used. The problem in social science, of course, lies not in the lack of variables to measure but in their multiplicity and variety. How to select, combine, or summarize them is where the real problem lies. Whatever theoretical weight this argument may have, it has little relevance here. A quick perusal of the following pages will show that the level of measurement required for many of our models is very rudimentary—often involving no more than simple classification by attributes. Further, as our later discussions of measuring social mobility and labour turnover will show, a model is a necessary prerequisite for defining a suitable scale of measurement and not vice versa.

A second objection arises from the complexity of social phenomena. This is often expressed by saying that social situations are far too complicated to allow mathematical study and that to ignore this fact is to be led into dangerous oversimplification. The premiss of this objection must be accepted. Social phenomena are often exceedingly complex and our models are bound to be simplifications. Even if it is allowed that our model need only reproduce the relevant features of the real process the problem remains formidable. However, we would argue that there is no alternative to simplification. The basic limiting factor is not the mathematical apparatus available but the ability of the human mind to grasp a complex situation. There is no point in building models whose ramifications are beyond our comprehension. Perhaps the only safeguard against oversimplification is to use a battery of models instead of a single one. Any particular model will be a special case of the more complex model which would be needed to achieve complete realism. Greater confidence can be placed in conclusions which are common to several special cases than in those applicable to one arbitrarily selected model. In many places the strategy dictated by these considerations will be evident. Instead of making 'realistic' assumptions and accepting the complexity which goes with them we have made two, or more, 'unrealistic' but simple assumptions. Where possible these have been selected as extremes between which the true situation must lie. If the conclusions about a particular question are similar we may apply them to the real-life situation without abandoning the framework of thought provided by the simple model.

The presence of computers leads to a third and more subtle objection to the model-solving aspects of the subject. A parallel situation is found in the theory of queues, where a great volume of effort has been devoted to the solution of special models. It is argued that, from a practical point of view, the results can be obtained more easily and rapidly by simulation on a computer. The same argument applies in the social sciences. If our objective is simply to obtain a quick answer in a specific situation, simulation will usually be the best method of attack. In applied social science this will be the norm and the practitioner in the field will find that his interest is in the model-building aspects of the book. On the other hand, if our primary interest in stochastic models is to gain insight into the workings of social phenomena, simulation is less satisfactory. In essence, we are then searching for general solutions whereas simulation provides us with solutions in special cases only. The economy and clarity of a conclusion conveyed by a simple formula is such that the effort of problem-solving is well worth while even if we have to be content with an approximation.

It is necessary to distinguish clearly between two senses in which the term 'simulation' is currently used. In the remarks made above we were speaking of simulating individual behaviour. In a stochastic model this involves determining each individual's changes of state according to the realization of random events as prescribed by the probabilistic assumptions of the model. By this means we generate artificial data on an individual basis of the kind that would be observed in practice. The term 'simulation' is also used by some writers to refer to the calculation of the expected values of random variables associated with the model. Such calculations tell us, in an average sense, what would happen if the model were allowed to operate and in this sense may be said to simulate the process. They provide no information about the stochastic behaviour of the system.

Extensive use has been made of simulation techniques, especially of the latter kind, in the study of social phenomena by Forrester and his colleagues at the Massachusetts Institute of Technology. Their work on industrial, urban and world dynamics is well known, and one of its main aims is to gain insight into the behaviour of very complex systems. Our approach is complementary to theirs, though less ambitious, in that it seeks to achieve the same end through mathematical analysis.

A major objection has been raised against much stochastic modelling, particularly by sociologists and economists, who feel that many models in use are devoid of real social or economic content. In economics, for example, they would find a model which purports to explain industrial size distribution in terms of a 'law of proportionate effect' unconvincing because it has no place for those economic variables which economists believe influence size. Too much of what is relevant, they would argue, is swept up into the stochastic component of the model. This amounts to abandoning the search for true explanation of how economic forces interact in the real world. There is much force in this argument but, to some extent, it betrays a misunderstanding of what stochastic modelling aims to achieve. It is rather like a physicist complaining that the treatment of coin-tossing experiments in elementary books on probability takes no account of the turbulence or viscosity of the air, the degree of spin imparted to the coin, or the nature of the surface on to which it falls. All of these variables are relevant in the sense that they influence the outcome. However, the manner in which they interact to produce that outcome is such that none of them has predictive value. One can therefore describe the distribution of outcomes in terms of a single parameter—the probability of a head—which, in a sense, summarizes the combined effect of all the causal variables. Where prior knowledge about a variable does have predictive value it should certainly appear in a stochastic model so that its effect can be quantified. Otherwise it has to be included in the 'error' part of the model in recognition that its effects cannot be disentangled from those of many other variables, known and unknown. The growth of firms doubtless depends on interest rates, foreign competition, the level of unemployment, the exchange rate, and a multitude of other factors, but their individual effects often cannot be identified. However, there is nothing to be gained, from the point of view of describing aggregate behaviour, by making them explicit in the model. This does not preclude, of course, the study of the detailed effect of such variables at the individual level. The two approaches are complementary.

The same point can be made somewhat differently. It is widely recognized that chaotic behaviour at one level produces order when aggregates are considered. Indeed this is the whole basis for applying probability theory. What is equally true is that the operation of deterministic processes can lead to apparently random effects. The generation of pseudo-random numbers in computers is a good example. The sequence of digits is completely determined by a mathematical formula, but in order to deduce what the formula was from the output would require an extremely long series. For practical purposes relatively short series are adequate and these contain only a negligible amount of information about the law of their generation. In the same way, to treat aspects of human or organizational behaviour as random is not the same as saying that there are no causal influences at work. We are merely acknowledging that our data are too limited to offer any hope of quantifying their individual effects.

1.4 PREREQUISITES, TERMINOLOGY, AND NOTATION

Prerequisites

Many social scientists will find that the mathematical level of the book causes them difficulty. With this in mind the discussion has been kept as elementary as possible and the arguments are developed informally. It should therefore be possible to glean some idea of the nature of the models and the uses to which they can be put by reading between the formulae. However, to obtain a real grasp of the subject there is no alternative to wrestling with the mathematics. As far as possible the treatment is self-contained but the various Complements sections will lead to further material on the mathematics and probability theory. A useful compendium of mathematical and statistical methods and results is provided by the *Handbook of Applicable Mathematics* (6 vols.) (Ed. W. Ledermann, Wiley) to which some specific references are given later. One of the guide-books to the main volumes, (Bartholomew, 1981) is expressly designed for social scientists and it covers, in outline, much of the material contained in this book. The non-mathematical reader must appreciate that mathematics is a very condensed language and hence be prepared to spend the time necessary to extract the meaning of simple formulae. He would be well advised to compute numerical examples and draw graphs until the meaning conveyed by a formula is clear to him. In some cases this has already been done in the text but, to conserve space, we have usually been content to point out only the most important results of the analysis. The ability to write computer programs in a language such as BASIC, FORTRAN, or ALGOL would be a great advantage for those who wish to fully explore the models.

Terminology

The line of development in this book is from the simple to the complex. This applies, broadly speaking, both to the mathematical theory and to the social processes. Some of the terminology used may be unfamiliar or liable to cause confusion because of the different connotations which some terms have in different disciplines.

The *state* of a system is a fundamental concept in the theory of stochastic processes. In social applications this term may be applied both to an individual or to the system as a whole. Thus, we shall speak of someone as being 'in state i' meaning that they belong to the class, category, or group designated by 'i'. When applied to a system the term refers, usually, to the numbers of individuals in the various categories. In the diffusion of news, for example, the numbers actively spreading, not yet informed, and so on constitute a description of the state of the system at a given time. The appropriate choice of a set of states, in both senses, is a key element in the art of model-building.

Social systems will be classified as *closed* or *open*. These terms appear to have various connotations in sociology. They are used here in their direct and obvious sense. The membership of a closed system does not change over time; no members leave and none are admitted. An open system has both gains and losses. Systems with losses but no gains are regarded as closed by adopting the convention that leavers move into a residual category within the system. All open systems can be formally converted into closed systems by this kind of device, but there seems to be no advantage in doing so.

Time may be treated as *discrete* or *continuous*. 'Real' time is continuous, but for modelling purposes it may be more convenient to suppose that changes of state can only take place at fixed points in time. This may be more realistic as when, for example, changes take place in educational systems at the beginning of terms or academic years. Additionally, a discrete model may lead to more tractable equations and so be acceptable as an approximation to the real process.

The *Markov property* is fundamental to virtually all our models. Where it does not arise naturally its advantages are so great that it is worthwhile attempting to redefine the problem in such a way that it can be assumed. Roughly speaking, the Markov property requires that a knowledge of the current state of the system provides all the information relevant to predicting its future. At first sight this is a highly restrictive assumption, but the art of model-building is to so define the states that any relevant history of the process is incorporated in the definition of the current state.

Notation

Notation is often a source of difficulty to those who lack mathematical experience. The problem is compounded when material from diverse sources has to be blended into a single coherent account. While it is clearly desirable to have a consistent notation throughout, it is equally desirable to conform, as far as possible, with what has become standard notation in different fields of application. These two desiderata are often in conflict and a compromise has to be made. Even then, and when the Greek alphabet is included with upper- and lower-case letters, there are not enough symbols to go round. We have adopted the following strategy. Fundamental quantities which occur throughout the book are denoted by the same symbol throughout. Where possible they are the initial letter of the quantity designated. The most common of these are as follows:

T denotes time measured from the origin of the process and it may be discrete or continuous.

t and τ also refer to time, but measured from some other origin.

n refers to a number of people. A single subscript, as in n_i, means the number in a given state, i; where these numbers change with time the number at time T is denoted by $n_i(T)$; a double subscript, as in n_{ij}, indicates a flow from state i to state j in an interval of time or at a point of time, this interval or point being identified by adding the time in brackets.

N denotes the total number of people in a system.

p and q are used in a similar way to n except that they refer to probabilities (or proportions); thus p_{ij} is the probability of an individual moving from state i to state j.

k is the total number of states in a system.

$g, h, i, j, m, r,$ and s are often used as subscripts indexing the states of a system, occasionally with primes attached; i and j are used only in this connection.

θ is used, with subscripts, for the eigenvalues (latent roots) of a matrix. This avoids confusion with λ which was used in earlier editions for both eigenvalues and hazard functions. It is now used only in the latter sense.

Some other symbols have different meanings in different contexts, but these have been arranged to minimize the risk of confusion. A few letters such as X, x, u, and v are general-purpose symbols which have only a local meaning.

$f(\cdot)$ denotes a probability density function; $F(\cdot)$ is the corresponding distribution function and $G(\cdot) = 1 - F(\cdot)$, in some applications called the survivor function.

Matrices and vectors are denoted in one of two ways. The most common is an upper-case bold letter for a matrix and lower-case bold for a vector. The alternative method for a matrix is to write the (i, j)th element in brackets, thus $\{p_{ij}\}$; a vector would be written as $\{p_i\}$. We only use the second notation where it is necessary to be explicit about what the elements are. Sometimes we shall meet matrices where non-zero elements occur only on the main diagonal. These will be written either as diag $\{p_i\}$ or $[\mathbf{p}]_d$. In the second case the diagonal elements are those of the vector $\mathbf{p}$. Here, $\mathbf{1}$ denotes a row vector of 1's and $\mathbf{e}_i$ is a vector with 1 in the ith position and zeros elsewhere. The dimensions of matrices are almost always omitted as they can easily be deduced from the context. Vectors are written as rows and transposition of vectors and matrices is denoted by a prime.

Logarithms to the base e are written 'ln'; 'det' means 'determinant of'. The notation $X \sim B(\cdot\;\cdot)$, Poiss $(\cdot)$ or $N(\cdot\;\cdot)$ is a shorthand way of saying that X (or any random variable) has a binomial, Poisson, or normal distribution with parameter values in the positions marked with dots.

The expectation of a random variable is denoted by an 'E' in front of the variable or by a bar as in $\bar{n}(T)$; var (X) means variance of X and cov(X, Y) the covariance of X and Y.

Care needs to be taken with the symbols R and S which each have more than one meaning. In upper-case italic R refers to a number of recruits. In lower case, $\mathbf{r} = (r_1, r_2, \ldots r_k)$ is a vector of recruitment proportions. An upper-case bold-face $\mathbf{R}$ denotes the vector of transition rates with elements denoted by r_{ij}. Without subscripts, r also has a purely local use as an index in Chapter 7. The main use of S is in connection with transition probabilities for vacancy chains in Chapter 8 but, following standard practice, it is also used as an ordinary transition matrix in Chapter 4 and as the parameter of a Laplace transform or generating function.

CHAPTER 2

Models for Mobility in Closed Social Systems

2.1 INTRODUCTION

The most rudimentary way of quantifying the characteristics of human populations is by classifying their members on the basis of some personal attribute. Thus we may classify voters according to the way they vote, families according to where they reside, or workers by their occupation. When we begin to study the dynamics of social processes it is thus natural to start by looking at the movement of people between categories. Since such moves are largely unpredictable at the individual level it is necessary for a model to describe the mechanism of movement in probabilistic terms. We shall begin this task in the present chapter for discrete time systems which are closed. A discrete time system is one in which changes of state take place only at fixed points in time. It is not necessary that these points be equidistant though they often will be. A closed system is one in which there are no gains or losses. Both of these restrictions will be relaxed in later chapters.

A system such as we have described may be characterized by two sets of quantities which will be called the 'stocks' and 'flows'. The stocks at any particular time are the numbers of people in each category. The flows are the numbers of people (or other entities) who move between pairs of categories over an interval of time. There are very many processes studied in the social sciences which can be regarded in this way. We have already mentioned some and others are listed in the Complements section at the end of Chapter 3. The very diversity of these potential applications poses a problem for our exposition. From a mathematical point of view the most economical way to proceed is to keep the discussion at a fairly abstract level. This kind of treatment is found in most formal textbook expositions of stochastic processes, but it tends to place mathematical tidiness before practical relevance. At the other extreme one could consider each field of application separately. However, this would lead to much repetition and it would obscure the structural similarities which underly many diverse phenomena. We shall therefore aim to steer a middle course. Most of our models (here and in later chapters) will be firmly rooted in one particular application. However, we shall not hesitate to branch out into other spheres if they serve to demonstrate novel features. Thus, in this chapter we shall begin by focusing on social mobility which is one of the oldest and most interesting processes to have been modelled stochastically. Later, however, when we come to interactive Markov chain models, it will be more natural to describe them in the context of attitude change and voting behaviour.

The earliest paper in which social mobility was viewed as a stochastic process appears to be that of Prais (1955a). Since then there has grown up a large literature to which we refer later. A distinction has to be made between intergenerational mobility and intragenerational mobility. The former refers to changes of social class from father to son through succeeding generations. Here the generation provides a natural discrete time unit; this is essentially an ordinal variable and we make no assumption about its magnitude in terms of years. Intragenerational mobility refers to changes of class which take place during an individual's life span. One may reasonably speak of this as social mobility if the classes chosen reflect social status, and this usage is common in the literature. We prefer to call it occupational, or labour, mobility because it is usually more directly concerned with occupations rather than the status which derives from them. A characteristic feature of occupational mobility is that there is no natural time interval between changes of state. Moves can take place at any time and therefore such processes are more appropriately modelled in continuous time. We shall do this in Chapter 4 where it will be shown that, if such a process is only observed at fixed intervals of time, it can sometimes be treated as if it were genuinely discrete. This means that the discrete time models of this chapter have a wider relevance than might at first sight appear. In a similar way we shall see that some of the open systems treated in Chapter 3 can be treated as if they were closed.

2.2   THE MARKOV MODEL FOR SOCIAL MOBILITY

The basic model

First we consider a very simple model for the development of a single family line and then, in later sections, investigate the consequences of removing its more unrealistic features. The fundamental requirement in a model is that it must specify the way in which changes in social class occur. We shall assume that these are governed by transition probabilities which are independent of time. Let p_{ij} denote the probability that the son of a father in class i is in class j; since the system is closed

$$\sum_{j=1}^{k} p_{ij} = 1,$$

where k is the number of classes. We denote by $\mathbf{P}$ the matrix of transition probabilities. If we consider only family lines in which each father has exactly one son the class history of the family will be a Markov chain. By regarding society as composed of such family lines we could make deductions about the changing structure of society. In practice the requirement that each father shall have exactly one son is not met. As a result some lines become extinct and others branch. However, in a population whose size remains constant over a period of time, each father will have *on average* one son. We may expect our results for the simple model to apply in an average sense in such an actual society. Later we shall place this reasoning on a firmer footing and show that this expectation is fulfilled.

Suppose that the probability that the initial progenitor of a family line is in class j at time zero is $p_j(0)$. Let the probability that the line is in class j at time $T(T = 1, 2, 3, \dots)$ be $p_j(T)$. The probabilities $\{p_j(T)\}$ can then be computed recursively from the fact that

$$p_j(T+1) = \sum_{i=1}^{k} p_i(T)p_{ij} \qquad (j = 1, 2, \dots, k). \tag{2.1}$$

In matrix notation these equations may be written as

$$\mathbf{p}(T+1) = \mathbf{p}(T)\mathbf{P}, \tag{2.2}$$

where $\mathbf{p}(T) = (p_1(T), p_2(T), \dots, p_k(T))$. Repeated application of (2.2) gives

$$\mathbf{p}(T) = \mathbf{p}(0)\,\mathbf{P}^T. \tag{2.3}$$

The elements of $\mathbf{p}(T)$ may also be interpreted as the expected proportions of the population in the various classes at time T. If the original classes of the family lines are known the vector $\mathbf{p}(0)$ would then represent the initial class structure.

The matrix $\mathbf{P}^T$ plays a fundamental role in the theory of Markov chains. It can be used to obtain the 'state' probabilities from (2.3), but its elements also have a direct probabilistic interpretation. Let $p_{ij}^{(T)}$ denote the (i, j)th element in $\mathbf{P}^T$, then (2.3) may be written

$$p_j(T) = \sum_{i=1}^{k} p_i(0)p_{ij}^{(T)} \qquad (j = 1, 2, \dots, k). \tag{2.4}$$

It is clear from this representation that $p_{ij}^{(T)}$ is the probability that a family line goes from class i to class j in T generations. The case $i = j$ is of special interest as the probabilities $p_{ii}^{(T)}$ can be made the basis of measures of mobility.

In many applications the population has been in existence for many generations so that the 'present' state corresponds to a large value of T. It is therefore of considerable practical interest to investigate the behaviour of the probabilities $\{p_i(T)\}$ and $\{p_{ij}^{(T)}\}$ as T tends to infinity. It is shown in the general theory of Markov chains that this limiting behaviour depends on the structure of the matrix $\mathbf{P}$. Provided that the matrix $\mathbf{P}$ is *regular* it may be shown that these probabilities all approach limits as T tends to infinity. A regular (finite) Markov chain is one in which it is possible to be in any state (class) after some number, T, of generations, no matter what the initial state. More precisely, a necessary and sufficient condition for the chain to be regular is that all of the elements of $\mathbf{P}^T$ are non-zero for some T. All transition matrices which are likely to occur in the present context are regular but, later in the book, we shall meet examples which do not possess this property.

With the existence of the limits assured it is a straightforward matter to calculate them. Thus if we write $\lim_{T \to \infty} p_j(T) = p_j$ it follows from (2.2) that the limiting structure must satisfy

$$\mathbf{p} = \mathbf{p}\mathbf{P} \tag{2.5}$$

with

$$\sum_{j=1}^{k} p_j = 1.$$

The limiting structure, or distribution, can thus be obtained by solving a set of simultaneous equations. An important property of the solution is that it does not depend on the initial state of the system. Since, by our assumptions, each family line extant will have reached the equilibrium given by (2.5), the vector $\mathbf{p}$ gives the expected structure of the population at the present time. If this structure is all that can be observed we have no means of reconstructing the transition matrix. Neither, in fact, can we deduce $\mathbf{P}$ from two consecutive observed structures $\mathbf{p}(T)$ and $\mathbf{p}(T+1)$, although White (1963) and Matras (1967) have considered what incomplete information can be obtained in these circumstances. The limiting value of $\mathbf{P}^T$, denoted by $\mathbf{P}^\infty$, can be deduced from (2.3). It must satisfy

$$\mathbf{p} = \mathbf{p}(0)\mathbf{P}^\infty$$

which can only be so if

$$\mathbf{P}^\infty = \begin{Bmatrix} p_1 & p_2 & \cdots & p_k \\ p_1 & p_2 & \cdots & p_k \\ \vdots & \vdots & & \vdots \\ p_1 & p_2 & \cdots & p_k \end{Bmatrix} \tag{2.6}$$

which implies that

$$\lim_{T \to \infty} p_{ij}^{(T)} = p_j. \tag{2.7}$$

The foregoing analysis shows that if our model provides an adequate description of actual societies then their future development depends only on their initial structure and the transition matrix. Of these two features the initial distribution has a diminishing influence as time passes. In the long run, therefore, the structure of the society is determined by its transition matrix. This conclusion implies that the study of mobility must be centred upon the transition probabilities. In particular, *measures* of mobility should be functions of the elements of $\mathbf{P}$. Before pursuing this proposal we may test the adequacy of the model using actual data on mobility.

Several empirical studies of mobility have been published which give sufficient data to provide a partial test of the theory. One of these due to Glass and Hall (Glass, 1954) is based on a random sample of 3,500 pairs of fathers and sons in Britain. A second study made by Rogoff (1953) is based on data from marriage licence applications in Marion County, Indiana. Rogoff obtained data for two periods; one from 1905 to 1912, with a sample size of 10,253, and a second from 1938 to 1941, when the sample size was 9,892. Further data relating to Denmark are given in Svalagosta (1959).

The data obtained by Glass and Hall were used by Prais (1955a) and much of the material in this section is taken from his paper. Glass and Hall classified the members of their sample according to the seven occupational groups listed in Table 2.1, which also gives the estimated transition probabilities.

Table 2.1   Estimated transition probabilities for England and Wales in 1949 (Glass and Hall's data)

	Son's class						
Father's class	1	2	3	4	5	6	7
1. Professional and higher administrative	0.388	0.146	0.202	0.062	0.140	0.047	0.015
2. Managerial and executive	0.107	0.267	0.227	0.120	0.206	0.053	0.020
3. Higher grade supervisory and non-manual	0.035	0.101	0.188	0.191	0.357	0.067	0.061
4. Lower grade supervisory and non-manual	0.021	0.039	0.112	0.212	0.430	0.124	0.062
5. Skilled manual and routine non-manual	0.009	0.024	0.075	0.123	0.473	0.171	0.125
6. Semi-skilled manual	0.000	0.013	0.041	0.088	0.391	0.312	0.155
7. Unskilled manual	0.000	0.008	0.036	0.083	0.364	0.235	0.274

The first two columns of Table 2.2 give the class structure of the population in two succeeding generations. If the Markov model is adequate and if the society has reached equilibrium we would expect these distributions to be the same, apart from sampling fluctuations. We would also expect them both to agree with the equilibrium distribution obtained from (2.5). Prais (1955a) made the calculations necessary for this comparison and his results are given in Table 2.2.

The differences between the three distributions are not large although there does appear to be a shift towards the lower classes as we move across the table. If this trend is genuine and not merely the result of sampling fluctuations it might be taken to indicate that the process had not reached equilibrium. Another possible explanation is discussed below. A complete answer to the question of sampling error is not available, but a first step towards the solution of the problem is

Table 2.2 *Actual and equilibrium distributions of the social classes in England and Wales (1949), estimated from Glass and Hall's data*

Class	Fathers	Sons	Predicted equilibrium
1	0.037	0.029	0.023
2	0.043	0.046	0.042
3	0.098	0.094	0.088
4	0.148	0.131	0.127
5	0.432	0.409	0.409
6	0.131	0.170	0.182
7	0.111	0.121	0.129

provided by the sampling theory derived later. Although we cannot obtain a complete test of the model using these data it does appear that there is a broad compatibility between the data and the predictions of the theory.

Rogoff's data lead to similar conclusions about the applicability of Markov chain theory to social mobility. She obtained information at two dates separated by 30 years and so we can see whether there had been any significant change in the transition probabilities during that period. The two transition matrices are given in Table 2.3.

Table 2.3 *Transition probabilities estimated from Rogoff's data for Marion County, Indiana*[a]

	1905–12			1938–41		
	1	2	3	1	2	3
1. Non-manual	0.594	0.396	0.009	0.622	0.375	0.003
2. Manual	0.211	0.782	0.007	0.274	0.721	0.005
3. Farm	0.252	0.641	0.108	0.265	0.694	0.042

[a] From *Finite Markov Chains* by John G. Kemeny and J. Laurie Snell © 1960 by D. Van Nostrand Company. Reprinted by permission of the publisher.

We have followed Kemeny and Snell (1976) in adopting a coarse grouping for the classes in place of Rogoff's very fine breakdown. Allowing for sampling error, it appears that changes did take place in the transition matrix over the period in question. However, the changes are not large and suggest that we shall not be involved in gross error if we treat them as constant over moderately short periods.

Neither set of data allows us to make a direct test of the Markov property. This property requires that a son's class should depend only on that of his father and not on that of his grandfather. To test this assumption we need records of family history over at least three generations. Indirect support for the assumption is provided by the close agreement between the equilibrium class structure predicted by Markov theory and the observed class structure.

There are theoretical grounds for believing that the Markov property will not hold exactly for a social mobility process. These grounds arise from the fact that the class boundaries are drawn arbitrarily. Thus, for example, we could subdivide the seven categories used by Glass and Hall. Alternatively, some classes could be amalgamated to give a smaller number of categories. It is known (see Kemeny and Snell, 1976, Chapter 6) that if the states of a Markov chain are pooled then the new chain does not, in general, have the Markov property. In the present context this means that we cannot arbitrarily rearrange the classes and retain the Markov property. Even if there is one system of classification for which the property holds this may not be the one which we happen to have chosen. However, as we have pointed out in Chapter 1, it is sufficient if our model embodies the main features of the process without being correct in every detail. To summarize: although the assumptions on which the theory depends are not completely realistic the model is sufficiently near to reality to justify its further use and development.

Time reversal in the Markov chain model

The basic difference equation of the Markov model for intergenerational mobility arose by considering the expected class in the next generation conditional upon present class. By this means it has proved possible to predict future class structures and to deduce their limiting form. It is sometimes interesting to proceed in the reverse direction and consider what happens if we try to go backwards in time. The relevant transition probabilities are then $Pr\{$family line now in class i *came from* class $j\} = \tilde{p}_{ij}$, say. These probabilities are simply related to the forward transition probabilities as follows. Let $X(T)$ temporarily denote the number of the class in which a particular family line is at time T. Then

$$Pr\{X(T+1) = j, X(T) = i\} = Pr\{X(T) = i\}Pr\{X(T+1) = j \,|\, X(T) = i\}$$
$$= Pr\{X(T+1) = j\}Pr\{X(T) = i \,|\, X(T+1) = j\}.$$

Equating the two alternative forms of the right-hand side,

$$Pr\{X(T) = i \,|\, X(T+1) = j\} = p_{ij}p_i(T)/p_j(T+1) = \tilde{p}_{ji}. \qquad (2.8)$$

Although the reversed process is Markovian it is not in general time-homogeneous. However, if the system has reached the steady state, $p_i(T)$ and $p_j(T+1)$ will have their limiting values. In that case, exchanging i and j,

$$\tilde{p}_{ij} = p_{ji}p_j/p_i \qquad (i, j = 1, 2, \ldots, k). \qquad (2.9)$$

It may happen that the backward and forward transition matrices turn out to be the same, that is $\tilde{\mathbf{P}} = \mathbf{P}$. This means that the process appears the same in whichever direction it is viewed in time: we shall call such a process, reversible.

One consequence of reversibility is that if we consider someone in class i then the chance that his son is in class j is the same as the chance that his father was in class j. Yet another way of expressing this property is in terms of the expected numbers of changes of class between one generation and the next. Suppose that we have a population of N family lines; then the expected number of changes from class i to class j from one generation to the next will be

$$Np_i p_{ij}.$$

The expected flow in the opposite direction is

$$Np_j p_{ji}.$$

Under the reversibility condition, obtained by putting $\tilde{p}_{ij} = p_{ij}$ in (2.9), these two flows are equal. Reversibility therefore implies an equal exchange between classes. Since the system is in equilibrium there is no *net* change in the class sizes from one generation to the next. But reversibility requires something much stronger; it is as though a family line can only move to a new class if there is a compensatory move in the reverse direction. This does not seem very plausible in the case of social mobility but, as Kemeny and Snell (1976) noted, it does seem to be the case for the Glass and Hall data. The point can be examined by computing the matrix with (i, j)th element equal to $p_i p_{ij}$; this matrix will be symmetric for a reversible process. If we form the matrix for Glass and Hall's data given in Tables 2.1 and 2.2 we obtain

$$\begin{pmatrix} 0.0089 & 0.0034 & 0.0046 & 0.0014 & 0.0032 & 0.0011 & 0.0003 \\ 0.0045 & 0.0112 & 0.0095 & 0.0050 & 0.0087 & 0.0022 & 0.0008 \\ 0.0031 & 0.0089 & 0.0165 & 0.0168 & 0.0314 & 0.0059 & 0.0054 \\ 0.0027 & 0.0050 & 0.0142 & 0.0269 & 0.0546 & 0.0157 & 0.0079 \\ 0.0037 & 0.0098 & 0.0307 & 0.0503 & 0.1935 & 0.0699 & 0.0511 \\ 0.0000 & 0.0024 & 0.0075 & 0.0160 & 0.0712 & 0.0568 & 0.0282 \\ 0.0000 & 0.0010 & 0.0046 & 0.0107 & 0.0470 & 0.0303 & 0.0353 \end{pmatrix}$$

The near symmetry of this matrix implies that there is almost equal exchange between all pairs of class. It is difficult to think of any sociological reason why this should be so, and the matter obviously requires further investigation.

The variability of the class sizes

The theory developed so far enables us to calculate expected numbers, or proportions, in each class at any time in the future. So far we have no means of determining the variances and covariances of our predictions. A method for obtaining the moments and product moments of the class numbers has been given by Pollard (1966), and we shall now describe its application to this problem.

Let N denote the number of family lines in the population; this remains constant through time. Let the size of the jth class at time T be $n_j(T)$ and let the number of transitions between class i and class j between T and $T+1$ be $n_{ij}(T)$. It follows from the definitions that

$$n_j(T+1) = \sum_{i=1}^{k} n_{ij}(T) \qquad (j = 1, 2, \ldots, k). \tag{2.10}$$

If we take expectations on each side of this equation we arrive at (2.1) because $En_j(T+1) = Np_j(T+1)$ and $En_{ij}(T) = Np_i(T)p_{ij}$. Consider the covariance of $n_j(T+1)$ with $n_l(T+1)$. It will simplify the presentation of the theory if we adopt the convention that $\operatorname{cov}(x_i, x_j) \equiv \operatorname{var}(x_i)$ if $i = j$. We then have

$$\begin{aligned}
\operatorname{cov}&\{n_j(T+1), n_l(T+1)\} \\
&= E\{n_j(T+1)n_l(T+1)\} - En_j(T+1)En_l(T+1) \\
&= \sum_{i=1}^{k} \sum_{i'=1}^{k} [E\{n_{ij}(T)n_{i'l}(T) - En_{ij}(T)En_{i'l}(T)\}]
\end{aligned} \tag{2.11}$$

by (2.10). In order to evaluate the expectations in (2.11), we make use of a well-known result about conditional expectations to the effect that $E(x) = E_X(x|X)$. In the present case we obtain the expectations conditional upon $n_i(T)$. These follow from the fact that, given $n_i(T)$, $n_{ij}(T)$ $(j = 1, 2, \ldots, k)$ are multinomially distributed with probabilities $p_{ij}(j = 1, 2, \ldots, k)$. Hence,

$$E\{n_{ij}(T)|n_i(T)\} = n_i(T)p_{ij} \tag{2.12}$$

and

$$\begin{aligned}
E\{n_{ij}(T)n_{i'l}(T)|n_i(T), n_{i'}(T)\} &= n_i(T)n_{i'}(T)p_{ij}p_{i'l} \qquad (i \neq i'), \\
E\{n_{ij}(T)n_{il}(T)|n_i(T)\} &= n_i(T)\{n_i(T)-1\}p_{ij}p_{il} + \delta_{jl}n_i(T)p_{ij},
\end{aligned} \tag{2.13}$$

where $\delta_{jl} = 1$ if $j = l$ and is zero otherwise. The unconditional expectations are now obtained from (2.12) and (2.13) by taking the expectations of the right-hand sides with respect to $n_i(T)$. Substituting these expressions in (2.11) then gives

$$\begin{aligned}
\operatorname{cov}\{n_j(T+1), n_l(T+1)\} &= \sum_{i=1}^{k} \sum_{i'=1}^{k} p_{ij}p_{i'l} \operatorname{cov}\{n_i(T), n_{i'}(T)\} \\
&\quad + \sum_{i=1}^{k} (\delta_{jl}p_{ij} - p_{ij}p_{il})En_i(T).
\end{aligned} \tag{2.14}$$

We have thus obtained a recurrence relation between the expectations and covariances at time T and the covariances at time $T+1$. Since the covariances at $T = 0$ are zero the complete set can be computed from (2.14) and (2.1). If we require them, the same method can be extended to yield the higher moments and product moments.

Since the expectations and covariances at $T+1$ are linear functions of the corresponding quantities at T we can express the relationship in matrix notation. To do this we introduce the vector of means and covariances and denote it by $\mu(T)$. In this vector we first list the k means followed by the k^2 covariances in dictionary order, that is

$$\boldsymbol{\mu} = [E(n_1), \ldots, E(n_k), \text{cov}(n_1, n_1), \text{cov}(n_1, n_2), \ldots, \text{cov}(n_1, n_k),$$
$$\text{cov}(n_2, n_1), \text{cov}(n_2, n_2), \ldots, \text{cov}(n_2, n_k), \ldots, \text{cov}(n_k, n_k)],$$

where we have omitted the arguments of the n's and $\boldsymbol{\mu}$ for brevity. There is some redundancy in this listing because, for example, the covariance between n_1 and n_2 appears as $\text{cov}(n_1, n_2)$ and as $\text{cov}(n_2, n_1)$, but by retaining this the symmetry of the expressions is preserved. Equations (2.1) and (2.14) may now be combined and written in the form:

$$\boldsymbol{\mu}(T+1) = \boldsymbol{\mu}(T)\boldsymbol{\Pi}, \tag{2.15}$$

where the elements of the $k(k+1) \times k(k+1)$ matrix $\boldsymbol{\Pi}$ are functions of the p_{ij}'s. The matrix $\boldsymbol{\Pi}$ may be partitioned as follows:

$$\begin{array}{c|c} \mathbf{P} & \mathbf{X} \\ \hline \mathbf{O} & \mathbf{Y} \end{array},$$

where $\mathbf{P}$ is the $k \times k$ transition matrix, $\mathbf{O}$ is a $k^2 \times k$ zero matrix, $\mathbf{X}$ is a $k \times k^2$ matrix with elements of the form $\delta_{jl} p_{ij} - p_{ij} p_{il}$ and $\mathbf{Y}$ is a $k^2 \times k^2$ matrix with elements of the form $p_{ij} p_{i'l}$.† In the case of $\mathbf{X}$ and $\mathbf{Y}$, (j, l) indexes the column and i or (i, i') the row.

Equation (2.15) is a generalization of (2.2) and it can be used in the same way. It is no longer true that $\boldsymbol{\Pi}$ is a stochastic matrix because it has negative elements but the rows still sum to one. This fact can be used to deduce that the elements of $\boldsymbol{\mu}(T)$ tend to limits as T tends to infinity. Consequently, these limits must satisfy the equations

$$\left. \begin{array}{c} \boldsymbol{\mu}(\infty) = \boldsymbol{\mu}(\infty)\boldsymbol{\Pi} \\ \sum_{i=1}^{k} \bar{n}_i(\infty) = N \end{array} \right\}, \tag{2.16}$$

where here, and subsequently, the bar denotes the expectation of the random variable. A computer program is available to generate the vector $\boldsymbol{\mu}(T)$ for all T. We shall give some results of calculations in Chapter 3 for a more general model. We can get an idea of the kind of solution obtained in the present application by considering a special case.

Suppose that we have what Prais calls a perfectly mobile society with $p_{ij} \doteq p_j (j = 1, 2, \ldots, k)$. Equation (2.14) then simplifies to give

$$\text{cov}\{n_j(T+1), n_l(T+1)\}$$
$$= N\delta_{jl} p_j - N p_j p_l + p_j p_l \sum_{i=1}^{k} \sum_{i'=1}^{k} \text{cov}\{n_i(T), n_{i'}(T)\}. \tag{2.17}$$

---

† $\mathbf{Y}$ is the *direct matrix product* $\mathbf{P} \times \mathbf{P}$. The direct matrix product (or Kronecker product) can be used to facilitate the solution of the more general problem of finding the higher moments and product moments of the $\{n_j\}$. See Pollard (1966).

Now

$$\sum_{i=1}^{k} \sum_{i'=1}^{k} \text{cov}\{n_i(T), n_{i'}(T)\} = E\left\{\sum_{i=1}^{k} (n_i - \bar{n}_i)\right\}^2 = 0$$

so that the last term in (2.17) vanishes. The part which remains will be recognized as giving the variances and covariances of the multinomial distribution. In fact, it is easy to see directly that $\{n_j(T)\}$ has a multinomial distribution with parameters N and $p_j(j = 1, 2, \ldots, k)$ for all T. The class distribution at time $T+1$ is independent of that at time T. Hence we may regard the N family lines as being allocated independently to classes with probabilities $p_j(j = 1, 2, \ldots, k)$, which is the condition for the distribution to be multinomial. In this case, therefore, the standard error of the predicted proportion p_j will be $\{p_j(1 - p_j)/N\}^{\frac{1}{2}}$.

The underlying assumption in the derivation of these results is that the individuals moving out of a given class are distributed according to a multinomial distribution over all classes. The probabilities governing these flows are those in the row of the transition matrix corresponding to the class in question. As far as the second moments are concerned, this assumption is embodied in the second part of (2.13) when $i = i'$, and this leads on to the result in (2.14). Wynn and Sales (1973a) have considered a more general error structure in which the covariances between any pair of flows out of a given class are allowed to be arbitrary subject to the overall requirement that the total flow is fixed. Thus Wynn and Sales (1973a) suppose that

$$\text{cov}\{n_{ij}(T), n_{i'l}(T)\} = 0 \quad \text{if } i \neq i' \text{ (as before)}$$

and

$$\text{cov}\{n_{ij}(T), n_{il}(T)\} = \alpha_{ijl} n_i(T) \qquad (j, l = 1, 2, \ldots, k) \quad (2.18)$$

conditional on $n_i(T)$. The α's are constrained by the fact that, for fixed $n_i(T)$,

$$\sum_{j=1}^{k} n_{ij}(T) = n_i(T)$$

and hence by

$$\text{var} \sum_{j=1}^{k} n_{ij}(T) = \sum_{j=1}^{k} \sum_{l=1}^{k} \text{cov}\{n_{ij}(T), n_{il}(T)\}$$

$$= \sum_{j=1}^{k} \sum_{l=1}^{k} \alpha_{ijl} = \text{var}\{n_i(T)\} = 0. \quad (2.19)$$

Subject to this restraint the values of $\{\alpha_{ijl}\}$ can have any values, except that when $j = l$ the covariance becomes a variance and so must be positive. In the multinomial case

$$\alpha_{ijl} = -p_{ij} p_{il} \qquad \text{if } j \neq l$$
$$= p_{ij}(1 - p_{ij}) \qquad \text{if } j = l. \quad (2.20)$$

The argument leading to (2.14) still holds, but the last term on the right-hand side has to be replaced by

$$\sum_{i=1}^{k} \alpha_{ijl} En_i(T).$$

In the matrix version of the equations the submatrix $\mathbf{X}$ becomes a $k \times k^2$ matrix with elements $\{\alpha_{ijl}\}$. Wynn and Sales (1973a) established the existence of the limiting form (2.16) and discussed the estimation of the errors; they applied the method to data relating to occupational mobility.

2.3　THE MEASUREMENT OF MOBILITY

It is often convenient to have a scalar measure of mobility which enables us both to compare different societies and to chart changes in mobility within the same society. Numerous descriptive measures of mobility have been devised for empirical work. Some examples are listed in Matras (1960b) and a full discussion is given by Boudon (1973). A useful review of measures is contained in Bibby (1975). It is not our intention to give a full account of this topic but rather to show what bearing stochastic theory has upon the problem of measurement. We shall assume that mobility is adequately described by a Markov model and then ask how, on that assumption, mobility should be measured.

Since movement is determined completely by the transition matrix $\mathbf{P}$ all of the information about mobility is contained in its elements. Our problem, stated formally, is thus to map the set of transition matrices on to some convenient interval so as to order them according to the degree of mobility which they represent.

The issues have been greatly clarified by two theoretical discussions by Shorrocks (1978) and Sommers and Conlisk (1979). The following discussion draws heavily upon their work, but we shall set it in the context of a distinction in the meanings of the word 'mobility', which appears to have been overlooked hitherto. The point at issue can be brought out by considering the special case $k = 2$. Consider the following three matrices:

$$\text{(a)} \begin{pmatrix} 1 & 0 \\ 0 & 1 \end{pmatrix}, \quad \text{(b)} \begin{pmatrix} p & 1-p \\ p & 1-p \end{pmatrix}, \quad \text{(c)} \begin{pmatrix} 0 & 1 \\ 1 & 0 \end{pmatrix}, \quad (0 < p < 1).$$

At one end of the scale there is no ambiguity. Matrix (a) corresponds to a society in which no movement takes place. It is clear that this matrix will have to be placed at one end of our scale of measurement. The distinction we are seeking to make becomes clear when we consider the other extreme. Here there are two possibilities represented by (b) and (c) above. Matrix (b) corresponds to what Prais (1955a) called a perfectly mobile society. Its distinguishing feature is that the son's class does not depend on that of his father. For general k any matrix having identical rows will have the same property. If we are concerned to measure the degree to which a son's class is constrained by that of his father then (a) and (b) are clearly at opposite ends of the scale. Case (c) then belongs with (a) because, as with (a), the son's class is completely determined by that of his father. It is this kind of mobility which is often of substantive interest because it has to do with the extent to which a son's opportunities are limited by the accident of birth. It would be less confusing if this kind of mobility were referred to as the degree of intergeneration dependence, or more generally, as temporal dependence.

The second sense in which the word 'mobility' is used is in referring to the amount of movement which takes place. If that is what we wish to measure then matrix (c) represents the opposite extreme to (a). In one case everyone moves and in the other no one moves. In practice this distinction is blurred, and made less important, by the fact that actual transition matrices almost always lie somewhere between (a) and (b). Nevertheless, when setting up criteria by which indices of mobility should be judged, as Shorrocks (1978) does, it is important to be clear about which kind of mobility one is aiming to measure.

A second feature of mobility tables which is often overlooked is that the classes may, or may not, be ordered according to status. If the classes are so ordered a move between adjacent classes may carry less weight than one between two widely separated classes. This consideration arises when we are trying to measure the degree of movement but not in the case of generation dependence.

Several indices, which are otherwise distinct, coincide when $k = 2$. We shall therefore begin with the general case, noting as we go the effect of specializing to $k = 2$.

Measures of generation dependence

We first consider how to measure the extent to which a son's class depends on that of his father. A method of doing this is suggested by considering the spectral representation of the transition matrix. If $\mathbf{P}$ is a stochastic matrix then it can be expressed in the form:

$$\mathbf{P} = \sum_{r=1}^{k} \theta_r \mathbf{A}_r. \qquad (2.21)$$

The matrices $\{\mathbf{A}_r\}$ are known as the spectral set and they have the following properties:

$$\mathbf{A}_r \mathbf{A}_s = \mathbf{0}, \quad \text{if } r \neq s; \qquad \mathbf{A}_r \mathbf{A}_s = \mathbf{A}_r \quad \text{if } r = s; \qquad \sum_{r=1}^{k} \mathbf{A}_r = \mathbf{I}. \qquad (2.22)$$

The coefficients $\{\theta_r\}$ are the eigenvalues of $\mathbf{P}$ and since $\mathbf{P}$ is a stochastic matrix one eigenvalue is unity and the remainder are less than or equal to one in modulus. We may therefore order them so that $1 = \theta_1 \geq |\theta_2| \geq \ldots \geq |\theta_k|$ and write (2.21) as

$$\mathbf{P} = \mathbf{1}'\mathbf{p} + \sum_{r=2}^{k} \theta_r \mathbf{A}_r \qquad (2.23)$$

where $\mathbf{1'p} = A_1$ and $\mathbf{1} = (1, 1, \ldots 1)$. The last step follows from the fact that (2.8) and (2.23) yield

$$\lim_{T \to \infty} \mathbf{P}^T = A_1$$

and we know from (2.7) that this limiting matrix has identical rows with elements equal to the steady-state structure for $\mathbf{P}$.

Inspection of (2.23) shows that if $\theta_2 = \theta_3 = \ldots = \theta_k = 0$ then $\mathbf{P}$ would have identical rows indicating a total absence of generation dependence. The degree of dependence can therefore be measured by how much $\mathbf{P}$ differs from $\mathbf{1'p}$. One way of measuring this distance is by constructing some suitable function of the θ's. Three such measures have been proposed which we now consider in turn.

The first, proposed by both Shorrocks (1978) and Sommers and Conlisk (1979), is based on the second largest, in absolute value, of the θ's. Let us denote this by $\theta_{\max}$, then the measure is

$$\mu_1(\mathbf{P}) = |\theta_{\max}| \tag{2.24}$$

This ranges between 0, when there is no dependence, and 1. The greater the value of $\mu_1(\mathbf{P})$ the stronger is the dependence of the son's class on the father's and hence the smaller the degree of social mobility *in that sense*. This measure can be given a variety of interpretations. Shorrocks (1978), for example, shows that it is a function of what he calls the half-life which, in this context, is the number of generations for the distance (suitably defined) between the present structure and the steady state to be halved. A disadvantage of $\mu_1(\mathbf{P})$, to which we shall return, is that the second largest eigenvalue of $\mathbf{P}$ could be imaginary.

A second measure can be based on the average of the θ's, thus:

$$\mu_2(\mathbf{P}) = \frac{1}{k-1} \sum_{r=2}^{k} \theta_r. \tag{2.25}$$

Since the sum of the eigenvalues of $\mathbf{P}$ is equal to the sum of its diagonal elements we can write

$$\mu_2(\mathbf{P}) = \left(\frac{\text{trace}(\mathbf{P}) - 1}{k-1} \right), \tag{2.26}$$

where trace $(\mathbf{P}) = \sum_{r=1}^{k} p_{rr}$. Again, this is zero when there is no generation dependence and rises to one in the case of complete immobility. The measure can also take on negative values, the minimum occurring when every son's class is different from his father's. The sign of $\mu_2(\mathbf{P})$ can thus be used to distinguish between what might be called 'positive' and 'negative' dependence.

Instead of taking the arithmetic mean of the θ's we could use the geometric mean of their moduli. This gives the measure

$$\mu_3(\mathbf{P}) = |\theta_2 \theta_3 \ldots \theta_k|^{1/(k-1)}. \tag{2.27}$$

The product of all the eigenvalues is equal to the determinant of $\mathbf{P}$ (remember that $\theta_1 = 1$), hence we may write

$$\mu_3(\mathbf{P}) = |\mathbf{P}|^{1/(k-1)}. \tag{2.28}$$

Apart from a possible difference of sign, all of these measures are equivalent when $k = 2$. For $k > 2$, $\mu_3(\mathbf{P})$ suffers from the disadvantage that it attains its minimum value of zero if any two rows of $\mathbf{P}$ are identical. It cannot therefore distinguish between generation independence for a single pair of classes and for the complete set.

Sommers and Conlisk (1979) suggested a variant of $\mu_1(\mathbf{P})$ which avoids the complication arising from imaginary eigenvalues. They proposed to replace $\mathbf{P}$ by

$$\mathbf{P}^* = \tfrac{1}{2}(\mathbf{P} + \mathbf{\Pi}^{-1}\mathbf{P'}\mathbf{\Pi}), \tag{2.29}$$

where $\mathbf{\Pi} = \mathrm{diag}\{p_i\}$. It is easily verified that $\mathbf{P}^*$ has the same limiting structure as $\mathbf{P}$. It is also reversible because it satisfies the condition

$$\mathbf{\Pi}\mathbf{P}^* = (\mathbf{P}^*)'\mathbf{\Pi}. \tag{2.30}$$

The advantage of $\mathbf{P}^*$ is that its eigenvalues are always real. Sommers and Conlisk (1979) give several arguments in favour of the modified measure but, from a practical point of view, it is hardly necessary. We have already noted that the reversibility condition is almost satisfied for Glass and Hall's data and so it is not surprising to find that $\mu_2(\mathbf{P}) = 0.505$ and $\mu_2(\mathbf{P}^*) = 0.506$. In 19 other cases for which the two measures were computed by Sommers and Conlisk the difference between them was minimal.

Measures of movement

The simplest measure of the amount of movement is the expected proportion of family lines which change class from one generation to the next. Between T and $T+1$ this is

$$\sum_{i=1}^{k} \sum_{\substack{j=1 \\ j \neq i}}^{k} p_i(T)p_{ij}$$

However, this violates our requirement that the measure shall depend only on $\mathbf{P}$. As $\mathbf{p}(T)$ changes with T the value of the measure would depend on T even if $\mathbf{P}$ remained constant. To avoid this we can compute the measure for a system in equilibrium by replacing $p_i(T)$ by p_i. This gives

$$m_1(\mathbf{P}) = \sum_{i=1}^{k} \sum_{\substack{j=1 \\ j \neq i}}^{k} p_i p_{ij} = 1 - \sum_{i=1}^{k} p_i p_{ii}. \tag{2.31}$$

This measure treats all transitions alike and so takes no account of any ordering of the classes. If the classes are ordered it seems reasonable to give more weight to

moves between distant classes than between those close together. Since a move
from class 1 to class 3, say, can be thought of as made up of two moves—from 1 to
2 and from 2 to 3—we propose to use as weights the number of class boundaries
crossed. This suggests basing our measure on

$$\sum_{i=1}^{k} \sum_{j=1}^{k} p_i p_{ij} |i-j|$$

which is the expected number of class boundaries crossed from one time to the
next when the system is in its steady state. Since it is convenient to confine the
index to the interval $(0, 1)$ we shall divide this expression by its greatest attainable
value. The greatest amount of movement occurs when every family line moves, at
the first step, to one of the extreme classes. Thereafter, they will alternate between
the two extremes crossing $(k - 1)$ class boundaries in the process. Hence we define

$$m_2(\mathbf{P}) = \frac{1}{k-1} \sum_{i=1}^{k} \sum_{j=1}^{k} p_i p_{ij} |i-j|. \tag{2.32}$$

A society in which there is no generation dependence does not yield a unique
value of either measure—and there is no reason why it should. If $k = 2$, $m_1(\mathbf{P})$ and
$m_2(\mathbf{P})$ are equivalent and it is then easy to show that, in the absence of generation
dependence,

$$m_2(\mathbf{P}) = p_{12}(1 - p_{12}) \le \tfrac{1}{4}.$$

The following example illustrates the use of $m_1(\mathbf{P})$ and $m_2(\mathbf{P})$ for $k = 3$ and the
difference between them. Suppose that population I has transition matrix

$$\begin{pmatrix} 0.4 & 0.3 & 0.3 \\ 0.2 & 0.5 & 0.3 \\ 0.2 & 0.2 & 0.6 \end{pmatrix}.$$

It may easily be verified that the limiting structure is $(7/28, 9/28, 12/28)$;
population II has transition matrix

$$\begin{pmatrix} 0.5 & 0.3 & 0.2 \\ 0.2 & 0.5 & 0.3 \\ 0.2 & 0.3 & 0.5 \end{pmatrix}$$

with limiting structure $(16/56, 21/56, 19/56)$. For population I

$$m_1(\mathbf{P}) = 1 - \frac{1}{28}\left[7(0.4) + 9(0.5) + 12(0.6) \right] = 0.4821$$

$$m_2(\mathbf{P}) = \frac{1}{56}\left[7\{0.3 + 2(0.3)\} + 9\{0.2 + 0.3\} + 12\{2(0.2) + 0.2\} \right]$$

$$= \frac{18}{56} = 0.3214$$

and for population II

$$m_1(\mathbf{P}) = 1 - \frac{1}{56}\left[\,16(0.5) + 21(0.5) + 19(0.5)\,\right] = \tfrac{1}{2}(0.5)$$

$$m_2(\mathbf{P}) = \frac{1}{112}\left[\,16\{0.3 + 2(0.2)\} + 21\{0.2 + 0.3\} + 19\{2(0.2) + 0.3\}\,\right]$$

$$= \frac{35}{112} = 0.3125.$$

Mobility in population I is thus marginally higher than in population II as measured by $m_2(\mathbf{P})$, but the order is reversed with $m_1(\mathbf{P})$. This reflects the greater degree of movement to the extremes in population I.

One could generalize (2.32) by introducing arbitrary weights in place of $|i-j|$, but then the simple interpretation in terms of the number of class boundaries crossed would be lost. Nevertheless, such a step would establish a link with another class of measures based on correlation ideas. Sommers and Conlisk (1979) first supposed that the classes could be located on a scale of status. Thus let x_i be the status score for all members of class i $(i = 1, 2, \ldots k)$. Then a possible measure of the degree of movement is the correlation coefficient between father's and son's class. A high positive correlation would involve little movement, whereas a high negative correlation would indicate a high degree of movement. This measure has the interesting and useful feature that the absence of generation dependence always corresponds to the zero point on this scale. This is not surprising because, in an even more obvious sense, it measures the dependence of son's class on father's.

Without loss of generality we may scale the x's so that they have zero mean and unit variance, and again we suppose the system to be in its steady state. The correlation coefficient is then

$$\rho = \sum_{i=1}^{k} \sum_{j=1}^{k} p_i p_{ij} x_i x_j \qquad (2.33)$$

since the number of family lines with father in i and son in j is proportional to $p_i p_{ij}$. Using the identity $(x_i - x_j)^2 = x_i^2 + x_j^2 - 2x_i x_j$ we have

$$\rho = 1 - \frac{1}{2} \sum_{i=1}^{k} \sum_{j=1}^{k} p_i p_{ij} (x_i - x_j)^2. \qquad (2.34)$$

If we now compare (2.34) with (2.32) we see that the only essential change is the replacement of $|i-j|$ by $(x_i - x_j)^2$. Here ρ is thus a natural generalization of $m_2(\mathbf{P})$. The problem with using ρ, of course, is that of finding a suitable scale on which to measure social status. To avoid this Sommers and Conlisk (1979) suggest that the x's should be chosen so as to maximize (2.34) subject to the requirement that they shall have zero mean and unit variance. If the classes are ordered such that $x_1 \le x_2 \le \ldots \le x_k$ we could introduce this additional constraint into the maximization drawing on the techniques of isotonic regression (see Barlow and co-workers, 1972). The appeal of ρ is somewhat diminished by the relative difficulty of the calculations and the rather involved interpretation of the maximized correlation.

Other approaches

Any measure which attempts to summarize the contents of a transition matrix in a single number is bound to result in oversimplification. A more detailed picture of the process can be obtained if we replace the single number by a set of numbers. In doing this, of course, we sacrifice simplicity. Prais (1955a) made some suggestions about how to enlarge the set of measures. One of these is based on the durations of stay in each class. The distribution of length of stay in class i, under the Markov model, is geometric with mean $\mu_i = 1/(1 - p_{ii})$. In a mobile society these will be shorter than in one which is immobile. In order to judge whether a mean is large or small we need some standard of comparison and Prais proposed using the means for the corresponding society with no generation dependence. That is, he compared the mean lengths of stay for the society with transition matrix $\mathbf{P}$ with those for the society with matrix $\mathbf{1}'\mathbf{p}$. For Glass and Hall's data the ratios of the former to the latter varied between 1.11 for class 4 and 1.59 in class 1.

2.4 SOME GENERALIZATIONS OF THE MARKOV MODEL

Time-dependent models

In this and the following sections we shall examine the effect of relaxing some of the assumptions on which the foregoing model depends. The first of these is that the transition probabilities are time-homogeneous. Rogoff's data suggest that this assumption may be reasonable in some societies, but it is of interest to point out that the generalization is straightforward.

Suppose that the transition matrix for the Tth generation is $\mathbf{P}(T)$. Equation (2.2) still holds with $\mathbf{P}$ replaced by $\mathbf{P}(T)$. Solving this equation recursively then gives, in place of (2.3),

$$\mathbf{p}(T) = \mathbf{p}(0) \prod_{i=1}^{T} \mathbf{P}(i). \qquad (2.35)$$

It is thus possible to investigate the effect of any specified changes in the transition matrix. In general, there will not be an equilibrium class structure and the complexity of the model makes it difficult to devise useful measures of mobility. However, one property of the simple Markov chain model is retained under this generalization. As T increases, the effect of the initial structure, $\mathbf{p}(0)$, on $\mathbf{p}(T)$ diminishes and, in the limit, vanishes. This condition rests on a general result about products of positive matrices to the effect that

$$\lim_{T \to \infty} \prod_{i=1}^{T} \mathbf{P}(i)$$

is a matrix with identical rows. Very wide conditions under which this holds are given in Hajnal (1976) who also gives references to earlier work. The result is true, in particular, if the elements of all the transition matrices are all strictly positive. A point to note is that if any member of the sequence $\{\mathbf{P}(T)\}$ has identical rows then the effect of $\mathbf{p}(0)$ and all prior members of the sequence on $\mathbf{p}(T)$ is wiped out.

A special kind of time dependence was envisaged by Prais (1955a). He postulated that the discrepancy between the class structure for fathers and sons revealed by Table 2.2 might be due to a change in the definitions of the classes. Thus, suppose that, in the past, the process was time-homogeneous and had attained an equilibrium structure like that observed for fathers. The changes which occur in the next generation are then supposed to take place in two parts. The first is a change of 'true' class governed by the time-homogeneous matrix $\mathbf{P}$; the second is a change in 'apparent' class resulting from a change in the system of classification. If the second transition has an associated matrix $\mathbf{R}(T)$ then the observed transition matrix would be $\mathbf{PR}(T)$, if it were possible to estimate $\mathbf{R}(T)$ from census or other information, it would be possible, in turn, to estimate $\mathbf{P}$ and so predict the development of the process. Prais (1955a) showed that a matrix $\mathbf{R}(T)$ could be constructed that would account for the observed difference in the class distributions of fathers and sons in Glass and Hall's data.

An apparent dependence of the transition matrix on time can arise due to heterogeneity in the population or to the non-Markovian character of the transitions, as will shortly appear. Hodge (1966), who discussed the time-dependent model given above, obtained results illustrating this point by investigating the n-step transition matrix.

A model allowing for a differential birth-rate

The most restrictive and unrealistic of our assumptions is that each father has exactly one son. We have used a rough argument to suggest that our model will be a reasonable approximation if each father has, on average, one son. In this section we shall develop the theory necessary to place these remarks on a firmer footing.

Suppose that the distribution of the number of sons born to a member of the jth class is $P_j(i)$, $(i = 0, 1, 2, \ldots)$ and assume that these probabilities are time-homogeneous. The dependence of the distribution on j enables us to introduce differential birth-rates between classes. We assume also that family sizes are independent Let the mean number of male offspring in the jth class be denoted by v_j. Under this model the population size will not remain constant from one generation to the next so we must work in terms of class numbers instead of proportions. Using a simple conditional probability argument based on conditional expectations we have

$$\bar{n}_j(T+1) = \sum_{i=1}^{k} \bar{n}_i(T)v_i p_{ij} \qquad (j = 1, 2, \ldots, k; T \geq 0), \qquad (2.36)$$

where $\bar{n}_i(0) = n_i(0)$ is the initial number in class i. This equation generates the expected class sizes in a manner analogous to (2.1). In order to make comparisons with the simple model we shall find it necessary to revert to proportions, writing

$$p_j(T) = \bar{n}_j(T) \bigg/ \sum_{i=1}^{k} \bar{n}_i(T) \qquad (j = 1, 2, \ldots, k).$$

It is not strictly accurate to use this notation because the ratio of expected values is not, in general, equal to the expected value of the proportion. However, the distinction is not important for the heuristic reasoning which follows and we shall refer to $p_j(T)$ as defined above as an expected proportion.

If we divide both sides of (2.36) by

$$\sum_{i=1}^{k} \bar{n}_i(T)$$

and use the fact that

$$\sum_{i=1}^{k} \bar{n}_i(T+1) = \sum_{i=1}^{k} \bar{n}_i(T)v_i$$

we obtain the following expression:

$$p_j(T+1) = \sum_{i=1}^{k} p_i(T)v_i p_{ij}/\bar{v}(T) \qquad (j = 1, 2, \ldots, k) \qquad (2.37)$$

where

$$\bar{v}(T) = \sum_{i=1}^{k} p_i(T)v_i \bigg/ \sum_{i=1}^{k} p_i(T).$$

This equation was given by Matras (1960b). When the classes have the same birth-rate, v_i is independent of i and (2.37) reduces to (2.1). We therefore conclude that, as far as expectations are concerned, the theory developed at the beginning of this section is still applicable when the single family line is replaced by a more realistic branching process.

The effect of a differential birth-rate can most easily be seen by writing $p_j(T)$ in the following form:

$$p_j(T) = \sum_{i=1}^{k} p_i(T-1)v_i p_{ij} \bigg/ \sum_{i=1}^{k} p_i(T-1)v_i. \qquad (2.38)$$

Consider first the perfectly mobile society in which $p_{ij} = p_j$. In this case (2.38) gives

$$p_j(T) = p_j.$$

Expressed in words, this means that the social structure of a perfectly mobile society is unaffected by a differential birth-rate between classes. This conclusion is intuitively obvious. By contrast the birth-rates are of crucial importance in an immobile society. In that case $p_{ij} = 1$ if $i = j$ and zero otherwise. From (2.36) we find

$$En_j(T+1) = v_j En_j(T)$$
$$= v_j^{T+1} n_j(0) \qquad (j = 1, 2, \ldots, k; \; T \geq 0).$$

The corresponding expression for $p_j(T)$ is

$$p_j(T) = v_j^T n_j(0) \left/ \sum_{i=1}^{k} v_i^T n_i(0) \right. \qquad (j = 1, 2, \ldots, k). \tag{2.39}$$

It is clear from the form of this equation that the class with the largest birth-rate will eventually dominate the population. In the limit as T tends to infinity we shall have $p_j(\infty) = 1$ if j refers to the class with the highest birth-rate and zero otherwise. For degrees of mobility intermediate between the two extremes which we have considered we would expect to find some tendency for classes with high birth-rates to increase in size relative to the others. Some idea of the extent to which this is possible can be deduced from (2.38). The right-hand side of this equation is a weighted average of the probabilities $p_{ij}(j = 1, 2, \ldots, k)$ with positive weights. Hence, it follows that

$$\min_i p_{ij} \leq p_j(T) \leq \max_i p_{ij} \qquad (j = 1, 2, \ldots, k) \tag{2.40}$$

for all T. Thus, however much the birth-rates of the classes may differ, the class structure is bounded by the inequalities (2.40). For example, using the data of Table 2.1, the expected proportion in class 5 (skilled manual and routine non-manual) can never fall below 0.140 or exceed 0.473 as long as the model remains valid.

The theory presented above is clearly capable of further development. Matrix methods have been used in demographic work for many years. Matras (1967) has pointed out that these can easily be adapted to include changes in social class as well as changes in the age structure of a population.

Heterogeneity in the transition probabilities

The Markov model of mobility provides a probabilistic description of the progress of a single family line. If all family lines have the same transition probabilities the theory will also describe the aggregate behaviour of a society, as we noted when analysing empirical data on mobility. However, in spite of the modest success we had in fitting the model, it seems rather unlikely that all family lines would have the same transition matrix. It is therefore of interest to see what the effects of differing matrices would be on the aggregate behaviour of the system. The same problem arises in intragenerational mobility, and it is in that connection that most of the work has been done. A basic reference is McFarland (1970); his work was followed by a numerical investigation of the effects of heterogeneity on the limiting structure by Morrison and co-workers (1971).

One way in which the effects of heterogeneity show themselves is in the structure of the T-step transition matrix. In a homogeneous Markov population this will be the Tth power of the one-step matrix. The nature of the departure from this expectation is the subject of the following analysis.

Suppose that a proportion $x_i(h)$ of the members of class i at time zero move according to the transition matrix $\mathbf{P}(h)$ $(h = 1, 2, 3, \ldots)$. Then the expected proportion of the whole population in class i who will move to class j between time 0 and time 1 is

$$p_{ij} = \sum_h x_i(h)p_{ij}(h) \qquad (i, j = 1, 2, \ldots, k). \tag{2.41a}$$

In matrix notation this may be written

$$\mathbf{P} = \sum_h \mathbf{X}(h)\mathbf{P}(h), \tag{2.41b}$$

where $\mathbf{X}(h)$ is a square diagonal matrix with (i, i)th element equal to $x_i(h)$. After T steps those who have the transition matrix $\mathbf{P}(h)$ will have moved according to $\mathbf{P}^T(h)$, and therefore the overall transition matrix for the T-step transition will be

$$\mathbf{P}^{(T)} = \sum_h \mathbf{X}(h)\mathbf{P}^T(h). \tag{2.42}$$

This matrix will approach a limit as $T \to \infty$ because each of the matrices $\mathbf{P}^T(h)$ does so. But unlike the single Markov chain the rows of $\mathbf{P}^{(\infty)}$ will not, in general, be equal because of the unequal weighting introduced by the matrices $\{\mathbf{X}(h)\}$.

If we were only able to observe the one-step matrix given by (2.41b) then the Markov model would lead us to expect that $\mathbf{P}^{(T)}$ would be

$$\mathbf{P}^T = \left\{ \sum_h \mathbf{X}(h)\mathbf{P}(h) \right\}^T. \tag{2.43}$$

The question which now arises has to do with how (2.43) differs from (2.42). A partial answer can be found numerically. To take a simple example, suppose that $k = 3$ and that h takes just two values with

$$\mathbf{P}(1) = \begin{pmatrix} 0.7 & 0.2 & 0.1 \\ 0.2 & 0.6 & 0.2 \\ 0.2 & 0.3 & 0.5 \end{pmatrix}, \qquad \mathbf{P}(2) = \begin{pmatrix} 0.5 & 0.3 & 0.2 \\ 0.3 & 0.4 & 0.3 \\ 0.3 & 0.4 & 0.3 \end{pmatrix}.$$

Further, suppose that $x_i(h) = \frac{1}{2}(h = 1, 2; i = 1, 2, \ldots, k)$. Then

$$\mathbf{P} = \frac{1}{2}[\mathbf{P}(1) + \mathbf{P}(2)] = \begin{pmatrix} 0.60 & 0.25 & 0.15 \\ 0.25 & 0.50 & 0.25 \\ 0.25 & 0.35 & 0.40 \end{pmatrix}$$

and

$$\mathbf{P}^2 = \begin{pmatrix} 0.460 & 0.328 & 0.213 \\ 0.338 & 0.400 & 0.263 \\ 0.338 & 0.378 & 0.285 \end{pmatrix}, \quad \mathbf{P}^4 = \begin{pmatrix} 0.394 & 0.362 & 0.244 \\ 0.379 & 0.370 & 0.252 \\ 0.379 & 0.369 & 0.252 \end{pmatrix}.$$

The two- and four-step matrices calculated from (2.42) for this example are

$$\mathbf{P}^{(2)} = \begin{pmatrix} 0.475 & 0.320 & 0.205 \\ 0.330 & 0.415 & 0.255 \\ 0.330 & 0.370 & 0.300 \end{pmatrix}, \quad \mathbf{P}^{(4)} = \begin{pmatrix} 0.407 & 0.357 & 0.236 \\ 0.375 & 0.375 & 0.250 \\ 0.375 & 0.371 & 0.254 \end{pmatrix}.$$

For this particular population the Markov model underestimates the proportions that will remain in the same grade over both two and four time periods. Is this true in general or is it peculiar to this example?

We shall show that this characteristic is typical but not universal and try to give some insight into the features of the matrices $\{\mathbf{P}(h)\}$ which produce this phenomenon.

To simplify the analysis let $x_i(h) = x(h)$ for all i; it will be sufficient to consider the case $T = 2$. We shall compare the (j, j)th element of (2.42) with that of (2.43). From (2.42)

$$p_{jj}^{(2)} = \sum_h x(h) \sum_{l=1}^{k} p_{jl}(h)p_{lj}(h) \tag{2.44}$$

and from (2.43)

$$p_{jj}^{2*} = \sum_{l=1}^{k} \left\{ \sum_h x(h)p_{jl}(h) \right\} \left\{ \sum_h x(h)p_{lj}(h) \right\} \tag{2.45}$$

where p_{jj}^{2*} denotes the (j, j)th element of $\mathbf{P}^2$. The difference $p_{jj}^{(2)} - p_{jj}^{2*}$ may thus be written:

$$\sum_{l=1}^{k} \sum_h x(h) \left\{ p_{jl}(h) - \sum_m x(m)p_{jl}(m) \right\} \left\{ p_{lj}(h) - \sum_m x(m)p_{lj}(m) \right\}$$

$$= \sum_{l=1}^{k} \text{cov}(p_{jl}, p_{lj}), \tag{2.46}$$

where the covariances are calculated with respect to the distribution $\{\mathbf{x}(m)\}$. The conditions under which the diagonal elements of $\mathbf{P}^{(2)}$ will be larger than those of $\mathbf{P}^2$ are thus related to the correlation between the flows in opposite directions between each pair of classes. Thus if a large flow (proportionately speaking) from class j to l tends to be accompanied by a large flow in the reverse direction for most pairs of classes the phenomenon observed will occur. One set of conditions under which (2.46) is certainly positive is when the matrices are symmetric, that is with $p_{jl}(h) = p_{lj}(h)$ $(j, l = 1, 2, \ldots, k)$. Matrices having this property are called doubly stochastic because their columns as well as their rows add up to one. The limiting structure for such matrices has equal proportions in each category. Another example, which arises in the next section, occurs when $p_{jl}(h) = p_{lj}(h)p_l(h)/p_j(h)$ $(l, j = 1, 2, \ldots, k)$ where $p_l(h)$ and $p_j(h)$ are the limiting probabilities associated with $\mathbf{P}(h)$; the Glass and Hall matrix very nearly satisfies this condition.

An example in which the diagonal elements do not differ in this systematic way is provided by

$$\mathbf{P}(1) = \begin{pmatrix} 0.5 & 0.4 & 0.1 \\ 0.2 & 0.5 & 0.3 \\ 0.2 & 0.3 & 0.5 \end{pmatrix}, \quad \mathbf{P}(2) = \begin{pmatrix} 0.5 & 0.1 & 0.4 \\ 0.3 & 0.5 & 0.2 \\ 0.3 & 0.2 & 0.5 \end{pmatrix}$$

with $x_i(h) = \frac{1}{2}(h = 1, 2; i = 1, 2, 3)$. In this case

$$\mathbf{P} = \begin{pmatrix} 0.50 & 0.25 & 0.25 \\ 0.25 & 0.50 & 0.25 \\ 0.25 & 0.25 & 0.50 \end{pmatrix} \quad \text{and} \quad \mathbf{P}^2 = \begin{pmatrix} 0.375 & 0.312 & 0.312 \\ 0.312 & 0.375 & 0.312 \\ 0.312 & 0.312 & 0.375 \end{pmatrix}.$$

But

$$\mathbf{P}^{(2)} = \frac{1}{2}\begin{pmatrix} 0.35 & 0.43 & 0.22 \\ 0.26 & 0.42 & 0.32 \\ 0.26 & 0.38 & 0.36 \end{pmatrix} + \frac{1}{2}\begin{pmatrix} 0.40 & 0.18 & 0.42 \\ 0.36 & 0.32 & 0.32 \\ 0.36 & 0.23 & 0.41 \end{pmatrix}$$

$$= \begin{pmatrix} 0.375 & 0.305 & 0.320 \\ 0.310 & 0.370 & 0.320 \\ 0.310 & 0.305 & 0.385 \end{pmatrix}.$$

Even with matrices having such different structures as the two in this example, the two-step matrices do not differ very much. The reader will find it quite difficult to construct examples in which $\mathbf{P}^{(2)}$ has smaller diagonal elements than $\mathbf{P}^2$.

One particular special case of the model incorporating differing transition matrices has received a great deal of attention. This is the so-called 'mover–stayer' model which arises if $h = 1, 2$ and $\mathbf{P}(1) = \mathbf{I}$, the unit matrix, and $\mathbf{P}(2) = \mathbf{P}_M$, an arbitrary transition matrix. According to this model some members of the population never move, whereas the remainder change state as in the Markov model. This is perhaps more plausible when considering occupational mobility or migration rather than social class and we shall meet a continuous time version in Chapter 5. The model has been fitted to labour mobility data by Blumen and co-workers (1955) and to British mobility data by Wynn and Sales (1973b). McCall (1971) has used it as a model of income dynamics.

Intuitively, one would expect to observe clustering on the main diagonal of the

T-step matrix in the mover–stayer model because a certain proportion of each class never move. Whether or not this is the case can be investigated by a method introduced by Singer and Spilerman (1977b and 1979). They adopt a mathematically more tractable definition of clustering than the one implicit in our analysis given above. We were concerned with whether all the elements on the diagonal of $\mathbf{P}^{(T)}$ exceeded those of $\mathbf{P}^T$. Singer and Spilerman (1977b and 1979) define clustering to occur whenever trace $\mathbf{P}^{(T)} \geq$ trace $\mathbf{P}^T$; that is the sums, rather than the individual diagonal elements, have to satisfy the inequality. This is obviously a weaker requirement since the trace inequality can hold even when there are reversals in individual elements. Thus in the example above trace $\mathbf{P}^{(2)} = 1.130$ and trace $\mathbf{P}^2 = 1.125$. This would show evidence of clustering according to Singer and Spilerman's definition, whereas we have used it as a counter-example because a reversal occurred in the second position. Nevertheless, if all the diagonal elements of $\mathbf{P}^{(T)}$ exceed those of $\mathbf{P}^T$ then the trace inequality will hold and an understanding of the conditions under which the latter holds will certainly give insight into the phenomenon of clustering.

The point of adopting the trace inequality for clustering is that the trace of $\mathbf{P}^T$ is equal to the sum of the Tth powers of its eigenvalues. This fact enables us to translate the clustering criterion into a statement about eigenvalues. To illustrate the potential of the method we consider the special case of the mover–stayer model in which the same proportion, s, of each class are stayers. We then have

$$\mathbf{P} = s\mathbf{I} + (1 - s)\mathbf{P}_M \qquad (2.47)$$

It follows that the eigenvalues of $\mathbf{P}$ are $s + (1 - s)\theta_i$ ($i = 1, 2, \ldots, k$), where $\{\theta_i\}$ are the eigenvalues of $\mathbf{P}_M$. Hence

$$\text{trace } \mathbf{P}^T = \sum_{i=1}^{k} \{s + (1 - s)\theta_i\}^T \qquad (2.48)$$

For the mover–stayer model, the actual T-step transition matrix is

$$\mathbf{P}^{(T)} = s\mathbf{I} + (1 - s)\mathbf{P}_M^T \qquad (2.49)$$

for which

$$\text{trace } \mathbf{P}^{(T)} = \sum_{i=1}^{k} \{s + (1 - s)\theta_i^T\}. \qquad (2.50)$$

The trace inequality therefore holds if

$$\sum_{i=1}^{k} \left[s + (1 - s)\theta_i^T - \{s + (1 - s)\theta_i\}^T \right] > 0. \qquad (2.51)$$

A sufficient condition for this inequality to hold is that each term in the sum is positive. If θ_i is positive, real, and not equal to one this result follows by Jensen's inequality. Transition matrices with real, positive eigenvalues appear to be fairly common in practice, and thus we might expect clustering to occur with the mover–stayer model.

The trace inequality may, of course, hold when some of the eigenvalues are negative or imaginary and Singer and Spilerman (1977b and 1979) have investigated the position in some detail. They were also able to show that the trace inequality could be reversed as the following example shows. Suppose that $k = 4$ and let

$$\mathbf{P}_M = \begin{pmatrix} 0 & \frac{1}{4} & \frac{3}{4} & 0 \\ \frac{1}{4} & 0 & 0 & \frac{3}{4} \\ \frac{1}{2} & 0 & 0 & \frac{1}{2} \\ 0 & \frac{1}{2} & \frac{1}{2} & 0 \end{pmatrix}$$

Then for $s < 0.342$ there will be under-prediction of the diagonal elements.

We conclude from the foregoing analysis that there is no simple and certain way of diagnosing heterogeneity by inspection of $\mathbf{P}^{(T)}$ and $\mathbf{P}^T$. An excess of probability on the diagonal of $\mathbf{P}^{(T)}$ certainly suggests the strong possibility of heterogeneity in the population. However, we shall see later that this is not the only possible explanation and the counter-examples warn us that heterogeneity can be present even when there is no clustering on the diagonal.

Cumulative inertia models

We have already seen how heterogeneity in the transition matrix could explain the occurrence of larger values on the main diagonal of a T-step transition matrix than predicted by Markov theory. This is not the only type of generalization which is capable of producing this phenomenon. An alternative explanation is provided by a model due originally to McGinnis (1968). He introduced the idea of 'cumulative inertia', which describes the situation when an individual's chance of changing class declines with increasing length of stay in that class. Empirical evidence in support of such a hypothesis has been provided by Myers and co-workers (1967), Morrison (1967), and Land (1969), for the case of intragenerational mobility. It has some plausibility in the intergenerational case also. At first sight this assumption destroys the Markov property since the chance of moving now has to depend on past history as well as the present state. This difficulty can be circumvented by redefining the states in a manner which is widely used in the theory of stochastic processes.

Instead of defining the states of the Markov chain in terms of classes (or occupational groups) they are now defined in terms of both class and length of stay in the class. Length of stay is measured in units of the time interval between changes of state and the members of a given class are subclassified according to their length of stay. Each state is thus described by a pair of numbers—the class and the duration of service. The process defined on this set of states is now a Markov chain because we have made length of stay part of the description of the current state. There is nothing in the specification so far to prevent an indefinitely long stay in a given grade, which means that we must allow for an infinite number of states within each class. This takes us out of the realm of finite Markov chains but, at the theoretical level, this presents no insuperable problems. In practice, it is convenient to work with finite Markov chains having a fairly small number of states. Henry and co-workers (1971) suggested a modification of the McGinnis model which achieves this objective. Each class is divided into groups according to length of service, as before, but now all length of service groups above some chosen level are combined into a single group. Thus, for example, the length of service categories might be one, two, three, four, and 'more than four' years.

A transition matrix for such a modified model would have the following form, where X stands for a non-zero entry:

```
 0 X . . . . 0 | X 0 . . . . 0 | X 0 . . . . 0
 . 0 X       . | X 0 . . . . 0 | X 0 . . . . 0
 .   .   .   . |   .     .   . |   .     .   .
 .     . 0 X . |   .     .   . |   .     .   .
 0 . . . . 0 X | X 0 . . . . 0 | X 0 . . . . 0
---------------+---------------+---------------
 X 0 . . . . 0 | 0 X . . . . 0 | X 0 . . . . 0
 X 0 . . . . 0 | . 0 X       . | X 0 . . . . 0
 .   .   .   . |   .     .   . |   .     .   .
 .   .   .   . |     . 0 X . . |   .     .   .
 X 0 . . . . 0 | 0 . . . . 0 X | X 0 . . . . 0
---------------+---------------+---------------
 X 0 . . . . 0 | X 0 . . . . 0 | 0 X . . . . 0
 X 0 . . . . 0 | X 0 . . . . 0 | . 0 X       .
 .   .   .   . |   .     .   . |   .     .   .
 .   .   .   . |   .     .   . |         . 0 X
 X 0 . . . . 0 | X 0 . . . . 0 | 0 . . . . 0 X
```

In a typical row we have the probabilities that an individual who has been in a given class for a certain length of time will move to another state. He either stays in the same class and moves up to the next higher length of service group or he moves to the lowest length of service category of another class. The exception to this pattern is in the last length of service category of each class. In this case someone remaining in the class stays in the same category and so there is a non-zero entry on the diagonal. Various assumptions are possible about the value of these entries. They can be unity, meaning that a person reaching that length of stay remains in the grade for ever; they can be zero if a person must move on reaching that level; or they can have some intermediate value. This last possibility is equivalent to a constant staying probability for all lengths of service beyond the lower limit of the final category. McGinnis' postulate of cumulative inertia requires that the super-diagonal elements in each diagonal submatrix form a non-decreasing sequence. The idea of incorporating the effect of increasing seniority into the Markov model in this way is widely used in manpower planning, using an extended version of the model to be discussed in Chapter 3. For further discussion of this point see, also, Bartholomew (1971).

The long-run behaviour of the cumulative inertia model depends on the structure of the extended transition matrix. If absorption in any class is possible (that is if there is a non-zero chance that someone in a class will never move out) the ultimate structure will depend on the initial structure. The theory of absorbing Markov chains can be used to deal with such cases. In the contrary case the matrix is regular and a limiting structure exists which is independent of the initial structure.

In order to demonstrate that cumulative inertia can produce larger diagonal elements in the T-step transition matrix than would be predicted on Markov theory we consider a simple example. In a two-class system the members of each class are divided up according to whether they have less or more than one year of service. The four states of the system are then (class 1, < 1 year's service), (class 1, > 1 year's service), (class 2, < 1 year's service) and (class 2, > 1 year's service). Suppose that the transition matrix for annual movements is

$$\mathbf{P} = \begin{pmatrix} 0 & 0.6 & 0.4 & 0 \\ 0 & 0.8 & 0.2 & 0 \\ \hline 0.3 & 0 & 0 & 0.7 \\ 0.1 & 0 & 0 & 0.9 \end{pmatrix}. \tag{2.52}$$

The dotted lines indicate the class boundaries. This matrix has the same form as the general case shown above, and the chance of staying in a class also increases as length of stay increases. Suppose further that, initially, there are equal numbers in each state. Then someone observing the transitions over one time period, but ignoring length of stay, will record the one-step matrix for the two classes as

$$\mathbf{P}^* = \begin{pmatrix} \frac{1}{2}(0.6+0.8) & \frac{1}{2}(0.4+0.2) \\ \frac{1}{2}(0.3+0.1) & \frac{1}{2}(0.7+0.9) \end{pmatrix} = \begin{pmatrix} 0.7 & 0.3 \\ 0.2 & 0.8 \end{pmatrix}.$$

His estimate of the two-step matrix will then be

$$(\mathbf{P}^*)^2 = \begin{pmatrix} 0.55 & 0.45 \\ 0.30 & 0.70 \end{pmatrix}. \tag{2.53}$$

However, the actual transitions will be governed by (2.52) for which

$$\mathbf{P}^2 = \begin{pmatrix} 0.12 & 0.48 & 0.12 & 0.28 \\ 0.06 & 0.64 & 0.16 & 0.14 \\ \hline 0.07 & 0.18 & 0.12 & 0.63 \\ 0.07 & 0.06 & 0.04 & 0.81 \end{pmatrix}. \tag{2.54}$$

For someone who fails to distinguish the length of service categories the observed two-step matrix between classes will be

$$(\mathbf{P}^2)^* = \begin{pmatrix} \frac{1}{2}(0.12+0.48+0.06+0.64) & \frac{1}{2}(0.12+0.28+0.16+0.14) \\ \frac{1}{2}(0.07+0.18+0.09+0.06) & \frac{1}{2}(0.12+0.63+0.04+0.81) \end{pmatrix}$$

$$= \begin{pmatrix} 0.65 & 0.35 \\ 0.20 & 0.80 \end{pmatrix}. \tag{2.55}$$

When compared with (2.53) we see that the incorporation of cumulative inertia has the effect of producing the higher diagonal elements which the theory was designed to explain.

2.5 INTERACTIVE MARKOV CHAIN MODELS

The basic Markov model is essentially concerned with the behaviour of a single entity. The behaviour of the population is then deduced by assuming that individuals act independently of one another. It was this assumption which has enabled us to interpret probabilities as expected proportions. In practice, individuals interact by observing one another's behaviour and being influenced by what they see. The alleged influence of opinion polls on subsequent voting behaviour at elections is an example of a situation in which the perceived state of the population may influence the behaviour of individuals. Buying behaviour, fashions and the formation of popular attitudes are other fields where similar influences are at work. Our aim in this section is to obtain some understanding of the broad qualitative effects of interaction between individuals.

The class of models to be described appears to have been first suggested by Matras (1967). He proposed that the transition probabilities between T and $T+1$ should be made to depend on the numbers in the classes at time T. This is, perhaps, not very realistic in the context of social mobility, but it is easy to imagine in migration studies that the popularity of a region might depend on the size of its population. The term 'interactive' model is due to Conlisk (1976 and 1978) who developed much of the existing theory. Our treatment largely follows his.

In order to specify the class of models to be discussed we must start with $\mathbf{n}(T)$ the vector giving the numbers in the classes at time T. We then suppose that the elements of the transition matrix for the interval $(T, T+1)$ depend on the observed value of $\mathbf{n}(T)$. The expectation of $\mathbf{n}(T+1)$ given $\mathbf{n}(T)$ can then be written

$$E\{\mathbf{n}(T+1)|\mathbf{n}(T)\} = \mathbf{n}(T)\mathbf{P}\{\mathbf{n}(T)\}. \tag{2.56}$$

The unconditional expectation is therefore

$$\bar{\mathbf{n}}(T+1) = E\mathbf{n}(T)\mathbf{P}\{\mathbf{n}(T)\}. \tag{2.57}$$

The expectation on the right-hand side of (2.57) involves a non-linear function of $\mathbf{n}(T)$ and hence its determination is not a straightforward matter. Some progress can be made, but here we shall use the deterministic approximation to the model proposed by Conlisk (1976). This involves replacing $\mathbf{n}(T)$ in $\mathbf{P}\{\mathbf{n}(T)\}$ by $\bar{\mathbf{n}}(T)$. As an approximation to (2.57) we therefore have

$$\bar{\mathbf{n}}(T+1) = \bar{\mathbf{n}}(T)\mathbf{P}\{\bar{\mathbf{n}}(T)\}. \tag{2.58}$$

An alternative way of justifying this step is to suppose that individuals are influenced not by the actual numbers but by those expected. This is unrealistic, but in large populations we would expect the effect of the approximation to be slight. The question of the asymptotic convergence of the two cases has been investigated by Brummelle and Gerchak (1980) and Lehoczky (1980).

We shall now revert to expected proportions, writing $\mathbf{p}(T) = \bar{\mathbf{n}}(T)/N$, and develop the analysis from the equation in the form:

$$\mathbf{p}(T+1) = \mathbf{p}(T)\mathbf{P}\{\mathbf{p}(T)\}. \tag{2.59}$$

The first thing to notice about the process is that it is a time-dependent Markov chain. This enables us to deduce, for example, that the long-run structure will not depend on $\mathbf{p}(0)$ if all the matrices $\mathbf{P}\{\mathbf{p}(T)\}$ satisfy the appropriate conditions. Beyond this little can be achieved so we consider the analysis of (2.59) from first principles.

It is clear that some restrictions will have to be imposed on the function $\mathbf{P}\{\mathbf{p}(T)\}$ in order to ensure that it is a stochastic matrix. Once this is done there is no difficulty in computing the expected structures recursively. We would like to go beyond this and identify what kinds of behaviour are possible within the framework provided by (2.59). In particular it would be useful to know in what ways an interactive model differs from the simple Markov chain especially as $T \to \infty$.

One simple result which follows from (2.59) is that if $\mathbf{P}(\cdot)$ is a continuous, single-valued function then there exists at least one stationary structure satisfying the equation

$$\mathbf{p} = \mathbf{p}\mathbf{P}(\mathbf{p}). \tag{2.60}$$

This follows from the 'fixed point' theorem of Brouwer which says that in a mapping of a space into itself there is at least one point which remains unchanged. There may be many such points, of course, and the theorem says nothing about their stability. A stable equilibrium is one such that if the structure is slightly perturbed the system will eventually return to that structure. Otherwise a structure is unstable. For a regular Markov chain there is one stable equilibrium, but with interactive models both kinds of equilibrium may exist in varying numbers.

The full analysis of systems governed by (2.59) is complicated and much remains to be done. However, there are some interesting special cases for which a fairly complete analysis is possible. We shall give three examples, based on Conlisk's work, with a view to discovering the main qualitative characteristics of interactive systems.

An attraction/repulsion model

This is, perhaps, the simplest special case to be considered. It allows the attractiveness of a state to depend linearly on its current size; that is, the larger the number in the state, the greater the probability of moving to that state from anywhere else. The model takes the form:

$$p_{ij}\{\mathbf{p}(T)\} = a_{ij} + p_j(T)b_i \qquad (i, j = 1, 2, \ldots k). \qquad (2.61)$$

The parameters must be restricted to ensure that the probabilities are non-negative and add up to one by rows whatever $\mathbf{p}(T)$. The condition for positivity is

$$a_{ij} \geq 0, \qquad b_i > -\max_i a_{ij}.$$

The sum of the ith row is

$$\sum_{j=1}^{k} p_{ij}\{\mathbf{p}(T)\} = \sum_{j=1}^{k} a_{ij} + b_i = 1, \qquad (i = 1, 2, \ldots k). \qquad (2.62)$$

Using this equation we can eliminate b_i from the model and re-express the condition for positivity as

$$0 \leq \sum_{\substack{h=1 \\ h \neq i}}^{k} a_{ih} \leq 1, \qquad (i = 1, 2, \ldots k). \qquad (2.63)$$

If $b_i > 0$ (i.e. $\Sigma_{j=1}^{k} a_{ij} < 1$) we have a model in which the attraction of a class increases with its size. If $b_i < 0$, without violating (2.63), size has a deterrent effect.

In the special case $b_1 = b_2 = \ldots = b_k = b$ the model reduces to a Markov chain and hence can be handled by the methods of earlier sections. To see this we go back to (2.59) which becomes

$$\mathbf{p}(T+1) = \mathbf{p}(T)\{\mathbf{A} + b\mathbf{1}'\mathbf{p}(T)\} = \mathbf{p}(T)\{\mathbf{A} + b\mathbf{I}\} \qquad (2.64)$$

where $\mathbf{A} = \{a_{ij}\}$. The system thus behaves as if it had the constant transition matrix $\mathbf{P} = \mathbf{A} + b\mathbf{I}$.

In the general case we have

$$\mathbf{p}(T+1) = \mathbf{p}(T)\{\mathbf{A} + \mathbf{B}\mathbf{1}'\mathbf{p}(T)\} = \mathbf{p}(T)\{\mathbf{A} + (\mathbf{I} - \mathbf{A})\mathbf{1}'\mathbf{p}(T)\} \qquad (2.65)$$

where $\mathbf{B} = \text{diag}\{b_i\}$. In order to explore the characteristics of such a system, without undue mathematical complexity, we first consider the case $k = 2$. Here it suffices to consider $p_1(T)$ alone since $p_2(T) = 1 - p_1(T)$. From (2.65)

$$p_1(T+1) = a_{21} + p_1(T)\{1 + a_{11} - 2a_{21} - a_{22}\} + p_1^2(T)\{a_{22} + a_{21} - a_{12} - a_{11}\}. \qquad (2.66)$$

The behaviour of the system can be deduced by plotting $p_1(T+1)$ against $p_1(T)$ as in Figure 2.1. Obviously when $p_1(T) = 0$, $p_1(T+1) = a_{21}$ and when $p_1(T) = 1$,

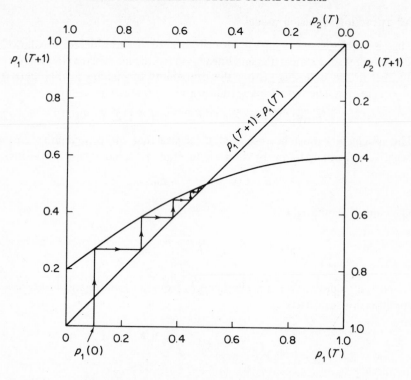

Figure 2.1 The graph of $\mathbf{p}(T+1) = \mathbf{p}(T)\mathbf{P}\{\mathbf{p}(T)\}$ illustrating the approach to the steady state when $k = 2$.

$p_1(T+1) = 1 - a_{12}$. The restrictions on $\mathbf{A}$ ensure that the whole curve lies in the unit square. Any stationary values must occur at points where the curve given by (2.66) cuts the diagonal $p_1(T+1) = p_1(T)$. The case illustrated has $\mathbf{A}$ given by

$$\mathbf{A} = \begin{pmatrix} 0.5 & 0.4 \\ 0.2 & 0.3 \end{pmatrix}$$

The transient behaviour of the system can also be deduced from such a graph as follows. Given $p_1(0)$ we can read off $p_1(1)$. Next, setting $T = 1$ we use the graph to find $p_1(2)$ and so on. This sequence can be traced out as a series of steps as illustrated on the figure; the arrows mark the direction of movement. It will be clear that the system will always converge on the point $(\frac{1}{2}, \frac{1}{2})$ whatever the starting structure. If $a_{21} = 0$ or $a_{12} = 1$ the stationary point will be $(1, 0)$ or $(1, 1)$ respectively. If $a_{21} = 1$ and $a_{12} = 0$ with $a_{11} = a_{22}$, say, then $p_1(T+1) = 1 - p_1(T)$ and the system goes into a limit cycle with $p_1(T)$ oscillating between $p_1(0)$ and $1 - p_1(0)$.

In the case of a Markov chain a similar analysis would make the curve on Figure 2.1 a straight line, but otherwise there is no essential difference between the two models. In general both approach a unique equilibrium which is independent of the starting structure.

The same conclusion applies for larger k. Conlisk (1976) has proved that there is a unique stationary vector $\mathbf{p}$ and he later showed (Conlisk, 1978) that the system will converge to $\mathbf{p}$ whatever the starting point provided that $b_i \geq 0$ (all i) and that $\mathbf{A}$ is primitive ($\mathbf{A}^T > 0$ for some T). We know that convergence also occurs with negative b's if $k > 2$, but no general proof is available in the general case. Positive b's arise in the 'attraction' version of the model. The stationary structure can be obtained as follows. Putting $\mathbf{p}(T + 1) = \mathbf{p}(T) = \mathbf{p}$ in (2.65) gives

$$\mathbf{p}(\mathbf{I} - \mathbf{A}) = \mathbf{p}(\mathbf{I} - \mathbf{A})\mathbf{1}'\mathbf{p}$$

which is satisfied if

$$\mathbf{p}\mathbf{A} = \mathbf{p}\mathbf{A}\mathbf{1}'\mathbf{p}. \tag{2.67}$$

Any vector $\mathbf{p}$ which also satisfies $\mathbf{p}\mathbf{A} = \theta\mathbf{p}$ for some θ will also satisfy (2.67). There is such a vector whenever θ is an eigenvalue of $\mathbf{A}$. The only one which meets the requirement of having all elements non-negative is that associated with the largest eigenvalue of $\mathbf{A}$. Hence the unique solution of (2.67) is the lefthand eigenvalue of $\mathbf{A}$, corresponding to the largest eigenvalue, scaled so that its elements sum to one.

A dissatisfaction model

The foregoing model assumes that individuals know (or can estimate) the numbers in all classes, including their own. It is easy to envisage circumstances in which each individual responds only to the size of his own class. Suppose, for example, that we consider the distribution of customers between suppliers of some commodity or service (garages, dentists, retail stores, etc.). If the number of customers using a given supplier increases then, in the short term, the quality of service is likely to decrease because a fixed service facility has to supply a larger number of customers. Poorer service is likely to increase the probability of the customer looking for another supplier. In other applications increasing size may have the opposite effect. A larger club or society will offer a wider range of facilities to its members and so make them less likely to go elsewhere. Both cases are covered by modifying the linear model of (2.61) so that only the size of the individual's own class affects the probability of moving. We do this by choosing

$$p_{ij} \propto a_{ij} \qquad (i \neq j)$$
$$p_{ii} \propto a_{ii} + b_i p_i(T). \tag{2.68}$$

The condition for positivity is $b_i \leq -a_{ii}(i = 1, 2, \ldots k)$ and the requirement that the row sums be unity gives

$$p_{ij} = \frac{a_{ij} + \delta_{ij}b_i p_i(T)}{\sum_{j=1}^{k} a_{ij} + b_i p_i(T)} \qquad (i, j = 1, 2, \ldots k), \tag{2.69}$$

where δ_{ij} is the Kronecker delta function. A negative b_i indicates that the larger the class the more likely a person is to leave it; the magnitude of b_i is thus an indicator of the unattractiveness of size.

When $k = 2$,

$$p_1(T+1) = p_1(T)\left(1 - \frac{a_{12}}{a_{11} + a_{12} + b_1 p_1(T)}\right)$$

$$+ \{1 - p_1(T)\}\left(\frac{a_{21}}{a_{21} + a_{22} + b_2\{1 - p_1(T)\}}\right) \qquad (2.70)$$

and any stationary value will thus satisfy

$$p a_{12}\{a_{21} + a_{22} + b_2(1-p)\} = (1-p)a_{21}\{a_{11} + a_{22} + b_1 p\}. \qquad (2.71)$$

It may easily be shown that this equation has only one root in $(0, 1)$ and a diagram like Figure 2.1 will show that the sequence $\{p_1(T)\}$ will converge to that limiting value from any starting value. To illustrate the model let

$$\mathbf{A} = \begin{pmatrix} 0.6 & 0.3 \\ 0.5 & 0.5 \end{pmatrix} \qquad \mathbf{b} = (-0.4, -0.4)$$

then

$$p_1(T+1) = (0.45 - 0.29p_1 + 0.20p_1^2 - 0.16p_1^3)/(0.54 + 0.12p_1 - 0.16p_1^2) \qquad (2.72)$$

where we have written $p_1 = p_1(T)$. The equilibrium value is $p = 0.5739$. Examination of the graph of $p_1(T+1)$ against $p_1(T)$ will show that if $p_1(0) > 0.5739$ then $p_1(1) < p_1(0)$ as we would expect from the deterrent effect of a negative b_1. After the first step $p_1(T)$ will always lie within the interval $(0.40, 0.83)$.

For general k, Conlisk (1976) was able to show that a unique equilibrium existed, though a proof that it will be reached from any starting point is lacking. He also gives a method of finding the equilibrium. To do this we first form a matrix $\mathbf{A}_0$ obtained by dividing the elements in each row of $\mathbf{A}$ by the corresponding row sum; $\mathbf{A}_0$ will thus be a stochastic matrix. If it is regular it will have a stationary structure which we denote by $\mathbf{a}_0$. The equilibrium vector, $\mathbf{p}$, required is then given by

$$p_i = A_i a_{0i}/(\mu A_i - b_i a_{0i}), \qquad (i = 1, 2, \ldots k), \qquad (2.73)$$

where $A_i = \sum_{j=1}^{k} a_{ij}$ and μ is the unique positive constant necessary to make $\sum_{i=1}^{k} p_i = 1$. When $k = 2$ it may be verified that this formula leads to the same answer as the direct method for the example given above. In general, μ has to be obtained by numerical methods.

A voting model

In the two interactive models we have investigated so far we have failed to find any marked difference from the simple Markov chain. In both cases the system typically approached a steady state independent of the initial structure. The computation of the path to equilibrium was slightly more complicated but there was no feature that was distinctive. We next turn to a voting model, also due to Conlisk (1976), which exhibits entirely new features.

The idea can be explained by reference to a system in which successive votes are taken and where, on each occasion, individuals are required to vote for one of k candidates, or propositions. After each vote has been taken individuals decide how to cast their next vote in the light of each candidate's current support. According to the model this takes place in two stages. At the first stage the individual considers whether or not to reconsider his position. The second stage only arises if a decision to change has already been made. He then has to decide which of the candidates will receive his vote in the next round. The main question now is whether the voting proportions will converge and, if so, what the final position will be. In Conlisk's model the relevant probabilities at the two stages are as follows:

Pr{individual in i decides not to reconsider} $= \alpha_i + \beta_i p_i \, (0 < \alpha_i < \alpha_i + \beta_i < 1)$
Pr{individual in i votes for j|decision to reconsider} $\propto p_j$.

Thus the individual is influenced by the existing support for his candidate in deciding whether to reconsider; having decided to change he tends to favour candidates who are already well supported. For such a system the transition probabilities are:

$$p_{ii}(\mathbf{p}(T)) = \alpha_i + \beta_i p_i(T) + \{1 - \alpha_i - \beta_i p_i(T)\} p_i(T)$$
$$p_{ij}(\mathbf{p}(T)) = \{1 - \alpha_i - \beta_i p_i(T)\} p_j(T) \qquad (i \neq j). \tag{2.74}$$

It is difficult to carry out a full analysis, but much can be learnt from the case $k = 2$. Here we have

$$p_1(T+1) = p_1(T)(1 - \beta_2 - \alpha_2 + \alpha_1) + p_1^2(T)\,(\beta_1 + 2\beta_2 - \alpha_1 + \alpha_2)$$
$$- (\beta_1 + \beta_2) p_1^3(T). \tag{2.75}$$

The points of equilibrium therefore satisfy the equation

$$(\beta_1 + \beta_2)p^3 - (\beta_1 + 2\beta_2 - \alpha_1 + \alpha_2)p^2 + (\beta_2 - \alpha_1 + \alpha_2)p = 0. \tag{2.76}$$

It is easily verified that $p = 0$ and $p = 1$ are both roots of this equation and hence that the remaining root is

$$p = (\beta_2 + \alpha_2 - \alpha_1)/(\beta_1 + \beta_2). \tag{2.77}$$

There are two cases to consider:

(a) if $\beta_2 + \alpha_2 > \alpha_1$ and if $\beta_1 + \alpha_1 > \alpha_2$, then $0 < p < 1$;
(b) otherwise there is no interior stationary point.

In case (a) the graph of $p_1(T+1)$ is as illustrated in the left-hand half of Figure 2.2. It appears from this diagram that the interior equilibrium is unstable. Unless

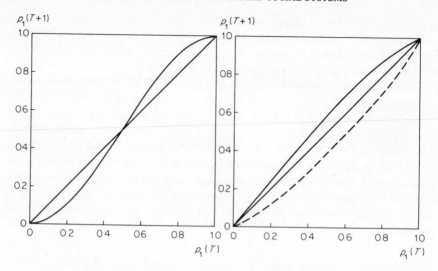

Figure 2.2   Behaviour of the voting model when $k = 2$.

$p_1(0) = p$ the structure will converge to one or other of the extremes. Unlike the earlier models the limiting state is here determined by the initial state and one candidate ultimately obtains all the votes. The final outcome depends solely on whether the initial vote for candidate 1 exceeds the critical value given by (2.77). If, for example, $\alpha_1 = \alpha_2$, $\beta_1 = \frac{1}{3}$, $\beta_2 = \frac{2}{3}$ the critical value of p is $\frac{2}{3}$. Thus candidate 1 needs to get at least two-thirds of the vote on the first round to become the eventual winner. The system is biased against him in the sense that candidate 2's initial supporters are more loyal (i.e. they have a larger β). In case (b) the curve will take one of the two forms shown on the right-hand part of Figure 2.2. The solid curve arises if $\beta_2 + \alpha_2 \leq \alpha_1$, in which case candidate 1 will always win as long as he gets at least one vote in the first round. If $\beta_1 + \alpha_1 \leq \alpha_2$ the broken curve applies and candidate 2 always wins. Case (b) cannot arise unless $\alpha_1 \neq \alpha_2$.

     Some light has been thrown on the case $k > 2$ by Conlisk (1976) though the situation remains obscure. Suppose we label the states such that $\alpha_1 \geq \alpha_2 \geq \ldots \geq \alpha_k$. Then Conlisk showed that $\mathbf{p} = \mathbf{e}_i$ ($i = 1, 2, \ldots k$) are all stationary structures ($\mathbf{e}_i$ is a vector with 1 in the ith position and zeros elsewhere). Further, $\mathbf{p} = \mathbf{e}_i$ is locally stable if $\alpha_i + \beta_i > \alpha_1$; this implies, in particular, that $\mathbf{p} = \mathbf{e}_1$ is always stable. If this were the only stable equilibrium (as in case (b) above) then candidate 1 could ultimately obtain all the votes provided only that $p_1(0) > 0$. This is not a complete account but it does show the possibility of converging on an extreme point. It leaves open the question of whether there are any stable interior equilibria, though we know they cannot exist when $k = 2$.

CHAPTER 3

Markov Models for Open Systems

3.1 INTRODUCTION

The social systems envisaged in Chapter 2 were closed. This meant either that no members moved in or out or that any losses were immediately made good. Our interest thus centred on the changing internal structure of the system. The assumption of a closed system is often reasonable for applications to social and occupational mobility, but there are other situations in which gains and losses are an important feature of the process. In this chapter, therefore, we shall discuss two generalizations of the closed Markov model in which gains and losses appear explicitly.

A feature common to both these models is the stochastic mechanism governing losses from the system. This will be accommodated by associating a time-homogeneous loss probability with each grade or stratum. The difference between the two models lies in the assumptions which govern the input. For clarity, we shall define the models by reference to two particular applications without intending thereby to restrict their applicability. References to applications in other fields are given in the Complements section at the end of the chapter.

In the first model the inflow at time T is either a known quantity or a realization of a known stochastic process. Under these circumstances the individual stock numbers will be random variables with distributions determined by the stochastic properties of all the flows. As an example of a situation in which such a model is appropriate we consider an educational system in which the strata represent successive years of study. The model we shall describe was used by Gani (1963) for the university system of Australia. He wished to predict the total enrolment in universities and the numbers of degrees to be awarded in the future. The system was the total university population in the country. The strata were the four undergraduate and three postgraduate years of study. At the end of each year three alternatives were available to any student. He could move into the next higher year by passing an examination; he could repeat the year if he failed the examination; or he could leave the university. Past data showed that it was reasonable to suppose that each type of transition had a fixed probability. The input to the system consisted of qualified students reaching the age for university entrance. This number could be predicted up to 18 years ahead from the known numbers in each age cohort. In practice there might well be constraints on the total size of the system, or its various parts, but we shall ignore them for the moment. Even if there were such constraints the model would still be useful for predicting the demand for university places. Other applications to educational systems are mentioned in the Complements.

The second model finds frequent application in the field of manpower planning. The states of the system then denote grades, age groups, or other relevant classifications of the employees in a firm. In this case we assume that the total size of the system is fixed rather than the total number of recruits. The recruitment needs are then determined by the losses, together with any change in the size of the system. If this version of the model is to be realistic there must always be a sufficient supply of recruits available to fill all vacancies which arise. In other words, the supply must exceed the demand. The version of the model with a fixed total size is due to Young and Almond (1961), who developed some theory and applied the model to the staff structure of a British university. They intended their model to be used for expanding systems but, with some modifications, it can be adapted for use with any pattern of change.

The fixed size model of Gani can also be useful in manpower planning when the recruitment flow is determined exogeneously. In practice this is not often the case because few systems operate with no constraint on total size. However, there are two circumstances in which it is meaningful to consider a model in which recruitment numbers are given. One is where the demand for labour exceeds the supply; in this case it is the supply which limits the input and this is determined externally. The other circumstance is when, as a planning exercise, we wish to explore the consequences of various patterns of recruitment. An example of this kind of application is given in Sales (1971) and Forbes (1971a).

In both versions of the model our main interest will be in the 'stocks' of people in the various 'grades'. In particular we shall seek to find the means and variances of the stock numbers and to gain insight into how the structure of the system is determined by the flow mechanism. Later, in Chapter 6, we shall look at the control problem where we proceed in the reverse direction. That is, we are given desired values for the stocks and have to find what flow probabilities will achieve them.

3.2 A MODEL FOR A SYSTEM WITH GIVEN INPUT

The basic model

We consider a population whose members are divided into k strata. In many applications these will be ranked according to seniority, but this feature will not be considered explicitly until later. As far as possible the notation of Chapter 2 will be retained. Thus $n_j(T)$ denotes the number of people in grade j at time T $(T = 0, 1, 2, \ldots)$. The initial grade sizes, $n_j(0)$ $(j = 1, 2, \ldots, k)$, are assumed to be given and we define

$$N(T) = \sum_{j=1}^{k} n_j(T).$$

For $T > 0$ the grade sizes are random variables and we shall be concerned mainly with their expectations. These will be denoted as before by placing a bar over the symbol representing the random variable; thus $\bar{n}_j(T)$ is the expected number in grade j at time T. The number of new entrants to the system at time T will be written as $R(T)$. As remarked above, this may be a constant or a random variable. In the latter case $R(T)$ is to be understood as the expected number of entrants. The assumptions and notation for the transitions between grades used in the model of social mobility are retained. A member of grade i moves to grade j with probability p_{ij}, but it is no longer true that

$$\sum_{j=1}^{k} p_{ij} = 1.$$

In general

$$\sum_{j=1}^{k} p_{ij} < 1$$

because, in an open system, transitions out of the system are possible. The probability of loss from the ith grade at time T is denoted by $p_{i,k+1}$ and we note that

$$p_{i,k+1} = 1 - \sum_{j=1}^{k} p_{ij}.$$

To complete the specification of the model we must say how new entrants are allocated to the various grades. In many applications all recruits are placed in the lowest grade but we shall make the more general assumption that a proportion p_{0j} enter the jth grade. Obviously

$$\sum_{j=1}^{k} p_{0j} = 1$$

and the distribution $\{p_{0j}\}$ will be referred to as the 'recruitment distribution'. An alternative assumption about entry to the system is that a new recruit is allocated to grade j with *probability* p_{0j}. The actual numbers entering the various grades will then be multinomially distributed instead of fixed. Our theory covers both cases as long as it refers only to expectations because it then makes no difference whether p_{0j} is regarded as an actual or an expected proportion. The difference becomes important for the discussion of variances and covariances given later.

The probabilities which specify the process can be conveniently set out in standard form as follows:

$$
\begin{array}{ccccc}
p_{01} & p_{02} & \cdots & p_{0k} & \\
\hline
p_{11} & p_{12} & \cdots & p_{1k} & p_{1,k+1} \\
p_{21} & p_{22} & \cdots & p_{2k} & p_{2,k+1} \\
\cdot & \cdot & & \cdot & \cdot \\
\cdot & \cdot & & \cdot & \cdot \\
\cdot & \cdot & & \cdot & \cdot \\
p_{k1} & p_{k2} & \cdots & p_{kk} & p_{k,k+1} \\
\hline
\end{array}
$$

As before, $\mathbf{P}$ will be the matrix with elements $\{p_{ij}\}$ ($i, j = 1, 2, \ldots, k$); $\mathbf{p}_0$ denotes the recruitment distribution and $\bar{\mathbf{n}}(T)$ the vector of expected grade sizes at T.

Now that the operation of the system has been specified it is easy to write down equations relating the expected grade sizes at successive points in time. A straightforward argument using conditional expectations gives

$$
\bar{n}_j(T+1) = \sum_{i=1}^{k} p_{ij}\bar{n}_i(T) + R(T+1)p_{0j} \qquad (T = 1, 2, 3, \ldots) \qquad (3.1a)
$$

$$
(j = 1, 2, \ldots, k)
$$

or

$$
\bar{\mathbf{n}}(T+1) = \bar{\mathbf{n}}(T)\mathbf{P} + R(T+1)\mathbf{p}_0. \qquad (3.1b)
$$

Note that the recruits are treated as though they all arrive in the system at $T+1$. Since $R(T)$ is known for all T this equation may be used to compute the expected grade sizes recursively. Repeated application of (3.1b) gives

$$
\bar{\mathbf{n}}(T) = \bar{\mathbf{n}}(0)\mathbf{P}^T + \mathbf{p}_0\left\{\sum_{\tau=0}^{T-1} R(T-\tau)\mathbf{P}^\tau\right\} \qquad (3.2)
$$

where $\mathbf{P}^0$ is again defined to be the unit matrix $\mathbf{I}$. The probabilities $\{p_{i,k+1}\}$ do not appear explicitly in these formulae but, because they are the complements of the row sums of $\mathbf{P}$, $\mathbf{P}^\tau$ and hence $\bar{\mathbf{n}}(T)$ depend upon their values. If $R(T)$ has a suitable mathematical form it may be possible to sum the matrix series appearing in (3.2) and so obtain the solution in closed form. This is the case if $R(T)$ is constant for all T or, more generally, if

$$
R(T) = Rx^T \qquad (R > 0; \ x > 0; \ T \geq 1). \qquad (3.3)
$$

We then have

$$
\bar{\mathbf{n}}(T) = \bar{\mathbf{n}}(0)\mathbf{P}^T + Rx\mathbf{p}_0(x\mathbf{I} - \mathbf{P})^{-1}(x^T\mathbf{I} - \mathbf{P}^T), \qquad (3.4)
$$

provided that the inverse of $x\mathbf{I} - \mathbf{P}$ exists. This will be so unless x is equal to any of the k eigenvalues of the matrix $\mathbf{P}$. For a matrix with non-negative elements and row sums strictly less than unity, the eigenvalues lie between 0 and 1. In particular $x = 1$ is not an eigenvalue and hence (3.4) applies for the case of constant input.

However, as we shall see below, the chief value of this form is for the light which it throws on the limiting behaviour of the system.

The analysis of the following sections is designed to yield general results about the form of the solutions and their implications for the social process in question.

For some purposes it is convenient to include in the stock vector the expected number who have left the system. If we denote this by $\bar{n}_{k+1}(T)$ we may define a set of augmented vectors and matrices as follows:

$$\bar{\mathbf{n}}_a(T) = (\bar{\mathbf{n}}(T), \bar{n}_{k+1}(T)), \qquad \mathbf{p}_{0a} = (\mathbf{p}_0, 0),$$

$$\mathbf{P}_a = \left(\begin{array}{c|c} \mathbf{P} & \mathbf{p}'_{k+1} \\ \hline 0 & 1 \end{array} \right).$$

The matrix $\mathbf{P}_a$ is a stochastic matrix with the form appropriate for a Markov chain with one absorbing state.

A repetition of the argument leading to (3.1b) gives

$$\bar{\mathbf{n}}_a(T+1) = \bar{\mathbf{n}}_a(T)\mathbf{P}_a + R(T+1)\mathbf{p}_{0a}. \tag{3.5}$$

This form of the equation is not particularly convenient in applications where our primary interest is in the sequence of stocks because the last element $\bar{n}_{k+1}(T)$ behaves in a different way from the others. In particular, it grows without limit even when the remaining elements approach limits. It is useful, however, for the study of cohort behaviour as we show below.

The spectral representation of $\bar{\mathbf{n}}(T)$

An alternative representation of $\bar{\mathbf{n}}(T)$ which has some advantages may be obtained by using a standard result in matrix theory known as Sylvester's theorem, according to which the Tth power of $\mathbf{P}$ can be expressed in the form

$$\mathbf{P}^T = \sum_{r=1}^{k} \theta_r^T \mathbf{A}_r, \tag{3.6}$$

where $\theta_1, \theta_2, \ldots \theta_k$ are the eigenvalues, assumed distinct, of $\mathbf{P}$. We have already met this result for a stochastic matrix in Section 2.3. If there are multiplicities among the eigenvalues, $\mathbf{P}^T$ can still be expressed in terms of powers of the eigenvalues, but the expressions are more complicated.

The matrices of the spectral set can also be found from

$$\mathbf{A}_r = \mathbf{b}'_r \mathbf{a}_r,$$

where $\mathbf{a}_r$ is the left eigenvector of $\mathbf{P}$ associated with θ_r, and $\mathbf{b}_r$ is the right eigenvector. (This means that $\mathbf{a}_r$ and $\mathbf{b}_r$ satisfy $\mathbf{a}_r\mathbf{P} = \theta_r\mathbf{a}_r$ and $\mathbf{P}\mathbf{b}_r' = \theta_r\mathbf{b}_r'$ respectively: it follows that $\mathbf{a}_r\mathbf{b}_s' = 0$ $(r \neq s)$ and the vectors are scaled so that $\mathbf{a}_r\mathbf{b}_r' = 1$.) Substituting into (3.6) we therefore have

$$\mathbf{P}^T = \sum_{r=1}^{k} \theta_r^T \mathbf{b}_r' \mathbf{a}_r$$

$$= (\mathbf{b}_1', \mathbf{b}_2', \ldots, \mathbf{b}_k')\mathbf{D}^T \begin{pmatrix} \mathbf{a}_1 \\ \mathbf{a}_2 \\ \vdots \\ \mathbf{a}_k \end{pmatrix}$$

where $\mathbf{D}^T$ is the diagonal matrix with elements $\{\theta_r^T\}$.

Since

$$\begin{pmatrix} \mathbf{a}_1 \\ \mathbf{a}_2 \\ \vdots \\ \mathbf{a}_k \end{pmatrix} (\mathbf{b}_1', \mathbf{b}_2', \ldots, \mathbf{b}_k') = \mathbf{I}$$

we may write:

$$\mathbf{P}^T = \mathbf{H}\mathbf{D}^T\mathbf{H}^{-1}, \tag{3.7}$$

where $\mathbf{H} = (\mathbf{b}_1', \mathbf{b}_2', \ldots, \mathbf{b}_k')$. We shall make use of this representation in Chapter 4.

If we substitute the expression for $\mathbf{P}^T$ given by (3.6) in (3.2) we shall have

$$\bar{\mathbf{n}}(T) = \sum_{r=1}^{k} \left[\theta_r^T \bar{\mathbf{n}}(0)\mathbf{A}_r + \left\{ \sum_{\tau=0}^{T-1} R(T-\tau)\theta_r^\tau \right\} \mathbf{p}_0 \mathbf{A}_r \right]. \tag{3.8}$$

This representation will be particularly useful if the series

$$\sum_{\tau=0}^{T-1} R(T-\tau)\theta_r^\tau$$

can be expressed in closed form. If this is possible $\bar{\mathbf{n}}(T)$ will be a sum of k terms for all T; it can be found without having to calculate the structure for intermediate values of T. Equation (3.8) is also a good starting point for the investigation of the limiting behaviour of $\bar{\mathbf{n}}(T)$. In general, the determination of the eigenvalues and the spectral set involves extensive calculations if k is large, but these are routine operations for which computer programs are available. However, the analysis is simplified considerably in certain special cases of practical interest. One such example arises when the organization is hierarchical in form with downward movement not permitted. We shall discuss this special case and illustrate the foregoing theory in the following section.

Hierarchical structure with no demotions

Let us suppose that the k grades are arranged in increasing order of seniority and assume that transitions within the organization are to a higher grade only. This was certainly the case in the application to the Australian university system and is sufficiently common to warrant special attention. The matrix $\mathbf{P}$ is now upper triangular, thus

$$
\mathbf{P} = \begin{pmatrix} p_{11} & p_{12} & \cdots & p_{1k} \\ 0 & p_{22} & \cdots & p_{2k} \\ \vdots & \vdots & & \vdots \\ 0 & 0 & \cdots & p_{kk} \end{pmatrix}.
$$

The eigenvalues of such a matrix are obtained at once as $\theta_r = p_{rr}$. Hence the eigenvalues will be distinct if the diagonal elements of $\mathbf{P}$ are distinct. In this case it is also possible to give an explicit expression for the elements of the matrices $\{\mathbf{A}_r\}$. If we denote the (i, j)th element of $\mathbf{A}_r$ by $A_{r, ij}$ then

$$
A_{r, ij} = \sum_{m=1}^{j-i} \left(\prod_{s=i}^{j-m} p_{s, s+m} \middle/ \prod_{\substack{s=i+m-1 \\ s \neq r}}^{j} (p_{rr} - p_{ss}) \right) \quad \left. \begin{cases} j > r > i \\ j \geq r > i \\ j > r \geq i \end{cases} \right\} \tag{3.9}
$$

$$
= 1 \quad \text{if} \quad i = r = j
$$
$$
= 0 \quad \text{otherwise}
$$

Substitution of the eigenvalues and the spectral set in (3.8) gives an expression from which $\bar{\mathbf{n}}(T)$ can be calculated.

Further simplification is possible if we place additional restraints on the kind of promotions which can occur. Suppose, for example, that promotion can only occur into the next highest grade so that $p_{s, s+m} = 0$ for $m > 1$. The first expression given in (3.9) can then be replaced by

$$
A_{r, ij} = \prod_{s=i}^{j-1} p_{s, s+1} \middle/ \prod_{\substack{s=i \\ s \neq r}}^{j} (p_{rr} - p_{ss}). \tag{3.10}
$$

If $R(T)$ has the geometric form Rx^T

$$
\bar{\mathbf{n}}(T) = \sum_{r=1}^{k} \left\{ p_{rr}^T \mathbf{n}(0) \mathbf{A}_r + Rx \left(\frac{x^T - p_{rr}^T}{x - p_{rr}} \right) \mathbf{p}_0 \mathbf{A}_r \right\} \quad (x \neq p_{rr}). \tag{3.11}
$$

It is thus clear that the diagonal elements of $\mathbf{P}$ and the value of x play a crucial role in determining the development of the process in time.

To illustrate the calculations required and to give some insight into the behaviour of hierarchical systems we shall discuss a numerical example with $k = 5$. The table of transition probabilities which we assume is set out in standard form below:

0.75	0.25	0	0	0	
0.65	0.20	0	0	0	0.15
0	0.70	0.15	0	0	0.15
0	0	0.75	0.15	0	0.10
0	0	0	0.85	0.10	0.05
0	0	0	0	0.95	0.05

The figures in this table have been chosen to reflect the kind of conditions which one might find in a typical management hierarchy. In an educational system the diagonal elements would tend to be much smaller. Three-quarters of new recruits enter the lowest grade and one-quarter the next lowest. The wastage probabilities, $\{p_{i,\,k+1}\}$, decrease as we move up the hierarchy since mobility between firms is usually more common at the lower levels. Loss from grade 5 would include retirements as well as transfers and so might well have a higher probability than the figure of 0.05 used in this example. The average time spent in grade i is $(1 - p_{ii})^{-1}$; in the present case this is between three and four years for the lowest three grades. The expected time to reach the highest grade is almost 17 years.

The matrix $\mathbf{P}$ is triangular and hence the eigenvalues are equal to the diagonal elements. It therefore remains to calculate the spectral set $\{\mathbf{A}_r\}$. To do this we have to substitute numerical values given above in (3.10). For example,

$$A_{1.11} = 1, \qquad A_{1.12} = p_{12}/(p_{11} - p_{22}) = (0.20)/(-0.05) = -4,$$
$$A_{1.13} = p_{12}p_{23}/(p_{11} - p_{22})(p_{11} - p_{33}) = (0.20)(0.15)/(-0.05)(-0.10) = 6,$$

etc. Written out in full the matrices are

$$\mathbf{A}_1 = \begin{pmatrix} 1 & -4 & 6 & -4.5 & 1.5 \\ 0 & 0 & 0 & 0 & 0 \\ 0 & 0 & 0 & 0 & 0 \\ 0 & 0 & 0 & 0 & 0 \\ 0 & 0 & 0 & 0 & 0 \end{pmatrix}$$

$$\mathbf{A}_2 = \begin{pmatrix} 0 & 4 & -12 & 12 & -4.8 \\ 0 & 1 & -3 & 3 & -1.2 \\ 0 & 0 & 0 & 0 & 0 \\ 0 & 0 & 0 & 0 & 0 \\ 0 & 0 & 0 & 0 & 0 \end{pmatrix}$$

$$\mathbf{A}_3 = \begin{pmatrix} 0 & 0 & 6 & -9 & 4.5 \\ 0 & 0 & 3 & -4.5 & 2.25 \\ 0 & 0 & 1 & -1.5 & 0.75 \\ 0 & 0 & 0 & 0 & 0 \\ 0 & 0 & 0 & 0 & 0 \end{pmatrix}$$

$$\mathbf{A}_4 = \begin{pmatrix} 0 & 0 & 0 & 1.5 & -1.5 \\ 0 & 0 & 0 & 1.5 & -1.5 \\ 0 & 0 & 0 & 1.5 & -1.5 \\ 0 & 0 & 0 & 1 & -1 \\ 0 & 0 & 0 & 0 & 0 \end{pmatrix}$$

$$\mathbf{A}_5 = \begin{pmatrix} 0 & 0 & 0 & 0 & 0.3 \\ 0 & 0 & 0 & 0 & 0.45 \\ 0 & 0 & 0 & 0 & 0.75 \\ 0 & 0 & 0 & 0 & 1 \\ 0 & 0 & 0 & 0 & 1 \end{pmatrix}.$$

Of the quantities appearing in (3.11) the initial grade structure $\bar{\mathbf{n}}(0)$ and the input sequence $\{R(T)\}$ remain to be specified. Let us assume that $\mathbf{n}(0) = N(0)$ (0.40, 0.30, 0.15, 0.10, 0.05) and $R(T) = R$. If we compute the total population size and the individual grade sizes at time $T(T > 0)$ as multiples of $N(0)$ we need only express R as a fraction of $N(0)$. For the illustrative calculations we have chosen $N(0) = 9.8333R$. The reason for this particular choice is that it makes $N(\infty)$ $= N(0)$. This ensures that there is no long-term trend in the overall size and so enables us to concentrate on the changes in structure which take place. Under the assumptions we have made

$$\bar{\mathbf{n}}(T) = \sum_{r=1}^{5} \left\{ p_{rr}^T \mathbf{n}(0) \mathbf{A}_r + R \left(\frac{1 - p_{rr}^T}{1 - p_{rr}} \right) \mathbf{p}_0 \mathbf{A}_r \right\} \tag{3.12}$$

and, in the limit,

$$\bar{\mathbf{n}}(\infty) = R \sum_{r=1}^{5} (1 - p_{rr})^{-1} \mathbf{p}_0 \mathbf{A}_r. \tag{3.13}$$

In the example the vectors $\{\mathbf{p}_0 \mathbf{A}_r\}$ are found to be

$$\mathbf{p}_0 \mathbf{A}_1 = (0.75 \quad -3.00 \quad 4.50 \quad -3.375 \quad 1.125),$$
$$\mathbf{p}_0 \mathbf{A}_2 = (0 \quad 3.25 \quad -9.75 \quad 9.75 \quad -3.90),$$
$$\mathbf{p}_0 \mathbf{A}_3 = (0 \quad 0 \quad 5.25 \quad -7.875 \quad 3.9375),$$
$$\mathbf{p}_0 \mathbf{A}_4 = (0 \quad 0 \quad 0 \quad 1.50 \quad -1.50),$$
$$\mathbf{p}_0 \mathbf{A}_5 = (0 \quad 0 \quad 0 \quad 0 \quad 0.3375).$$

Substitution in (3.13) thus gives the limiting structure as

$$\bar{\mathbf{n}}(\infty) = N(0)(0.218 \quad 0.230 \quad 0.138 \quad 0.138 \quad 0.275). \tag{3.14}$$

The difference between this and the initial structure is quite striking. The highest grade has increased more than fivefold in size and the sizes of the lower grades have become more nearly equal. This feature is primarily due to the long average stay of 20 years in grade 5. In order to counteract this excessive growth at the top we might be able to increase p_{rr} for $r < 5$ by reducing the promotion probabilities. This example illustrates the fact that a promotion policy which seems reasonable in itself may lead to undesirable consequences for the structure of the organization. In particular, it suggests that the pressure to grow at the top which is exhibited in many organizations may be the direct consequence of a fixed promotion policy.

To calculate the intermediate structures we must first calculate the vectors $\{\mathbf{n}(0)\mathbf{A}_r\}$. We omit the details and proceed directly to Table 3.1 giving values of $\bar{\mathbf{n}}(T)$ for selected values of T.

Table 3.1   *Values of $\bar{n}_j(T)/N(0)$ for the example discussea in the text*

				T		
Grade	0	1	2	5	10	∞
1	0.400	0.336	0.295	0.239	0.220	0.218
2	0.300	0.315	0.313	0.280	0.243	0.230
3	0.150	0.158	0.165	0.174	0.159	0.138
4	0.100	0.107	0.115	0.136	0.154	0.138
5	0.050	0.058	0.065	0.091	0.137	0.275
$\sum_j \bar{n}_j(T)/N(0)$	1.000	0.974	0.953	0.920	0.913	1.000

The approach to equilibrium exhibits a pattern which we have observed in other instances. In the lower grades the limiting expectations are attained relatively quickly, but the approach is very slow in grade 5. After ten years the expected size of grade 5 is still only half its equilibrium value while grades 1 and 2 have almost attained theirs. The total expected size of the population shows a steady decrease over the period up to ten years, but ultimately recovers its original value. The slowness of the approach to equilibrium in the upper grades and in the total size is accounted for by the term involving $(0.95)^T$ which occurs in (3.12). With a coefficient of -0.6864 its value does not become negligible until T is of the order of 100. In cases like this the limiting structure of the system is of little direct practical interest, although it does indicate the general direction of change. For some purposes it might be of greater interest to look at the expected proportions in each grade instead of the expected numbers. These can easily be obtained from Table 3.1, but they do not materially alter the general picture.

Limiting behaviour of $\bar{\mathbf{n}}(T)$

We have already considered the limiting behaviour of the model in a special case with constant input. Here we shall investigate the position in more detail. In an open system the question of the behaviour of the long-term structure is complicated by presence of the input $\{R(T)\}$. Intuition suggests that if this sequence settles down to a constant value R, say, then the sequence of expected stock vectors will also approach a steady state. If, on the other hand, the sequence $\{R(T)\}$ increases indefinitely then the stocks will likewise increase and there will be no steady state of the kind we met in Chapter 2. However, there remains the possibility that the relative sizes of the expected stocks will become stable. If we define the relative stock vector by

$$\mathbf{q}(T) = \bar{\mathbf{n}}(T)/\bar{\mathbf{n}}(T)\mathbf{l}',$$

where $\mathbf{l} = (1, 1, \ldots, 1)$ then it is pertinent to ask whether the sequence $\{\mathbf{q}(T)\}$ approaches a limit as $T \to \infty$. Similarly, if $R(T)$ decreases to zero with T, the stock vector will likewise vanish, in the limit, but it might still happen that $\mathbf{q}(T)$ had a non-zero limiting value.

In order to explore these possibilities we shall consider the special case $R(T) = Rx^T$ used above. This includes the constant input as a special case and also embraces increasing and decreasing input sequences.

If we substitute $R(T) = Rx^T$ in (3.2) and then divide through by x^T we find

$$\mathbf{n}(T)/x^T = \mathbf{n}(0)(\mathbf{P}^T/x^T) + \mathbf{p}_0 R \left\{ \sum_{\tau=0}^{T-1} (\mathbf{P}/x)^\tau \right\}. \tag{3.15}$$

We shall now let $T \to \infty$ on the right-hand side of this equation. Recalling the spectral representation of $\mathbf{P}^T$ given in (3.6), we deduce that

$$\mathbf{P}^T/x^T = \sum_{r=1}^{k} (\theta_r/x)^T \mathbf{A}_r.$$

If $x > \max_r \theta_r$ this expression vanishes in the limit. Under the same condition the series in the second term converges to $(\mathbf{I} - \mathbf{P}/x)^{-1}$ and hence

$$\lim_{T \to \infty} \bar{\mathbf{n}}(T)/x^T = R\mathbf{p}_0 x(x\mathbf{I} - \mathbf{P})^{-1}. \tag{3.16}$$

An alternative, but equivalent, expression follows from (3.8) namely that

$$\lim_{T \to \infty} \bar{\mathbf{n}}(T)/x^T = R\mathbf{p}_0 \sum_{r=1}^{k} (x - \theta_r)^{-1} \mathbf{A}_r. \tag{3.17}$$

Thus if $x > \theta = \max_r \theta_r$ we have shown that the expected stock numbers will grow geometrically in line with the input. The limiting relative stock numbers are clearly given by

$$\mathbf{q}(\infty) \propto \mathbf{p}_0 (x\mathbf{I} - \mathbf{P})^{-1} \tag{3.18}$$

the constant of proportionality being determined by the requirement that $\mathbf{q}(\infty)\mathbf{l}' = 1$. Notice that the limiting structure given by (3.18) does not depend on

$\mathbf{n}(0)$ or R. If x is very large then

$$\mathbf{q}(\infty) \sim \mathbf{p}_0$$

indicating that the structure is then dominated by the recruitment vector. The only case in which $\bar{\mathbf{n}}(T)$ has a limit is when $x = 1$, which implies that the input is constant.

The foregoing argument breaks down if $x \leq \theta$. For a matrix $\mathbf{P}$ it is known (see, for example, Cox and Miller, 1965, p. 120) that

$$0 < \theta \leq \max \sum_{j=1}^{k} p_{ij}$$

where the equality sign holds only if all the row sums are equal. The largest eigenvalue will thus be strictly less than one unless all the row sums of $\mathbf{P}$ are equal to one. This is impossible for an open system so we conclude that $\theta < 1$. The condition $x > \theta$ thus means that (3.18) will always hold when the input, and hence the total size, is expanding and also for a contracting input provided that the contraction is not too rapid. To find out what happens when $x \leq \theta$ we turn to (3.8) and divide through both sides by θ^T to give

$$\bar{\mathbf{n}}(T)/\theta^T = \sum_{r=1}^{k} \left[\mathbf{n}(0)\mathbf{A}_r(\theta_r/\theta)^T + \left\{ \sum_{\tau=0}^{T-1} R(x/\theta)^T(\theta_r/x)^\tau \right\} \mathbf{p}_0 \mathbf{A}_r \right]. \qquad (3.19)$$

Consider first the case $x < \theta$. As $T \to \infty$ every term in the first sum vanishes except the one for which $\theta_r = \theta$ so this part contributes $\mathbf{n}(0)\mathbf{A}_1$ if we label the eigenvalues so that the largest has the subscript 1. The sum in curly brackets is

$$R\left\{ \frac{(x/\theta)^T - (\theta_r/\theta)^T}{1 - (\theta/x)^T} \right\}.$$

In the limit this is zero except when $r = 1$, in which case it is

$$Rx/(\theta - x).$$

The second part of (3.19) thus contributes

$$\frac{Rx}{(\theta - x)} \mathbf{p}_0 \mathbf{A}_r$$

giving, finally,

$$\lim_{T \to \infty} \bar{\mathbf{n}}^{(T)}/\theta^T = \{ \mathbf{n}(0) + Rx\mathbf{p}_0/(\theta - x) \} \mathbf{A}_r. \qquad (3.20)$$

In this case, therefore, the stocks decline at a somewhat slower rate than the input.

This leaves the case $x = \theta$. The first term on the right hand side of (3.19) is unaffected, but the second becomes

$$\sum_{r=1}^{k} \sum_{\tau=0}^{T-1} R(\theta_r/\theta)^\tau \mathbf{p}_0 \mathbf{A}_r \quad \cdot \text{as } T \to \infty$$

The geometric series converges as $T \to \infty$ except when $r = 1$. In that case the sum is $R\mathbf{p}_0\mathbf{A}_1 T$. It therefore follows that

$$\lim_{T \to \infty} \frac{\bar{\mathbf{n}}(T)}{T\theta^T} = R\mathbf{p}_0\mathbf{A}_1. \tag{3.21}$$

The limiting form of $\mathbf{q}(T)$ is the same both when $x = \theta$ and when $x < \theta$. We have seen that the members of the spectral set may be written $\mathbf{A}_r = \mathbf{b}'_r\mathbf{a}_r$, where $\mathbf{a}_r$ and $\mathbf{b}_r$ are the left and right eigenvectors respectively, associated with θ_i. Substituting $\mathbf{A}_1 = \mathbf{b}'_1\mathbf{a}_1$ into (3.20) and (3.21) it follows, in both cases, that

$$\mathbf{q}(\infty) \propto \mathbf{a}_1. \tag{3.22}$$

We thus have two different forms of the steady-state structure corresponding to whether $x > \theta$ or $x \leq \theta$.

To illustrate these results we take the case of a hierarchical system with no demotions and promotion into the next higher grade for which we have already found the spectral set in (3.10). The eigenvalues are given by $\theta_r = p_{rr} (r = 1, 2, \ldots, k)$ and we further assume that $\mathbf{p}_0 = (1, 0, 0, \ldots, 0)$. It then follows from (3.17) and (3.10) that for a fixed input

$$\bar{n}_j(\infty) = R \prod_{r=1}^{j-1} p_{r,r+1} \Big/ \prod_{r=1}^{j} (1 - p_{rr}), \qquad (j = 1, 2, \ldots, k), \tag{3.23}$$

where the first product is defined to be one when $j = 1$. This formula clearly demonstrates how the stock numbers depend on the values of the staying probabilities $\{p_{rr}\}$ and, in particular, on their closeness to one. If $R(T) = Rx^T$, (3.17) yields

$$q_j(\infty) \propto \prod_{r=1}^{j-1} p_{r,r+1} \Big/ \prod_{r=1}^{j} (x - p_{rr}), \qquad (j = 1, 2, \ldots, k) \tag{3.24}$$

provided that $x > \max r p_{rr} = \theta$. In the contrary case reference to (3.20) or (3.21) and (3.8) gives

$$q_j(\infty) \propto \prod_{r=1}^{j-1} p_{r,r+1} \Big/ \prod_{r=1}^{j} (\theta - p_{rr}), \tag{3.25}$$

where the value of r for which $p_{rr} = \theta$ is omitted from the product in the denominator. Comparison with (3.23) shows that θ has replaced 1 in the denominator and that one factor is omitted.

Application to a cohort

If we put $R(T) = 0$, for all T, in the model with given input, the various formulae describe the movement of the original members through the system. The stock numbers at time T are now given by

$$\bar{\mathbf{n}}(T) = \mathbf{n}(0)\mathbf{P}^T \tag{3.26}$$

and they will clearly tend to zero as T tends to infinity. The importance of this special case lies in its application to a cohort and the information which this gives about career patterns.

A cohort is a group of individuals joining the organization at the same time. Suppose a group is of size N and that they all enter grade i. (Here, as elsewhere in this chapter, we use the word 'grade' instead of 'state' because the main application is in manpower planning, but the results have a wider application.) In this case

$$\mathbf{n}(0) = N\,\mathbf{e}_i$$

where $\mathbf{e}_i$ is a vector having a one in the ith position and zeros elsewhere. The expected distribution of the cohort over the grades at various stages in its history can then be obtained from (3.26). One variable of practical interest is the expected proportion of the original cohort who survive in the organization for different lengths of time. This can easily be obtained by summing the elements of $\bar{\mathbf{n}}(T)$ and dividing by N. This line of approach was developed by Young (1971) and will be pursued further in Chapter 7, along with a number of other questions relating to durations. For the present we shall concentrate on the determination of the means and, where possible, the variances of various quantities of relevance in career planning.

Our analysis depends on the theory of absorbing Markov chains, a good account of which can be found in Kemeny and Snell (1976). The $(k + 1)$th grade, consisting of all those who have left, can be treated as a single absorbing state, the remaining states being transient. The (stochastic) transition matrix of such a chain thus has the augmented form:

$$\mathbf{P}_{\mathrm{a}} = \left[\begin{array}{c|c} \mathbf{P} & \mathbf{p}'_{k+1} \\ \hline 0 & 1 \end{array} \right]$$

where $\mathbf{p}_{k+1} = (p_{1,\,k+1}, p_{2,\,k+1}, \ldots, p_{k,\,k+1})$. It turns out that a great deal of information about individual career histories can be deduced from the so-called fundamental matrix $(\mathbf{I} - \mathbf{P})^{-1}$. This matrix will occur again in Chapter 6 where it will be used to determine whether or not a given grade structure can be maintained.

Consider first a person who enters the system in grade i: what is the average length of time he will spend in grade j? To answer questions such as this we introduce random variables $\{X_{ij}^{(r)}\}$ defined as follows:

$$X_{ij}^{(r)} = 1 \text{ if an entrant to grade } i \text{ is in grade } j \text{ after } r \text{ time units}$$

$$= 0 \text{ otherwise } (i, j = 1, 2, \ldots, k; r = 0, 1, 2, \ldots).$$

The total time spent by such an individual in grade j is thus

$$X_{ij} = \sum_{r=0}^{\infty} X_{ij}^{(r)} \qquad (i, j = 1, 2, \ldots, k).$$

(Note that $X_{ij}^{(0)} = 0$ if $j \neq i$ and $= 1$ otherwise.) We first find the expectation of X_{ij} which is given by

$$E(X_{ij}) = \sum_{r=0}^{\infty} E(X_{ij}^{(r)}). \qquad (3.27)$$

It is well known from the general theory of Markov chains that

$$Pr\{X_{ij}^{(r)} = 1\} = p_{ij}^{(r)},$$

where $p_{ij}^{(r)}$ is the (i, j)th element of $\mathbf{P}^r$. Hence

$$E(X_{ij}^{(r)}) = p_{ij}^{(r)}$$

and therefore

$$E(X_{ij}) = \sum_{r=0}^{\infty} p_{ij}^{(r)}. \qquad (3.28)$$

If we introduce the matrix $\mathbf{X} = \{X_{ij}\}$, (3.28) yields

$$E(\mathbf{X}) = \sum_{r=0}^{\infty} \mathbf{P}^{(r)} = \sum_{r=0}^{\infty} \mathbf{P}^r = (\mathbf{I} - \mathbf{P})^{-1}. \qquad (3.29)$$

This result establishes the relationship between the fundamental matrix and the expected lengths of service in the grades. The expected stay of an entrant to grade i in the whole system is

$$E(X_i) = \sum_{j=1}^{k} E(X_{ij}) = d_i, \text{ say,} \qquad (3.30)$$

where d_i is thus the sum of the ith row of $(\mathbf{I} - \mathbf{P})^{-1}$. Notice that all of these expectations apply equally to a new entrant to the organization and to someone newly promoted from inside. This fact is an immediate consequence of the Markov assumption which treats all entrants to a grade alike, regardless of their past history. We might well doubt the validity of such an assumption in practice, but in organizations where all, or most, recruitment is at the bottom of a hierarchy the difficulty does not arise.

Stone (1972) has discussed the application of these results to a manpower system and has made the calculations for the five-grade matrix used above. For this matrix we find:

					Row totals
2.86	1.90	1.14	1.14	2.29	9.33
0	3.33	2.00	2.00	4.00	11.33
0	0	4.00	4.00	8.00	16.00
0	0	0	6.67	13.33	20.00
0	0	0	0	20.00	20.00

$$(\mathbf{I} - \mathbf{P})^{-1} =$$

Inspection of the row totals shows how the expectation of service increases as we move up the hierarchy. This result reflects the decrease of wastage with increasing seniority—once a recruit has survived the high wastage rates of the lowest grades his prospects of a long stay are improved.

The individual elements in the rows of $(I - P)^{-1}$ show how the total expected service is divided among the grades. Thus on entering grade 1 an individual will expect to spend 2.86 years in the first grade, 1.90 years in the second, and so on. On moving up to the second grade the pattern changes, reflecting the fact that the individual has survived through grade 1. The diagonal elements are the expected stays in each grade at the time of entry to that grade.

Care must be taken in interpreting the off-diagonal elements. For example, 2.29 years is the time an entrant to grade 1 expects to spend in grade 5. Of all entrants to grade 1 most will never reach grade 5 and so will contribute nothing to the average. By the time a successful entrant reaches the threshold of grade 5 his expectation has increased to 20 years. The set of numbers in the ith row can thus be thought of as a typical career expectation for an entrant to that grade but, like all averages, they conceal considerable variation.

We have mentioned the chance that an entrant at the bottom will reach the top. Such chances can be calculated from the fundamental matrix. Let π_{ij} denote the probability that an entrant to grade i spends some time in grade j before leaving. (In general, an individual may pass through j more than once; in a simple hierarchy with one-step promotion he can pass through it no more than once.) If μ_{ij} is the (i, j)th element of $(I - P)^{-1}$ then clearly,

$$\mu_{ij} = \pi_{ij}\mu_{ij} + (1 - \pi_{ij}) \times 0$$

or

$$\pi_{ij} = \mu_{ij}/\mu_{jj} \qquad (i, j = 1, 2, \ldots, k). \tag{3.31}$$

Thus to obtain the set of probabilities $\{\pi_{ij}\}$ we must divide the elements in each column of $(I - P)^{-1}$ by the diagonal element of that column. For the example,

$$\{\pi_{ij}\} = \begin{pmatrix} 1 & 0.57 & 0.29 & 0.17 & 0.11 \\ 0 & 1 & 0.50 & 0.30 & 0.20 \\ 0 & 0 & 1 & 0.60 & 0.40 \\ 0 & 0 & 0 & 1 & 0.67 \\ 0 & 0 & 0 & 0 & 1 \end{pmatrix}.$$

The diagonal elements must obviously be unity; the off-diagonal elements give the chance of reaching the grade corresponding to the column, given that we enter that corresponding to the row. Thus, for example, on entering grade 2 an individual has a 50 per cent chance of reaching the next grade, a 30 per cent chance of rising two grades and a 20 per cent chance of reaching the top.

In the case of a matrix like that just illustrated, in which P has non-zero elements, only on the diagonal and super-diagonal, $(I - P)^{-1}$ can be found algebraically (see equation 6.8). We then easily find

$$\pi_{ij} = \prod_{r=i}^{j-1} p_{r,r+1}/(1 - p_{r,r}) \qquad (j > i). \tag{3.32}$$

This result could have been deduced directly from the fact that $p_{r,r+1}/(1 - p_{r,r})$ is the probability of promotion from grade r, given that the individual does not stay. These formulas can, of course, be used in reverse to deduce the transition matrix required to achieve desired career prospects as specified by the π_{ij}'s or the μ_{ij}'s.

Another set of indices which may be of use in career planning are the probabilities of leaving from particular grades. Let a_{ij} be the probability that someone in grade i will leave when he is in grade j. The method is to introduce one absorbing state corresponding to the leavers from each grade so that the transition matrix of the chain has the form:

$$\begin{array}{c|c} \mathbf{P} & \mathbf{Q} \\ \hline \mathbf{O} & \mathbf{I} \end{array},$$

where $\mathbf{Q}$ has diagonal elements equal to those of the vector $\mathbf{p}_{k+1}$ and zeros elsewhere. This makes it a simple matter to carry out analyses like those discussed above for those who terminate their employment in a specified grade.

The method used to find the expected length of stay in each group (the μ_{ij}'s) can easily be extended to find the variances. Thus:

$$E(X_{ij}^2) = E\left(\sum_{r=0}^{\infty} X_{ij}^{(r)} \right)^2 = \sum_{r=0}^{\infty} \sum_{s=0}^{\infty} E(X_{ij}^{(r)} X_{ij}^{(s)}). \tag{3.33}$$

If $r = s$,

$$E(X_{ij}^{(r)})^2 = E(X_{ij}^{(r)}) = p_{ij}^{(r)}.$$

Otherwise

$$\begin{aligned} E(X_{ij}^{(r)} X_{ij}^{(s)}) &= Pr\{X_{ij}^{(r)} = 1, X_{ij}^{(s)} = 1\} \\ &= Pr\{X_{ij}^{(r)} = 1\} Pr\{X_{ij}^{(s)} = 1 | X_{ij}^{(r)} = 1\} \\ &= p_{ij}^{(r)} p_{jj}^{(s-r)} \text{ if } s \geq r. \end{aligned} \tag{3.34}$$

If $s < r$ we simply interchange r and s. Substituting in (3.33)

$$\begin{aligned} E(X_{ij}^2) &= \sum_{r=0}^{\infty} p_{ij}^{(r)} + 2 \sum_{r=0}^{\infty} \sum_{s=r+1}^{\infty} p_{ij}^{(r)} p_{jj}^{(s-r)} \\ &= \mu_{ij} + 2 \sum_{r=0}^{\infty} p_{ij}^{(r)} (p_{jj}^{(1)} + p_{jj}^{(2)} + \ldots) \\ &= \mu_{ij} + 2\mu_{ij}(\mu_{jj} - 1) = 2\mu_{ij}\mu_{jj} - \mu_{ij}. \end{aligned} \tag{3.35}$$

Hence

$$\begin{aligned} var(X_{ij}) &= 2\mu_{ij}\mu_{jj} - \mu_{ij} - \mu_{ij}^2 \\ &= \mu_{jj}^2(2\pi_{ij} - \pi_{ij}^2) - \mu_{jj}\pi_{ij} \qquad (i, j = 1, 2, \ldots, k). \end{aligned} \tag{3.36}$$

In the important special case when $i = j$, $\pi_{ii} = 1$ and

$$\operatorname{var}(X_{ii}) = \mu_{ii}^2 - \mu_{ii} = \mu_{ii}(\mu_{ii} - 1) \qquad (i = 1, 2, \ldots, k). \tag{3.37}$$

This result, in particular, shows that the lengths of stay in each grade will be highly variable. If μ_{ii} is fairly large the standard deviation of length of stay will be almost as large as the expectation.

Variances and covariances of the grade sizes

The theory for closed systems given in Chapter 2 can be generalized to include open systems. For an open system (2.10) must be replaced by

$$n_j(T+1) = \sum_{i=1}^{k} n_{ij}(T) + n_{0j}(T+1) \qquad (j = 1, 2, \ldots, k), \tag{3.38}$$

where $n_{0j}(T+1)$ is the number of new entrants to grade j at time $T+1$. On taking expectations of both sides of this equation we are led back to (3.1). If we assume that the input is stochastically independent of the internal movements of the system then

$$\operatorname{cov}\{n_j(T+1),\, n_l(T+1)\}$$

$$= \operatorname{cov}\left\{ \sum_{i=1}^{k} n_{ij}(T) \sum_{i=1}^{k} n_{il}(T) \right\} + \operatorname{cov}\{n_{0j}(T+1),\, n_{0l}(T+1)\}. \tag{3.39}$$

The first term on the right-hand side of (3.39) is essentially the same as the expression given in (2.11) of Chapter 2. Its evaluation can be carried through as in that chapter without any modification. All that we have to do to make the result of (2.15) of Chapter 2 applicable in the present case is to add on the last covariance term given in (3.39) above. Thus in matrix notation we have

$$\boldsymbol{\mu}(T+1) = \boldsymbol{\mu}(T)\boldsymbol{\Pi} + \boldsymbol{\mu}_0(T+1), \tag{3.40}$$

where $\boldsymbol{\mu}(T)$ and $\boldsymbol{\Pi}$ are as defined in Chapter 2. The first k elements of the vector $\boldsymbol{\mu}_0(T+1)$ are the expected numbers of entrants to each grade at time $T+1$ listed in ascending order; the remaining k^2 elements are the covariances of these numbers listed in dictionary order. In our examples the input will be time-homogeneous so we shall suppress the argument of $\boldsymbol{\mu}_0$ and write $\boldsymbol{\mu}_0$ instead of $\boldsymbol{\mu}_0(T+1)$. In this case (3.40) leads to

$$\boldsymbol{\mu}(T) = \boldsymbol{\mu}(0)\boldsymbol{\Pi}^T + \boldsymbol{\mu}_0(\mathbf{I} - \boldsymbol{\Pi})^{-1}(\mathbf{I} - \boldsymbol{\Pi}^T). \tag{3.41}$$

The inverse of $\mathbf{I} - \boldsymbol{\Pi}$ exists because the dominant eigenvalue of $\boldsymbol{\Pi}$ is strictly less than one. If we adopt the more general error structure discussed in Chapter 2 the same equation holds with appropriate changes in $\boldsymbol{\Pi}$.

The form of $\boldsymbol{\mu}_0$ will depend upon the stochastic nature of the input and the method of allocation to grades. We shall illustrate the point using the two cases mentioned at the beginning of the section. First suppose that the input $\{R(T)\}$ is a sequence of independently and identically distributed random variables and that, at time T, a proportion p_{0j} go to grade j. In this case the first k elements of μ_0 are $\bar{R}(T)p_{0j}$ $(j = 1, 2, \ldots, k)$, where $\bar{R}(T) = ER(T)$. The covariance elements take the form:

$$\text{var}\{R(T)\}p_{0j}p_{0l} \qquad (j, l = 1, 2, \ldots, k).$$

Our second assumption about the allocation of recruits was that each one was to be allocated to the jth grade with probability p_{0j}. In this case the k means in the vector μ_0 are $\bar{R}(T)p_{0j}$ $(j = 1, 2, \ldots, k)$ as before, but the covariances are now given by

$$\begin{aligned}
&\text{cov}\{n_{0j}(T), n_{0l}(T)\} \\
&= \text{var}\{R(T)\}p_{0j}p_{0l} + \bar{R}(T)(\delta_{jl}p_{0j} - p_{0j}p_{0l}) \qquad (i, j = 1, 2, \ldots, k).
\end{aligned} \qquad (3.42)$$

This result is obtained by first finding expectations conditional upon $R(T)$ and then averaging over the distribution of $R(T)$. Under either assumption about allocation the determination of the successive values of $\mu(T)$ is straightforward. Some examples are discussed below, but first we consider the limiting behaviour of the system.

If the input to the system is time-homogeneous, μ_0 does not depend on T. Under these circumstances we may investigate the limiting form of the vector $\mu(T)$ as $T \to \infty$. We already know that the limit exists for the first k elements of the vector and it may be shown that the same result holds for all of the elements. This being so it is clear from (3.40) that the limiting vector must satisfy

$$\mu(\infty) = \mu(\infty)\Pi + \mu_0$$

or

$$\mu(\infty) = \mu_0(\mathbf{I} - \Pi)^{-1}. \qquad (3.43)$$

Once the matrix $(\mathbf{I} - \Pi)^{-1}$ has been computed the effects of different kinds of input can easily be compared. If $R(T)$ is a fixed quantity R, then $\text{var}\{R(T)\} = 0$ and there is a substantial simplification in μ_0.

To illustrate the foregoing theory we shall present some calculations for two different kinds of input. In the first case we shall suppose that $R(T)$ is a fixed number, R, independent of T, and in the second that $\{R(T)\}$ is a sequence of independent Poisson variates with constant mean equal to R. These two assumptions represent fairly extreme degrees of variability in the input. We shall assume, for the purposes of this illustration, that each new recruit is allocated to grade j with probability p_{0j}. The covariance elements of μ_0 are obtained from (3.42), where $\text{var}\{R(T)\}$ is zero for constant input and equal to $\bar{R}$ for Poisson input. The calculations in Table 3.2 relate to the example discussed earlier in this section for which the expected grade sizes were given in Table 3.1. We have further supposed that the initial size of the organization was 590 and that $\bar{R} = 60$.

Table 3.2 Variance–covariance matrices for the grade sizes for the example of Section 3.2 with (a) fixed input and (b) Poisson input where, in each case, $N(0) = 9.8333\,R(T)$ and $\bar{R}(T) = 60$

T	Fixed input					Poisson input				
1	64.94	−41.93	0	0	0	98.69	−30.68	0	0	0
		86.18	−18.59	0	0		89.93	−18.59	0	0
			39.16	−9.96	0			39.16	−9.96	0
				18.81	−5.02				18.81	−5.02
					6.71					6.71
2	83.82	−47.68	−4.09	0	0	131.83	−26.92	−2.99	0	0
		115.16	−21.51	−1.95	0		125.25	−20.78	−1.95	0
			60.94	−12.81	−0.75			61.02	−12.81	−0.75
				31.86	−7.99				31.86	−7.99
					12.61					12.61
5	80.20	−38.20	−7.69	−1.22	−0.07	137.86	−5.70	−2.82	−0.67	−0.05
		124.69	−17.15	−5.07	−0.57		149.79	−11.25	−4.24	−0.52
			85.60	−11.20	−2.56			87.63	−10.85	−2.54
				58.36	−11.47				58.43	−11.46
					28.83					28.83
10	71.54	−34.71	−6.73	−1.51	−0.24	129.97	−0.19	−0.25	−0.21	−0.07
		111.69	−11.99	−4.68	−1.16		142.27	−1.53	−1.57	−0.65
			84.77	−6.72	−2.91			90.83	−4.20	−2.39
				79.43	−10.50				80.80	−10.18
					56.05					56.13
∞	70.13	−34.58	−6.58	−1.43	−0.24	128.57	0	0	0	0
		104.79	−11.10	−3.97	−1.05		135.71	0	0	0
			74.13	−4.35	−1.86			81.43	81.43	0
				76.84	−3.74				81.43	0
					154.50[a]					162.26
Expected grade sizes for $T = \infty$	128.6	135.7	81.4	81.4	162.3	128.6	135.7	81.4	81.4	162.3

[a] This is the value for $T = 100$, at which point the limiting value had not been attained. The calculations in this table were carried out by J. H. Pollard.

The equilibrium values are approached quite rapidly for the lower grades, but only very slowly for the highest grade. For both types of input there is considerable uncertainty in predictions of the size of the fifth grade in the distant future. The difference between the two kinds of input is most apparent in the lowest grades. It is not until $T = 5$ that the input has any effect on the top grade and thereafter it is only slight. In the lowest grade the effect of Poisson variability is to roughly double the variance as compared with fixed input. All of the covariances are either zero or negative, but in the Poisson case they vanish in the limit. This fact coupled with equality of the limiting means and variances suggests that the grade sizes are asymptotically distributed like independent Poisson variates. Pollard (1967) has proved this to be true in the general case when the input consists of an independent sequence of Poisson variates.

An alternative approach leading to the same result has been given by Staff and Vagholkar (1971) based on generating functions. In principle it may be used to find the joint distribution of the grade sizes as well as the moments, but in practice the manipulations required are so heavy as to make the method of very limited value. Staff and Vagholkar do give a few explicit results in very simple cases, but for most purposes the first- and second-order moments are sufficient and these can be derived more easily using Pollard's method. The one exception to this remark is when the inputs $\{R(T)\}$ are an independent sequence of Poisson random variables. In that case, as remarked above, the joint distribution of the grade sizes is asymptotically Poisson and we shall illustrate the method of generating functions by using it to derive this result.

Let $p_j(T - \tau)$ $(j = 1, 2, \ldots, k+1)$ be the probability that a person recruited at time τ is in grade j at time T. Those who have left are considered to be in a grade $k + 1$. Let the number of recruits at $T - \tau$ be $R(T - \tau) = X$. Then, conditional upon X, those recruited at time $T - \tau$ will be distributed multinomially over the $k + 1$ grades at time T according to the probabilities $\{p_j(T - \tau)\}$. The joint probability generating function of these numbers may thus be written

$$g_\tau(z_{1\tau}, z_{2\tau}, \ldots, z_{k+1,\tau}|X) = \left(\sum_{j=1}^{k+1} p_j(T - \tau) z_{j\tau} \right) X. \tag{3.44}$$

The unconditional generating function is the expectation of (3.44) with respect to the distribution of X. If X has the probability generating function $l(z)$ then the generating function required is

$$g_\tau(z_{1\tau}, z_{2\tau}, \ldots, z_{k+1,\tau}) = l \left(\sum_{j=1}^{k+1} p_j(T - \tau) z_{j\tau} \right). \tag{3.45}$$

The total number in the grades is composed of the survivors from the initial establishment together with the cohorts recruited at times $1, 2, \ldots, T - 1$. Since the $R(T)$'s are independent the total number of survivors from these cohorts, excluding the originals, will be obtained by taking the product of (3.45) over τ and putting $z_{j\tau} = z_j$ for all j. This gives the generating function

$$g(z_1, z_2, \ldots, z_{k+1}) = \prod_{\tau=1}^{T-1} l\left(\sum_{j=1}^{k+1} p_j(T-\tau)z_j\right) \qquad (3.46)$$

for the distribution of $\mathbf{n}(T) - \mathbf{n}(0)\mathbf{P}^T$. For large T, $\mathbf{P}^T$ will be negligible and so (3.46) may then be treated as the generating function of $\mathbf{n}(T)$. This amounts to saying that T must be large enough for almost all of the original members to have left.

If the input consists of a sequence of independent Poisson variates with mean R then

$$l(s) = e^{R(s-1)}.$$

Hence, in this case,

$$g(z_1, \ldots, z_{k+1}) = \exp\left[R\left\{\sum_{\tau=0}^{T-1}\sum_{j=1}^{k+1} p_j(T-\tau)z_j - 1\right\}\right]$$

$$= \exp\left\{\sum_{j=1}^{k+1}(z_j - 1)R\sum_{\tau=0}^{T-1} p_j(T-\tau)\right\}. \qquad (3.47)$$

This shows that the grade sizes, apart from original survivors, are distributed independently with expectation in the jth grade equal to

$$R\sum_{\tau=0}^{T-1} p_j(T-\tau) = R\sum_{\tau=1}^{T} p_j(\tau).$$

The probabilities $p_j(\tau)$ are obtained from

$$\mathbf{p}(\tau) = \mathbf{p}_0\mathbf{P}^{\tau-1} \qquad (3.48)$$

which is the distribution of the destination of an individual recruited at time $T - \tau$. In the limit as $\tau \to \infty$ the vector of expectations tends to

$$\bar{\mathbf{n}}(\infty) = R\mathbf{p}_0\sum_{\tau=1}^{\infty}\mathbf{P}^{\tau-1}$$

$$= R\mathbf{p}_0(\mathbf{I} - \mathbf{P})^{-1} \qquad (3.49)$$

which checks with the result obtained earlier for the general case. In the case of Poisson input we now see that this determines the complete distribution.

The method of derivation shows that the rate of approach to the asymptotic distribution is governed by how long it takes the original members to leave. It is easy to see that the result generalizes to the case when the mean input is a function of time, and this enables us to incorporate trends or oscillations into the input sequence. This remark is made precise in Mehlmann (1977a).

An example with a periodic transition matrix

So far we have assumed that the transition probabilities and the recruitment distribution do not depend on T. Without this assumption (3.1b) still holds, although the subsequent analysis becomes more complicated. In particular, there may be several limiting structures or none at all. An interesting example of a non-homogeneous process arose in Gani's study of student enrolment at Michigan State University. The academic year there consisted of three 'quarters' and transitions took place at the end of each quarter. It was not considered realistic to assume that the transition and recruitment probabilities would be the same in each quarter of a given year. However, it could be assumed that these probabilities would be the same in, say, the first quarter of one year as they had been in the first quarter of previous years. This requires us to make both $\mathbf{P}$ and $\mathbf{p}_0$ functions of T satisfying

$$\mathbf{P}(T+3) = \mathbf{P}(T)$$

and

$$\mathbf{p}_0(T+3) = \mathbf{p}_0(T) \qquad (T = 0, 1, 2).$$

Suppose that $T = 0$ refers to the first quarter of the first academic year and let

$$\left.\begin{array}{l} \mathbf{P}(3T+j) = \mathbf{P}_{j+1} \\ \mathbf{p}_0(3T+j) = \mathbf{p}_{0,j+1} \end{array}\right\} \qquad (j = 0, 1, 2; T = 0, 1, 2, \ldots).$$

Then $\mathbf{P}_j$ and $\mathbf{p}_{0j}$ relate to the j th quarter of any year. The grade structures may now be computed from the following difference equations:

$$\left.\begin{array}{l} \bar{\mathbf{n}}(3T+1) = \bar{\mathbf{n}}(3T)\mathbf{P}_1 + R(3T+1)\mathbf{p}_{02} \\ \bar{\mathbf{n}}(3T+2) = \bar{\mathbf{n}}(3T+1)\mathbf{P}_2 + R(3T+2)\mathbf{p}_{03} \\ \bar{\mathbf{n}}(3T+3) = \bar{\mathbf{n}}(3T+2)\mathbf{P}_3 + R(3T+3)\mathbf{p}_{01} \end{array}\right\} \qquad (T = 0, 1, 2, \ldots). \qquad (3.50)$$

These equations express the structure in a given quarter in terms of the structure in the previous quarter. For some purposes it is more convenient to relate the structures for the same quarter of succeeding years. Thus, for example, for the first quarter (3.50) gives

$$\bar{\mathbf{n}}(3T+3) = \bar{\mathbf{n}}(3T)\mathbf{P}_1\mathbf{P}_2\mathbf{P}_3 + R(3T+1)\mathbf{p}_{02}\mathbf{P}_2\mathbf{P}_3$$
$$+ R(3T+2)\mathbf{p}_{03}\mathbf{P}_3 + R(3T+3)\mathbf{P}_{01}. \qquad (3.51)$$

Similar expressions can be obtained for the second and third quarters.

The limiting behaviour of the system may be investigated by the methods of earlier sections. The case of greatest interest is when $R(T)$ is constant or tends to a limit. When this is so the limiting grade structure in the *first* quarter will be given by

$$\bar{\mathbf{n}}(\infty) = (\mathbf{I} - \mathbf{P}_1\mathbf{P}_2\mathbf{P}_3)^{-1}\{\mathbf{p}_{02}\mathbf{P}_2\mathbf{P}_3 + \mathbf{P}_{03}\mathbf{P}_3 + \mathbf{p}_{01}\}R \qquad (3.52)$$

where R is the limiting value of the input. The corresponding expressions for the second and third quarters can either be obtained by substituting back in (3.50) or by a repetition of the above argument.

3.3   A MODEL FOR AN EXPANDING SYSTEM WITH GIVEN SIZE

The equations of the model

The practical context in which the present model arose was described at the beginning of the chapter. The model differs from that described in the preceding section only in that the total size of the organization rather than the input is fixed. Instead of being given the sequence of inputs $\{R(T)\}$ we now have a sequence of total sizes $\{N(T)\}$. As before, this may be a sequence of given numbers or a realization of a known stochastic process. In the latter case the symbol $N(T)$ is to be interpreted as an expected size. The distinction is of no importance until we come to consider the distributions of grade sizes. Young and Almond (1961), who proposed the model, were concerned with expanding organizations and we shall, at first, follow them in this respect. Much of the theory is also applicable to the more general case of fluctuating or decreasing sequences.

Let $M(T)$ denote the increase in size which takes place between $T-1$ and T; thus $M(T) = N(T) - N(T-1)(T = 1, 2, \dots)$. Equation (3.1) remains valid but cannot be used as it stands because $\{R(T)\}$ is an unknown in this version of the problem. At any time the number of recruits must be sufficient to achieve the desired expansion and to replace losses from the system. The expected number of recruits required at time $T+1$ is thus

$$\bar{R}(T+1) = M(T+1) + \sum_{i=1}^{k} p_{i,k+1}\bar{n}_i(T) \qquad (T = 1, 2, \dots). \qquad (3.53)$$

By substituting $\bar{R}(T+1)$ for $R(T+1)$ in (3.1a), the difference equations for the expected grade sizes become

$$\bar{n}_j(T+1) = \sum_{i=1}^{k} (p_{ij} + p_{i,k+1}p_{0j})\bar{n}_i(T)$$

$$+ M(T+1)p_{0j} \qquad (j = 1, 2, \dots , k). \qquad (3.54a)$$

If we write $q_{ij} = p_{ij} + p_{i,k+1}p_{0j}$ this equation may be written in matrix notation as

$$\bar{\mathbf{n}}(T+1) = \bar{\mathbf{n}}(T)\mathbf{Q} + \mathbf{p}_0 M(T+1), \qquad (3.54b)$$

where $\mathbf{Q}$ is the matrix $\{q_{ij}\}$. This matrix equation has the same form as (3.1b) and leads to the result that

$$\bar{\mathbf{n}}(T) = \mathbf{n}(0)\mathbf{Q}^T + \mathbf{p}_0 \left\{ \sum_{\tau=0}^{T-1} M(T-\tau)\mathbf{Q}^\tau \right\}. \qquad (3.55)$$

The foregoing formulae provide all that is necessary for the numerical investigation of the expected structure. If we wish to study the form of $\bar{n}(T)$ by analytical methods two approaches are possible. One of these is to use the spectral representation for the power of a matrix as in Section 3.2, but the advantages of this method are not as great here as when the input was fixed. In that case we were able to obtain many explicit results for the special but important case when the matrix $\mathbf{P}$ was triangular. This was because the eigenvalues were then equal to the diagonal elements of the matrix. It is clear from inspection of the q_{ij}'s as defined above that $\mathbf{Q}$ will not normally be triangular and hence much of the simplicity of the method is lost. An alternative approach is to use the fact that $\mathbf{Q}$ is a stochastic matrix, a conclusion which follows by observing that

$$\sum_{j=1}^{k} q_{ij} = \sum_{j=1}^{k} p_{ij} + p_{i,k+1} \sum_{j=1}^{k} p_{0j} = 1 \qquad (i = 1, 2, \ldots, k).$$

This fact enables us to use known results about powers of stochastic matrices and, in particular, about their limiting behaviour. We shall use this method in the following sections.

The special case $M(T) = 0$, for all T, is particularly interesting. Our equations then relate to an open system of constant size, but they are identical in form to those used in the study of closed systems in Chapter 2. They thus provide a formal justification for our earlier remark that an open system in which gains and losses were equal could be treated as closed. Each person who leaves can be paired with a new entrant and the two changes treated as one. Thus a transition from grade i to grade j can either take place within the system or by loss from grade i and replacement to grade j with total probability $p_{ij} + p_{i,k+1}p_{0j}$.

The exact solution for geometric growth rate

In order to gain insight into the behaviour of expanding organizations we shall suppose the growth rate to be geometric. That is we suppose that

$$M(T) = Mx^{T} \qquad (T \geq 1) \tag{3.56}$$

where M is a positive constant and x is non-negative. By varying the values of M and x we can generate a considerable variety of growth patterns. The simplest case arises when there are only two strata and we shall consider this first.

When $k = 2$ the matrix $\mathbf{Q}$ can be written in the form:

$$\mathbf{Q} = \begin{pmatrix} 1-\alpha & \alpha \\ \beta & 1-\beta \end{pmatrix}$$

where $1 - \alpha = p_{11} + p_{01}p_{13}$ and $1 - \beta = p_{22} + p_{02}p_{23}$. A well-known result then gives

$$\mathbf{Q}^{\tau} = \frac{1}{\alpha + \beta} \left\{ \begin{pmatrix} \beta & \alpha \\ \beta & \alpha \end{pmatrix} + (1 - \alpha - \beta)^{\tau} \begin{pmatrix} \alpha & -\alpha \\ -\beta & \beta \end{pmatrix} \right\}. \tag{3.57}$$

Substituting for $\mathbf{Q}^\tau$ in (3.55), with $M(T-\tau) = Mx^{T-\tau}$, we find

$$\bar{\mathbf{n}}(T) = \frac{\mathbf{n}(0)}{\alpha+\beta}\left\{\begin{pmatrix} \beta & \alpha \\ \beta & \alpha \end{pmatrix} + (1-\alpha-\beta)^T\begin{pmatrix} \alpha & -\alpha \\ -\beta & \beta \end{pmatrix}\right\}$$
$$+ \frac{1}{\alpha+\beta}\sum_{\tau=0}^{T-1} Mx^{T-\tau}\mathbf{p}_0\left\{\begin{pmatrix} \beta & \alpha \\ \beta & \alpha \end{pmatrix} + (1-\alpha-\beta)^\tau\begin{pmatrix} \alpha & -\alpha \\ -\beta & \beta \end{pmatrix}\right\}. \qquad (3.58)$$

Summing the geometric series in (3.58) and using the facts that $n_1(0)+n_2(0) = N(0)$ and $p_{01}+p_{02} = 1$, the vector of expected grade sizes becomes

$$\bar{\mathbf{n}}(T) = \frac{1}{\alpha+\beta}\left[\left\{N(0) + M\frac{(x-x^{T+1})}{1-x}\right\}\right.$$
$$\times (\beta, \alpha) + (1-\alpha-\beta)^T(\alpha n_1(0) - \beta n_2(0), -\alpha n_1(0) + \beta n_2(0))$$
$$\left. + M\left\{\frac{x^T - (1-\alpha-\beta)^T}{1-(1-\alpha-\beta)x^{-1}}\right\}(\alpha p_{01} - \beta p_{02}, -\alpha p_{01} + \beta p_{02})\right] \qquad (3.59)$$

The total size of the organization at time T is

$$N(T) = N(0) + Mx(1-x^T)/(1-x). \qquad (3.60)$$

The variable, T, enters the expression for $\bar{\mathbf{n}}(T)$ through the terms in x^T and $(1-\alpha-\beta)^T$. Of these the latter exerts a diminishing influence as T increases because $1-\alpha-\beta < 1$. If $x > 1$ the terms in x^T become dominant, whereas if $x < 1$ they vanish in the limit. In the intermediate case where $x = 1$ the grade sizes increase without limit, but their relative values approach a limit as $T \to \infty$. In fact, for $x \le 1$ we have

$$\bar{\mathbf{n}}(T) \sim \frac{N(T)}{\alpha+\beta}(\beta, \alpha). \qquad (3.61)$$

In order to appreciate the full implications of (3.59) it is necessary to consider special cases. One such case is considered below; the reader will find it helpful to construct others.

The following illustration is for a two-grade system in which all recruits enter grade 1. There are no demotions and the probability of withdrawal is the same for each grade. The table of transition probabilities is

	1	0
$\frac{1}{2}$	$\frac{1}{4}$	$\frac{1}{4}$
0	$\frac{3}{4}$	$\frac{1}{4}$

For this example

$$1-\alpha = 1-\beta = \tfrac{3}{4}.$$

Substitution in (3.59) gives

$$\bar{\mathbf{n}}(T) = N(T)(\tfrac{1}{2}, \tfrac{1}{2}) + (\tfrac{1}{2})^{T+2}(n_1(0) - n_2(0), n_2(0) - n_1(0))$$

$$+ Mx\left\{\frac{x^T - (\tfrac{1}{2})^T}{x - \tfrac{1}{2}}\right\}(\tfrac{1}{2}, -\tfrac{1}{2}). \tag{3.62}$$

If x is fairly small the approach to the limit will be rapid. For example, if $x = \tfrac{1}{2}$,

$$n_1(T) = \tfrac{1}{2}\{N(0) + M(1 - (\tfrac{1}{2})^T)\} + (\tfrac{1}{2})^{T+1}\{n_1(0) - n_2(0)\} + (\tfrac{1}{2})^{T+2}TM. \tag{3.63}$$

The difference between the initial and final structures and the size of M both affect the rate of approach to the limit but their effect is not great. If x is near to 1 the approach to the limit may be very slow, as we shall see below.

When $k > 2$ there is no simple expression for $\mathbf{Q}^T$ but similar conclusions can be drawn. If the growth rate is geometric the matrix series (3.55) can almost always be summed. It is easy to verify that

$$(x\mathbf{I} - \mathbf{Q})\left(\sum_{\tau=0}^{T-1} \mathbf{Q}x^{T-\tau}\right) = x(x^T\mathbf{I} - \mathbf{Q}^T)$$

and hence, if $x\mathbf{I} - \mathbf{Q}$ possesses an inverse,

$$\sum_{\tau=0}^{T-1} \mathbf{Q}^\tau x^{T-\tau} = x(x\mathbf{I} - \mathbf{Q})^{-1}(x^T\mathbf{I} - \mathbf{Q}^T).$$

Thus we have

$$\bar{\mathbf{n}}(T) = \mathbf{n}(0)\mathbf{Q}^T + Mx\mathbf{p}_0(x\mathbf{I} - \mathbf{Q})^{-1}(x^T\mathbf{I} - \mathbf{Q}^T). \tag{3.64}$$

The inverse of $x\mathbf{I} - \mathbf{Q}$ does not exist if the determinant of the matrix vanishes. This will happen whenever x is equal to an eigenvalue of the matrix $\mathbf{Q}$. For a stochastic matrix it is known that at least one eigenvalue is equal to unity and that the remainder lie between zero and one in modulus. Hence (3.64) holds for all $x > 1$ and when $x < 1$ for all but at most $k - 1$ values of x. For numerical work with $x < 1$ it is preferable to use the recursive formula of (3.54b). The usefulness of the representation of (3.64) is that it provides direct information about the transient behaviour of the system. This behaviour depends on the terms x^T and $\mathbf{Q}^T$. It is known from the general theory of Markov chains that, if $\mathbf{Q}$ is regular, $\mathbf{Q}^T$ tends to a limit as T increases. As in the case $k = 2$ the long-term behaviour of the system depends crucially on the magnitude of x.

In order to illustrate the theory we shall take $k = 5$ and use the same transition probabilities as in the example of Section 3.2. Let the rate of growth be

$$M(T) = \frac{1}{10}N(0)x^T$$

with $x = 0, \tfrac{1}{2}, 1$, and 2. Because the total size of the organization is changing we have tabulated (Table 3.3) the relative expected grade sizes given by

$$q_j(T) = \bar{n}_j(T)/N(T) \qquad (j = 1, 2, \ldots, k).$$

Table 3.3 *Percentage values of the expected grade sizes for an organization with geometric growth rate and transition probabilities as in Section 3.2*

T	x	1	2	3	4	5
		\multicolumn{5}{c}{$100q_i(T)$}				
0	0	40	30	15	10	5
	$\frac{1}{2}$	40	30	15	10	5
	1	40	30	15	10	5
	2	40	30	15	10	5
1	0	35.6	32.2	15.8	10.8	5.8
	$\frac{1}{2}$	37.4	31.8	15.0	10.2	5.5
	1	39.1	31.5	14.3	9.8	5.2
	2	42.1	31.0	13.1	9.0	4.8
2	0	32.5	32.8	16.6	11.5	6.5
	$\frac{1}{2}$	34.8	32.8	15.7	10.7	6.1
	1	38.4	32.4	14.2	9.6	5.4
	2	46.6	31.3	10.9	7.2	4.1
5	0	27.8	31.1	18.2	13.8	9.1
	$\frac{1}{2}$	29.2	32.0	17.7	12.8	8.3
	1	36.3	33.2	14.8	9.6	6.1
	2	59.6	31.6	5.4	2.2	1.3
10	0	24.8	27.5	17.7	16.2	13.8
	$\frac{1}{2}$	25.3	28.2	17.9	15.7	12.8
	1	33.8	32.4	15.8	10.6	7.4
	2	63.6	31.9	3.9	0.6	0.1
25	0	22.3	23.8	14.6	15.5	23.8
	$\frac{1}{2}$	22.4	23.9	14.7	15.6	23.3
	1	29.3	29.4	15.9	13.0	13.0
	2	63.8	31.9	3.8	0.5	0.0
50	0	21.8	23.0	13.8	13.9	27.3
	$\frac{1}{2}$	21.8	23.1	13.9	14.0	27.3
	1	26.2	26.8	15.2	13.6	18.1
	2	63.8	31.9	3.8	0.5	0.0

When $x = 0$ there is no expansion and the limit is approached rather slowly, especially in the highest grade. This is due mainly to the big difference between the initial and limiting structures. When $x = \frac{1}{2}$ the ultimate increase in total size is 10 per cent, but the rate of approach to the limiting structure is hardly affected. The case $x = 1$ shows a very slow approach to the limit, though here again the lower grades attain their limiting values more quickly than the higher ones. In this example grade 5 is little more than half its limiting value after 50 years. Under this kind of expansion the structure would not achieve equilibrium in any period likely to be of practical interest. By contrast, when $x = 2$, the limit is reached quite rapidly, although in this case it is a different limit. Few organizations would be able to maintain such a rapid rate of growth long enough for this final structure to be of real importance.

The limiting structure

Expressing (3.46b) in terms of the q's we have

$$q(T+1)\{1 + M(T+1)/N(T)\} = q(T)Q + p_0 M(T+1)/N(T). \qquad (3.65)$$

The possibility of there being a limiting structure q therefore depends on the behaviour of $M(T+1)/N(T)$, in particular it must tend to a limit as T increases. Let

$$\lim_{T \to \infty} M(T+1)/N(T) = c$$

then the limiting structure, if it exists, satisfies

$$q(1+c) = qQ + p_0 c \qquad (3.66)$$

and hence

$$q = c p_0 \{I(1+c) - Q\}^{-1} \qquad \text{if } c > 0. \qquad (3.67)$$

If $c = 0$ the limiting structure is precisely that for a closed Markov chain model with transition matrix Q. When the growth rate is geometric with $M(T) = Mx^T$, $c = 0$ if and only if $x \le 1$, and we observed in Table 3.1 the approach to the same limit in the three cases $x = 0, \frac{1}{2}, 1$. When $x > 1$, $c = x - 1$, and in that case the limiting structure is given by

$$q \propto p_0 \{Ix - Q\}^{-1} \qquad (3.68)$$

as may be deduced directly from (3.64). When $x = 2$ this yields

$$100q = (63.8, 31.9, 3.8, 0.5, 0.0)$$

which is identical with the last line of Table 3.3. The existence of the limit for $q(T)$ when $\lim_{T \to \infty} M(T+1)/N(T) = c$ can be deduced by writing (3.65) in the form

$$q(T+1) = q(T) P(T) + p_0 R(T+1),$$

where

$$\mathbf{P}(T) = \mathbf{Q}\{1 + M(T+1)/N(T)\}^{-1} \quad \text{and} \quad R(T+1) = M(T+1)/N(T+1)$$

and then appealing to the theory of products of positive matrices.

The results of this section can thus be summarized by saying that if $M(T) \sim Mx^T$ the steady-state structure is the same as in a fixed size system if $x \le 1$. Otherwise it is given by (3.68). If x becomes very large we have $\mathbf{q} = \mathbf{p}_0$, a result which could have been anticipated by noticing that with such a rapid rate of growth most members of the system would be new recruits.

Variances and covariances of the grade sizes

The method used to obtain the variances and covariances for the model with given input can be applied here with only minor modifications. If we start from (3.38) we can no longer assume that the input is independent of the internal movements, because the latter influence the wastage which, in turn, determines the input. To the two terms on the right-hand side of (3.39) we must therefore add terms involving the covariances between the $n_{0j}(T+1)$'s and the internal flows. However, the fairly heavy algebra can be circumvented by arguing as follows. Each loss is directly responsible for a recruitment; if the loss occurs from grade i the probability that the consequential recruitment is in grade j is $p_{i,k+1}\,p_{0j}$. The replacement of a leaver can thus be treated as if it were an internal transfer. The total probability of a move (of any kind) out of grade i which results in an addition to grade j is thus

$$q_{ij} = p_{ij} + p_{i,k+1}\,p_{0j}. \tag{3.69}$$

This is the same point as we made in another connection on page 73.

If $M(T) = 0$ for all T we can therefore use the theory developed in Chapter 2 for the case of a closed system. The vector of expectations, variances, and covariances will thus be given by

$$\boldsymbol{\mu}(T+1) = \boldsymbol{\mu}(T)\boldsymbol{\Pi}' \tag{3.70}$$

where the elements of $\boldsymbol{\Pi}'$ are the same functions of the q_{ij}'s as $\boldsymbol{\Pi}$ in (2.15) is of the p_{ij}'s.

When the system is expanding there are additional recruits who fill the vacancies created by expansion. If these are allocated to grades independently of the replacements and with the same probabilities $\{p_{0j}\}$, then these flows will be independent of all other flows and so their contribution to the variances and covariances can simply be added to (3.70). Let the vector of expectations, variances, and covariances attributable to the expansion be $\boldsymbol{\mu}_0'(T+1)$. Then

$$\boldsymbol{\mu}_0'(T+1) = M(T+1)\{p_{01}, p_{02}, \ldots, p_{0k}, p_{01}(1-p_{01}), -p_{01}p_{02} \cdots$$

$$\cdots p_{0k}(1-p_{0k})\}, \tag{3.71}$$

the covariance of the flows into grades i and j being

$$\delta_{ij} p_{0j} - p_{0j} p_{0l} \qquad (i,j = 1, 2, \ldots, k).$$

The total expression for the difference equation is thus

$$\mu(T+1) = \mu(T)\Pi' + \mu_0'(T+1). \qquad (3.72)$$

The variances and covariances can therefore be computed using the same computer program as for the fixed input, but with Π replaced by Π' and $R(T+1)$ by $M(T+1)$.

Contracting systems

The argument leading to (3.54a) and (3.54b) may break down if $M(T+1)$ is negative for any T. If this happens it may be that the expected number of losses, given by

$$\sum_{i=1}^{k} p_{i,k+1}\, \bar{n}_i(T),$$

is not sufficient to achieve the reduction in size implied by $M(T+1)$. In such circumstances $\bar{R}(T+1)$ as given by (3.53) will be negative. Since the model as described so far makes no provision for dealing with redundancies such an occurrence would bring the calculations to a halt.

If we wish to allow for redundancies the model must be extended. Two possibilities are as follows. One is to specify a redundancy vector $\mathbf{s} = (s_1, s_2, \ldots, s_k)$ whose elements are the proportions of $|R(T)|$ to be declared redundant in each grade. Instead of adding the vector $R(T)\mathbf{p}_0$ to $\mathbf{n}(T)\mathbf{P}$ we must then subtract $|R(T)|\mathbf{s}$. The basic difference equation for the expected grade sizes now takes one of two forms according to the sign of $R(T)$ as follows.

$$\bar{n}_j(T+1) = \sum_{i=1}^{k} (p_{ij} + p_{i,k+1} p_{0j})\bar{n}_i(T) + M(T+1)p_{0j}$$

$$(j = 1, 2, \ldots, k) \quad (3.73a)$$

if $\bar{R}(T+1) \geq 0$ or

$$\bar{n}_j(T+1) = \sum_{i=1}^{k} (p_{ij} + p_{i,k+1} s_j)\bar{n}_i(T) + M(T+1)s_j,$$

$$(j = 1, 2, \ldots, k) \quad (3.73b)$$

if $\bar{R}(T+1) < 0$. Note that $M(T+1)$ will certainly be negative in (3.73b) and may be negative in (3.73a). We must therefore alternate between (3.73a) and (3.73b) according to whether the current situation requires recruitment or redundancy. If the vector $\mathbf{s}$ is chosen to be the same as $\mathbf{p}_0$ it is clear that the two equations become identical and we can use the original equation (3.54b) as if it were valid for any $M(T+1)$. In doing this, however, we must check at each step that none of the elements of $\bar{\mathbf{n}}(T+1)$ has become negative. If negative elements do occur further progress can only be made by choosing a new $\mathbf{s}$.

An alternative method, which avoids the possibility of negative stock sizes, is to compute the number of redundancies as a proportion of the stocks after wastage and promotion have taken place. In the absence of recruitment or redundancy the expected size of the jth grade at $T+1$ is

$$\bar{n}'_j(T+1) = \sum_{i=1}^{k} p_{ij}\bar{n}_i(T) \qquad (j = 1,2,\ldots,k).$$

Redundancies will be necessary if

$$\sum_{j=1}^{k} \bar{n}'_j(T+1) > N(T+1)$$

and the rule proposed is to reduce each $\bar{n}'_j(T+1)$ by the same proportion. If this proportion is denoted by x then it must satisfy

$$x \sum_{j=1}^{k} \bar{n}'_j(T+1) = N(T+1)$$

or

$$x = N(T+1) \Big/ \sum_{i=1}^{k} (1 - p_{i,k+1})\bar{n}_i(T). \tag{3.74}$$

If this fraction is applied to each grade the resulting stock numbers will be

$$\bar{n}_j(T+1) = N(T+1) \sum_{i=1}^{k} p_{ij}\bar{n}_i(T) \Big/ \left\{ N(T) - \sum_{i=1}^{k} p_{i,k+1}\bar{n}_i(T) \right\}$$

$$(j = 1,2,\ldots,k) \quad (3.75a)$$

or

$$\mathbf{n}(T+1) = \mathbf{n}(T)\mathbf{P}\{1 + M(T)/N(T)\}/\{1 - W(T)\}, \tag{3.75b}$$

where $W(T) = \bar{\mathbf{n}}(T)\mathbf{p}'_{k+1}$. In this version of the model the last equation would be used whenever $\bar{R}(T+1)$ as given by (3.53) turned out to be negative.

3.4.   COMPLEMENTS TO CHAPTERS 2 AND 3

Our models draw extensively on the theory of discrete time Markov chains. Accounts of that subject, at a variety of levels, will be found in almost all modern texts in probability theory and stochastic processes. Feller (1968), Moran (1968), Cox and Miller (1965), and Karlin and Taylor (1975) are four examples. There are also a few books devoted entirely to Markov chains. A basic reference at an advanced level is Chung (1967), but Kemeny and Snell (1976) (reprint of 1960 publication) still provides one of the clearest accounts of the theory for chains with a finite number of states. More recently, Isaacson and Madsen (1976) provide a good coverage including a treatment of time-dependent chains.

The mathematics on which the theory depends is largely concerned with the theory of matrices and their associated eigenvalues and eigenvectors. Again, there are many sources; basic references are Frazer and co-workers (1946), the two-volume work by Gantmacher (1964), and Lancaster (1969). Since transition matrices belong to the wider class of positive matrices the book by Seneta (1973) is particularly useful.

Open and closed Markov models have found applications in almost all branches of social science and the literature is voluminous. The following list, by subject area, is intended as an introduction for readers who wish to pursue the subject within their own field of interest. It is by no means exhaustive and the subject classification is necessarily somewhat arbitrary. In particular the dividing line between geography and economics and between geography and demography is very difficult to draw. Readers whose interests lie in these areas should also consult adjacent categories. Publications are referred to by author's name and year; the full references are in the Bibliography. Some papers mentioned in the text of the two chapters are not repeated here.

Education and manpower planning

Applications to manpower planning can be approached through Bartholomew and Forbes (1979) which contains a full bibliography. A somewhat broader but less comprehensive bibliography which also covers education planning and national manpower planning is given in Bartholomew (1976b). Occupational mobility merges on the one hand with social mobility and on the other with manpower planning. Papers not mentioned elsewhere which fall within this area are Hodge (1966) and Stewman (1975a, 1975b, 1976). Education planners have made considerable use of Markov models for which see Gani (1963), Clough and McReynolds (1966), Kamat (1968b and c), Thonstad (1969), Armitage and co-workers (1969), Armitage and co-workers (1970), Menges and Elstermann (1971), Johnstone and Philp (1973), Moore (1975) (a reply to Johnstone and Philp), and Britney (1975).

Demography

Demographers have frequently used models which are in essence the same as Markov chains. More recently this fact has tended to become more explicit as the following papers will show: Matras (1960a, b, 1967), Rogers (1968), Sykes (1969), Krishnan (1977b), Feeney (1973), Feichtinger (1972, 1973), Rees and Wilson (1973), Joseph (1974), and Salkin and co-workers (1975).

Geography

Spatial analysis has come to occupy a prominent place in contemporary geographical studies. This is partly concerned with movement of population, but also includes questions of land use and industrial location and size. The following list spans this broad spectrum of topics: Clark (1965), Tarver and Gurley (1965), Morrison (1967), Drewett (1969), Brown (1970), Long (1970), Bell (1974), Berry (1971), Collins (1973), and Stafford (1977).

Economics

Applications here range very widely covering such things as income distribution, size of firms, flows of wool, stock price movements, market shares, and so forth. Some of these topics will be discussed in Chapter 7 where we deal with size distributions. A partial list in this area is as follows: Hart and Prais (1956), Adelman (1958), Preston and Bell (1961), Collins and Preston (1961), Dent (1967), Dryden (1969), Fielitz and Bhargava (1973), and Shorrocks (1976).

Health care planning

Markov models have found many applications in this field where the states are typically different kinds of provision. Examples will be found in Marshall and Goldhamer (1955), Navarro (1969), Meredith (1973), Anderson (1974), and Davies and co-workers (1975).

In addition to these major groupings we may note applications to voting behaviour (Anderson, 1954; Hawkes, 1969; Miller, 1972), interview communication (Hawes and Foley, 1973; Hedge and co-workers, 1978), accounting (Cyert and co-workers, 1962), dentistry (Lu, 1968), computer systems (Foley, 1967), linguistics (Miller, 1952), and interpersonal relations (Katz and Procter, 1959; Mayhew and Gray, 1971; Mayhew, 1972),

Social mobility is a much broader subject than our rather limited treatment of Chapter 2 suggests. A good introduction to the quantitative aspect is provided by Boudon (1973). Bibby (1975) is mainly concerned with measurement, but includes a useful general discussion with references. There is now a considerable collection of published social mobility tables. Some are mentioned in Bibby (1975) and a convenient listing of 20 national tables is in Table 1 of Sommers and Conlisk (1979) who give the sources in an appendix.

Shorrocks' (1978) treatment of the measurement of mobility proceeds by listing a set of properties which seem desirable. He then investigates how far they are mutually consistent. One of his main conclusions is that no measure exists which meets all his requirements. This appears to be a reflection of the fact that no distinction is made between intergeneration dependence and amount of movement. Shorrocks introduces, for example, the notion of period consistency which expresses the fact that a measure should order populations in the same way whatever the period of observation. Thus if two societies have transition matrices $\mathbf{P}_1$ and $\mathbf{P}_2$ and if for some measure μ, $\mu(\mathbf{P}_1) > \mu(\mathbf{P}_2)$, then we would want $\mu(\mathbf{P}_1^T) > \mu(\mathbf{P}_2^T)$ for all positive integers, T. The measure $m_2(\mathbf{P})$ does not possess this property. However, $m_2(\mathbf{P})$ was introduced as a measure of movement for which period consistency does not seem relevant. It is more concerned with what we have called intergeneration dependence.

In addition to the measures discussed in Chapter 2, Sommers and Conlisk (1979) discuss what they call a regression to the mean coefficient. They generalize the correlation coefficient measure, ρ, by considering the rth root of ρ calculated from $\mathbf{P}^r$, they also establish some inequalities for the various measures.

In our discussion of mobility measurement we have emphasized that everything depends on the realism of the Markov assumption. According to the model the flows produce changes in the structure and not vice versa. If, on the other hand, changes in structure generate the flows other considerations come into play. In this connection a distinction may be drawn between what is called *structural* and *pure* mobility. Structural mobility refers to the minimum amount of movement necessary to achieve a given structural change. Pure mobility is the excess of the actual movement over this minimum. Boudon (1973) and others have considered how to isolate and measure the pure component of mobility. In so far as our measures relate to a system in equilibrium—when, by definition, there is no change in structure—they are measures of pure mobility.

The phenomenon of clustering on the diagonal on the Tth power of a transition matrix has been extensively investigated by Singer and Spilerman (1977b). As we have noted, the situation is straightforward when the transition matrix has real positive eigenvalues and this seems to be the usual case in practice. What appears to be lacking is a characterization of matrices having real positive eigenvalues in terms which have substantive sociological meaning. If such could be found it might also help to illuminate the surprising phenomenon of reversibility which seems to be typical of mobility matrices, since a reversible matrix has real positive eigenvalues.

Since the second edition was published Feichtinger (1976) and Mehlmann (1977a) have given a more complete treatment of the limiting behaviour of open Markov models and this is reflected in the new accounts of these topics in Chapter 3. This is both simpler and more general than that originally given and it brings the treatment into line with that given later for the continuous time model.

Various attempts have been made to generalize Markov models in directions not covered in Chapters 2 and 3. Moya-Angeler (1976), for example, considers a model for education planning in which there are capacity constraints which place limits on the grade sizes. Once these limits are reached the surplus has to be relocated in other grades. This is akin to the bottleneck models introduced earlier by Armitage and co-workers (1969). With similar aims in view, Young and

Vassiliou (1974) allow the numbers promoted to depend not only on the stock from which they come—as in the Markov model—but also on the stock of the destination grade. These various generalizations lead to models which are intermediate between the Markov models of Chapter 3 and the renewal models to be discussed in Chapter 8. Schinnar and Stewman (1978) discuss a general class of models incorporating what they call 'duration memory'. These include the cumulative inertia effect as a special case, but also embrace heterogeneity and forward-looking memory effects.

The work described here on the variances and covariances of the stock numbers assumes that the parameters of the model are known. In practice this will rarely be the case. They will have to be estimated and this will introduce further errors into any predictions. Yet more variation will arise if the parameters are themselves subject to random variation. These points are investigated in Bartholomew (1975) where it is shown, for example, that if **P** is estimated from data over one time interval the variances and covariances of the stock numbers one step ahead are likely to be roughly doubled.

We have not discussed the statistical problems of estimating parameters of Markov models. The subject is treated in Bartholomew (1977a) and Bartholomew and Forbes (1979) and these will lead to other references. Particular care is required with social mobility tables where it is not always made clear how the population was sampled. One needs to know, for example, whether sons were questioned about their fathers or vice versa.

CHAPTER 4

Continuous Time Models for Closed Social Systems

4.1 THEORETICAL BACKGROUND

We now turn to the development of continuous time versions of the Markov chain models introduced in Chapter 2. The choice, in practice, between the discrete and continous versions of a model is partly a matter of realism and partly one of convenience. On grounds of realism, for example, one would usually want to model the movement of people between occupations or regions in continuous time, but in practice the computational advantages of treating time as discrete have often led to the choice of a discrete time model. Continuous time models are often more amenable to mathematical analysis and this may count in their favour even when realism calls for a discrete model. In order to bring out the close parallel between the two cases the pattern of development in this chapter follows that in Chapter 2 fairly closely. In particular, we shall derive the main theoretical results by limiting operations on formulae already obtained. Although this is not always the way in which a mathematician would want to proceed, it serves to emphasize the essential unity of the models.

In the discrete time Markov chain, changes of state take place at unit intervals of time. There are two ways of making the change to continuous time which we shall consider in turn. One is to allow the time interval between transitions to become smaller and smaller. Doing this alone is not sufficient because it merely speeds up the process by changing the time scale. In addition, therefore, we must allow the transition probabilities to change in such a way that the expected number of changes of state per unit time remains the same. In particular, the transition probabilities out of each state must decrease as the time interval between changes becomes shorter. This can be achieved as follows. Let the length of the intervals between transitions be δT and write:

$$p_{ij} = r_{ij}\delta T, (i \neq j); \qquad p_{ii} = 1 - \delta T \sum_{\substack{j=1 \\ j \neq i}}^{k} r_{ij}. \qquad (4.1)$$

Then as $\delta T \to 0$, the probability of a transition out of any state approaches zero, but the mean time interval between changes of state is constant. To see this we note that the mean number of steps in state i is $(1 - p_{ii})^{-1}$ (see p. 30); each step involves a stay of length δT so the mean length of stay (sojourn time) is

$$\delta T (1 - p_{ii})^{-1} = \left\{ \sum_{j \neq i} r_{ij} \right\}^{-1} = \lambda_i^{-1}, \text{ say.}$$

The stochastic process at which we arrive in the limit as $\delta T \to 0$ is a continuous time Markov chain. The p_{ij}'s given by (4.1) are often called infinitesimal transition probabilities. The r_{ij}'s are called transition *rates* or *intensities* and they measure the propensities to move between pairs of states. A time-dependent chain can be defined in which the transition rates are functions of time. Transition rates are frequently used in actuarial work and reliability studies under various names; perhaps the most familiar and expressive is the term 'force of mortality'. In our terminology this is the transition rate from the state 'life' to the state 'death'. Care must be taken not to confuse rates with probabilities; the relationship between the two is given by (4.1).

If we write $\mathbf{P}(\delta T)$ for the matrix of transition probabilities in (4.1) then we have

$$\mathbf{P}(\delta T) = \mathbf{I} + \delta T \mathbf{R} \tag{4.2}$$

where $\mathbf{R}$ is a matrix with the r_{ij}'s in the off-diagonal positions and with $r_{ii} = -\lambda_i = -\sum_{j \neq i} r_{ij}$ on the diagonal. It is convenient to define r_{ii} in this way in order to have the simple matrix representation of (4.2). The main theoretical results we need for applications can now be obtained by carrying out appropriate limiting operations on the corresponding results of Chapter 2.

Before turning to these we shall consider the second way of making the transition from discrete to continuous time. In the discrete time model changes of state take place at fixed intervals of time. We could generalize this by allowing them to take place at random intervals of time. Our process would then be defined by two stochastic processes. One would be a Markov chain determining what change of state occurs when the time comes for a change; the second would be a point process whose realization would give the times at which changes of state take place. The ordinary Markov chain then emerges as a special case when this point process degenerates into a set of points at fixed intervals. This way of extending the discrete time Markov model is, of course, much more general than the first method since it gives rise to as many models as there are point processes. We shall consider several process of this family during the course of this chapter and the next, but here we mention only one which happens to coincide with the continuous time Markov chain defined above.

Suppose that changes of state take place according to a Poisson process with rate λ. That is,

$$Pr\{\text{change takes place in } (T, T + \delta T)\} = \lambda \delta T$$

from which it follows that the probability of n changes of state taking place in

$(0, T)$ is the Poisson probability

$$P_n(T) = \frac{(\lambda T)^n}{n!} e^{-\lambda T}.$$

Further, let $\mathbf{P}$ be the transition probability matrix governing changes of state. We shall show below that such a process is a continuous time Markov chain with

$$\mathbf{R} = \lambda(\mathbf{P} - \mathbf{I}). \qquad (4.3)$$

The basic equations

Two sets of probabilities which we commonly wish to know are those denoted by $\mathbf{p}(T)$ and $\mathbf{P}^{(T)}$ in Chapter 2 (p. 15). The former, which are the probabilities of being found in the various states at time T, can easily be deduced from (2.2) as follows:

$$\mathbf{p}(T + \delta T) = \mathbf{p}(T)\,\mathbf{P}(\delta T) = \mathbf{p}(T)\,\{\mathbf{I} + \delta T\,\mathbf{R}\}$$

giving

$$\{\mathbf{p}(T + \delta T) - \mathbf{p}(T)\}/\delta T = \mathbf{p}(T)\mathbf{R}$$

which in the limit as $T \to \infty$ gives

$$\frac{\mathrm{d}}{\mathrm{d}T}\mathbf{p}(T) = \mathbf{p}(T)\mathbf{R}. \qquad (4.4)$$

Here, and subsequently, the differential operator is to be understood as applying to each element of the vector or matrix which follows. The form of the solution of (4.4) is suggested by the special case when R and $p(T)$ are scalars. In that case we know that the solution is

$$p(T) = p(0)e^{RT} = p(0) \sum_{i=0}^{\infty} (RT)^i/i!. \qquad (4.5)$$

In the matrix case it is easily verified that the series on the right-hand side of (4.5) satisfies (4.4) if we define $\mathbf{R}^0 = \mathbf{I}$. It may be shown that this is the solution required if the series converges, and that this is always so if k is finite. We may, therefore, write the solution as

$$\mathbf{p}(T) = \mathbf{p}(0) \exp\{\mathbf{R}T\}. \qquad (4.6)$$

The matrix $\mathbf{P}(T)$, whose (i, j)th element is the probability of moving from i to j between 0 and T, obviously satisfies

$$\mathbf{p}(T) = \mathbf{p}(0)\,\mathbf{P}(T) \qquad (4.7)$$

and so we deduce that

$$\mathbf{P}(T) = \exp\{\mathbf{R}T\}. \qquad (4.8)$$

Anticipating the needs of the Chapter 5, we shall work here with the vector of expected numbers $\bar{\mathbf{n}}(T)$, rather than $\mathbf{p}(T)$; thus $\bar{\mathbf{n}}(T) = \mathbf{n}(0)\,\mathbf{P}(T)$.

We are now in a position to verify (4.3). For the process defined there we have that

$$\mathbf{P}(T) = \sum_{n=0}^{\infty} P_n(T)\mathbf{P}^n$$

since $\mathbf{P}^n$ gives the transition probabilities *given that n transitions* have occurred. Substituting for $P_n(T)$ and noting that $\mathbf{I}e^{-\lambda T} = e^{-I\lambda T}$ we find

$$\mathbf{P}(T) = \sum_{n=0}^{\infty} \frac{(\lambda T)^n}{n!} e^{-\lambda T} \mathbf{P}^n = \exp\{\lambda(\mathbf{P}-\mathbf{I})T\}. \tag{4.9}$$

Comparison with (4.8) now yields (4.3).

Although the two ways of arriving at the continuous time Markov model are equivalent, there are practical reasons why one or other may be preferred in a particular application. These have greater significance when we generalize the model later in this chapter. The essential point concerns whether or not it is meaningful to talk about changes of category within a state. For example, in occupational mobility, the states of the system may be defined by industry. A person who changes their job, but within the same industry, has not made a change of state, but their behaviour can be distinguished from someone who has remained in the same job. In modelling such a process by a Markov chain, λ would represent the rate of changing *jobs* and $\mathbf{P}$ would be the transition probability matrix for industries. The non-zero elements on the diagonal of $\mathbf{P}$ would thus represent the probabilities of moving within the same industrial category. Note that it is necessary that λ should be the same for all states. In fitting the model to such a process one would want to express the results in terms of estimates of λ and $\mathbf{P}$.

In most applications, on the other hand, it has no meaning to speak of movements within the same state. This is certainly the case with the second of the examples mentioned in Section 4.2, and also with those in Chapter 5. In such applications the definition of the process in terms of the transition rates is more natural. However, even if this is the case, it may be instructive to reparameterize the model so as to focus attention separately on the two aspects—the durations in each state and the changes of state. This may be done as follows. The length of time a discrete Markov chain stays in state i has a geometric distribution with

$$Pr\{\text{sojourn time in } i = r\} = (1 - p_{ii}) p_{ii}^{r-1} \qquad (r = 1, 2, \ldots) \tag{4.10}$$

Under the limiting operations carried out at the beginning of this section this yields a sojourn time density which is exponential with parameter $\lambda_i = \sum_{j \neq i} r_{ij}$. The probability that the system changes from i to j $(i \neq j)$ given that a move out of i takes place is

$$m_{ij} = p_{ij}/(1 - p_{ii}) \qquad (i \neq j)$$

which, on substituting from (4.1) becomes

$$m_{ij} = r_{ij}/\lambda_i \qquad (i \neq j) \tag{4.11}$$

with $\sum_j m_{ij} = 1$. The process may thus be defined in terms of the rate parameters, λ_i, and the transition matrix $\mathbf{M}$ which has zeros on the main diagonal and m_{ij} in the (i, j)th position $(i \neq j)$. To summarize, (4.3) may be extended to give three equivalent parameterizations as follows:

$$\mathbf{R} = \lambda(\mathbf{P} - \mathbf{I}) = \Lambda(\mathbf{M} - \mathbf{I}) \tag{4.12}$$

where Λ is the diagonal matrix with λ_i in the ith position. This shows that if the sojourn time in i depends on i then the transition matrix must have zeros on the diagonal. Conversely, if the transition matrix does not have zeros on the diagonal the process cannot be Markovian unless the sojourn time distributions for all states are the same.

Solution of the basic equations

The direct determination of $\mathbf{P}(T)$ from (4.8) would involve the summation of k^2 infinite series whose terms would be obtained from the powers of $\mathbf{R}$. A better method is to use Sylvester's theorem (see Section 3.2) which enables us to express the solution as a finite series of k terms. In general the matrix $\mathbf{R}$ admits the spectral representation

$$\mathbf{R} = \sum_{i=1}^{k} \theta_i \mathbf{A}_i$$

where $\{\theta_i\}$ are the eigenvalues of $\mathbf{R}$ and $\{\mathbf{A}_i\}$ is the associated spectral set. The representation is only valid if all the θ's are distinct. It then follows from Sylvester's theorem that

$$\exp(\mathbf{R}T) = \sum_{i=1}^{k} e^{\theta_i T} \mathbf{A}_i \tag{4.13}$$
$$= \mathbf{P}(T),$$

using the matrix form of (4.5). Likewise, we have

$$\bar{\mathbf{n}}(T) = \sum_{i=1}^{k} e^{\theta_i T} \mathbf{n}(0) \mathbf{A}_i. \tag{4.14}$$

The behaviour of the solution, especially for large T, will clearly depend critically on the eigenvalues. Before proceeding to a discussion of the practical steps needed to obtain a complete solution we shall therefore obtain some general results about their values.

The eigenvalues are obtained by solving the equation

$$\begin{vmatrix} r_{11} - \theta & r_{21} & \cdot & \cdot & \cdot & \cdot & r_{k1} \\ r_{12} & r_{22} - \theta & & \cdot & \cdot & \cdot & r_{k2} \\ \vdots & \vdots & & & & & \vdots \\ r_{1k} & r_{2k} & \cdot & \cdot & \cdot & & r_{kk} - \theta \end{vmatrix} = 0. \tag{4.15}$$

The value of the determinant is unchanged if we replace the first row by a row whose elements are the column sums. Each element in the first row of the new determinant so formed is $-\theta$. The equation (4.15) is therefore always satisfied when $\theta = 0$. It is clear from (4.13) that all roots must be zero or negative otherwise there would be some T for which the transition probabilities did not lie between zero and one.

Since we are mainly interested in the vector $\bar{\mathbf{n}}(T)$ we shall consider the solution given by (4.14). It follows from the foregoing remarks that $\bar{n}_i(T)$'s can be expressed in the form

$$\bar{n}_i(T) = c_{i1} + \sum_{j=2}^{k} c_{ij} e^{\theta_j T} \qquad (i = 1, 2, \ldots, k) \qquad (4.16)$$

where $\theta_j < 0 (j = 2, 3, \ldots, k)$. This form is valid provided that all the θ's are distinct. It holds also, with a slight modification discussed below, if $\theta = 0$ is a multiple root. The coefficients $\{c_{ij}\}$ in (4.16) may be found by first determining the spectral set $\{\mathbf{A}_i\}$. An equivalent, but more direct approach, is the following. Substituting $\bar{\mathbf{n}}(T)$ from (4.16) into (4.4), we have

$$\sum_{j=1}^{k} c_{ij} \theta_j e^{\theta_j T} = \sum_{j=1}^{k} r_{ji} \sum_{h=1}^{k} c_{jh} e^{\theta_h T} \qquad (i = 1, 2, \ldots, k) \qquad (4.17)$$

where $\theta_1 = 0$. Equating coefficients of $e^{\theta_h T}$ we find that the c_{ij}'s must satisfy the equations

$$\sum_{j=1}^{k} r_{ji} c_{jh} = \theta_h c_{ih} \qquad (i = 1, 2, \ldots, k; h = 1, 2, \ldots, k). \qquad (4.18)$$

Although there are k^2 equations here for the same number of unknowns they are not independent. In fact, if we sum each side of (4.18) over i we obtain zero in each case. In order to determine the c's we therefore require k further equations. These arise from the necessity of ensuring that the initial conditions are satisfied. Thus, setting $T = 0$ in (4.16), we have

$$\sum_{j=1}^{k} c_{ij} = n_i(0) \qquad (i = 1, 2, \ldots, k). \qquad (4.19)$$

We shall solve the equations in the case of particular applications later.

The theory given above covers the case when all the θ's are distinct. If multiplicities occur among the roots the form of the solution can be determined by an appropriate limiting operation on (4.16). For example, suppose that $\theta_2 = \theta_3 = \theta$. The coefficients $\{c_{ij}\}$ are functions of the θ's and we must determine their limits as $\theta_2 \to \theta_3$. The terms involving θ_2 and θ_3 in the exponent require special attention. In a typical case we have

$$
\left.
\begin{aligned}
&\lim_{\theta_2 \to \theta_3 = \theta} \{c_{i2}e^{\theta_2 T} + c_{i3}e^{\theta_3 T}\} \\
&\quad = e^{\theta T} \lim_{\theta_2 \to \theta_3} \{c_{i2} + c_{i3}e^{(\theta_3 - \theta_2)T}\} \\
&\quad = e^{\theta T} \lim_{\theta_2 \to \theta_3} \{c_{i2} + c_{i3} + c_{i3}((\theta_3 - \theta_2)T + O(\theta_3 - \theta_2)^2)\}.
\end{aligned}
\right\}
\tag{4.20}
$$

Since the limit in curly brackets cannot be infinite in any meaningful problem, the pair of terms corresponding to $j = 2$ and $j = 3$ in (4.16) must be replaced by a single term of the form

$$
(d_{i2} + d_{i3}T)e^{\theta T}.
\tag{4.21}
$$

In general, if there is a root of multiplicity m with common value θ, the terms corresponding to that root in (4.16) must be replaced by the following expression:

$$
\sum_{j=2}^{m+1} d_{ij}T^{j-2}e^{\theta T}.
\tag{4.22}
$$

A particularly simple and important case occurs when the multiple root has the value zero. The general solution then has the form

$$
\bar{n}_i(T) = \sum_{j=1}^{m} d_{ij}T^{j-1} + \sum_{j=m+1}^{k} c_{ij}e^{\theta_j T} \qquad (i = 1, 2, \ldots, k).
\tag{4.23}
$$

It may be shown that, in this case, $d_{ij} = 0$ for $j > 1$ and all i. The necessity for this may be seen by considering the limit of $\bar{n}_i(T)$ as T tends to infinity. Under these conditions the second sum in (4.23) vanishes and the first sum will tend to $\pm \infty$ according to the sign of d_{im}. Since $\bar{n}_i(T)$ possesses a finite limit we conclude that $\bar{n}_i(T)$ must have the form

$$
\bar{n}_i(T) = d_{i1} + \sum_{j=m+1}^{k} c_{ij}e^{\theta_j T} \qquad (i = 1, 2, \ldots, k).
\tag{4.24}
$$

If we substitute this in (4.4) the equations for $\{d_{i1}\}$ and $\{c_{ij}\}$ are identical with those given in (4.18), if we delete those equations with $h \leq m - 1$ and replace c_{im} by d_{i1}.

Limiting behaviour

The limiting structure of a continuous time Markov process can be arrived at in a variety of ways. If we have found the full solution of the equations as described above we merely have to let $T \to \infty$. Referring to (4.16) and noting that the real parts of the eigenvalues for $j \neq 1$ are strictly negative, then

$$
\lim_{T \to \infty} n_i(T) = c_{i1} \qquad (i = 1, 2, \ldots, k).
\tag{4.25}
$$

In the case when $\theta = 0$ is a multiple root we have to replace c_{i1} by d_{i1}. If we are interested only in the limiting values (or do not have full information on the process) we can avoid the necessity of calculating all the c_{ij}'s. If a limiting state exists it is clear from (4.4) that the limiting vector $\bar{\mathbf{n}}(\infty)$ (or $\mathbf{p}(\infty)$) must satisfy

$$\bar{\mathbf{n}}(\infty)\mathbf{R} = \mathbf{0}. \tag{4.26}$$

These equations, together with $\bar{\mathbf{n}}(\infty)\mathbf{1}' = N$ can be solved in the usual way, but the solutions can also be expressed in two alternative ways by utilizing (4.12). Thus (4.26) is also equivalent to

$$\lambda\bar{\mathbf{n}}(\infty)\mathbf{P} = \lambda\bar{\mathbf{n}}(\infty) \tag{4.27}$$

which shows that the limiting structure is the same as for a discrete time model with transition matrix $\mathbf{P}$. A useful consequence is that the steady-state structure can be obtained without knowing the rate of change of state given by λ. From (4.12) we also have that $\bar{\mathbf{n}}(\infty)$ satisfies

$$\bar{\mathbf{n}}(\infty)\mathbf{\Lambda M} = \bar{\mathbf{n}}(\infty)\mathbf{\Lambda}. \tag{4.28}$$

Thus

$$\bar{\mathbf{n}}_j(\infty) \propto m_j/\lambda_j \qquad (j = 1, 2, \ldots, k), \tag{4.29}$$

where $\mathbf{m}$ is the limiting state vector associated with the matrix $\mathbf{M}$. We shall see later that this result generalizes almost immediately to semi-Markov processes.

4.2   EXAMPLES

Applications of closed Markov models in continuous time are rather uncommon outside the work of Coleman (1964a, b). In Chapter 5 we shall look in detail at two examples which involve open systems and so we shall defer the technical discussions until then. For the moment we shall also ignore the practical difficulties of fitting the models which arise from the fact that it is rarely possible to observe processes continuously; this topic is taken up in Section 4.3. Our attention here will be focused on the model-building aspects by discussing why and in what form a continuous time model might be considered appropriate for two kinds of social process.

Coleman (1964b) fitted a continuous time model to the social mobility data of Glass and Hall which we discussed in Chapter 2 and to some similar Danish data of Svalagosta (1959). In the Glass and Hall case this requires an assumption about the spacing of observations in time which does not correspond with the way the data were collected. Coleman's analysis proceeds on the assumption that each son starts life with his father's social class and is subsequently subject to the operation of time-homogeneous transition rates over a fixed interval of time. The class recorded for the son is that attained after that fixed interval of time—the same for all individuals. This would be a reasonable assumption if all the sons were questioned about their class at the same age. If their ages varied, as they did, some would have been exposed to risk of moving for a longer time than others and this would invalidate the fitting of the model to the aggregate data. This renders Coleman's application to these particular data rather doubtful, but the model is potentially useful for mobility studies in which changes of class are observed over fixed intervals of time.

Since the seven social classes in Glass and Hall's data are ranked in order of status or prestige Coleman argued that most transitions would be into adjacent classes. This would be plausible if one supposed that the classes arose from an underlying continuum along which individuals move in a continuous fashion. It is easy to think of examples of sudden changes of status, but these seem sufficiently rare to justify trying to fit the simpler model. The matrix **R** will be such that

$$r_{ij} = 0 \quad \text{if } |i-j| > 1 \qquad (i, j = 1, 2, \ldots, k).$$

If complete data are available one could check whether there ever were changes of more than one state and, if so, whether they were sufficiently rare to warrant fitting the model. If, as is more usual, the classes are recorded at two fixed points in time the presence of transitions between any pair of states is possible because there is no limit to how many transitions can take place in a finite interval. What one would expect to find would be that p_{ij} is a decreasing function of $|i-j|$ and this is what Coleman's model predicts. However, inspection of Table 2.1 shows that this is not the case and it is not surprising therefore that this model does not fit well. Once **R** has been estimated by methods to be discussed later, it is a relatively straightforward matter to compute expected sojourn times, the limiting structure, and so forth by the methods of the last section.

Coleman's main use of continuous time Markov models has been for the study of attitude change. A sample of individuals are allocated to categories at an initial point in time and then their subsequent changes of category are recorded. Ideally, one would want to observe the times at which changes took place but, in practice, it is easier simply to observe the new categories at fixed intervals of time. In either event the basic aim must be to estimate the transition rates, both for their own intrinsic interest and as a means of predicting the future attitude distribution and sojourn times. Such an analysis seems much better than the use of purely descriptive analyses which are often used by psychologists and others. For example, if we have a transition table for individuals showing their movement among a set of categories at two points in time we can compute a measure of correlation or association to show the extent to which their position has changed. Though this measures how much change has taken place it gives no insight into the structure of the underlying process nor does it provide any firm basis for prediction.

An interesting example of the use of the model for studying attitude change is given by Coleman (1961) who subsequently discussed it in Coleman (1964a); it has been further investigated in Singer and Spilerman (1977a). A sample of American high school youth were asked the following two questions in October 1957: (a) Are you a member of the 'leading crowd'? and (b) Do you agree that if a fellow wants to be part of the leading crowd round here, he sometimes has to go against his principles? Boys and girls were treated separately and each was allocated to one of the four possible response categories. The question was repeated in May 1958 which enabled the initial and final responses to be obtained for each individual. The model was then fitted to see whether attitude changes could be accounted for by the Markov model. The four categories of response were as follows:

1. Positive answer to both questions.
2. Positive answer to (a) and negative to (b).
3. Negative to (a) and positive to (b).
4. Negative answers to both.

Since two attitudes are involved Coleman argued that only one would change at once and hence that the rates r_{14}, r_{23}, r_{32}, r_{41} would be zero. The resulting restricted model fitted the data very well.

Singer and Spilerman (1977a) argue that the conclusions about the nature of attitude change emerge most clearly from the matrix $\mathbf{M}$ which gives the probabilities of transitions given that a transition takes place. For example, they estimate $\mathbf{M}$ for the 3260 girls to be

$$\mathbf{M} = \begin{pmatrix} 0 & 0.536 & 0.464 & 0 \\ 0.690 & 0 & 0 & 0.310 \\ 0.237 & 0 & 0 & 0.763 \\ 0 & 0.234 & 0.766 & 0 \end{pmatrix}$$

From this matrix, the most likely transitions can be immediately identified. For example, changes between 3 and 4 are likely in both directions, whereas those from 3 to 1 and 2 to 4 are much less common. Such figures seem to give a clearer insight into the substantive process than either the transition rates themselves or the raw data on transition numbers.

This is a second example in which the process is observed at only two points in time. We have bypassed the important question of how much can be learnt about the process by such occasional observation. The answer has many subtle features which we next consider.

4.3 DISCRETE OBSERVATION OF A CONTINUOUS PROCESS

For the practical reasons noted above we may only be able to observe a continuous time process at two or more fixed points in time. In the context of our discussion so far this raises a number of questions:

1. Is it possible to say whether the underlying process is Markovian?
2. If so how can its parameters be estimated?

To these we may add a third:

3. Under what conditions is it possible to treat a discretely observed continuous process as if it were really discrete?

This last question is prompted by the fact that the discrete time Markov model of Chapter 2 has often been fitted to processes which were really continuous. For example, models for regional migration have often treated individuals as if changes of residence were possible only at, say, annual, intervals. It is clearly desirable to know whether this is valid.

Let us deal with the last question first. If the system is observed at times 0 and T, let $\mathbf{P}(T)$ be the transition probability matrix for that interval. This can be directly estimated from the proportions of individuals who make the various transitions. In treating the process discretely we would predict the r-step matrix to be $\mathbf{P}^r(T)$. Further, since T is likely to have been fixed by administrative considerations unconnected with the process itself, we would want to get an answer which was consistent with those from other choices. That is, we would want

$$\mathbf{P}(rT) = [\mathbf{P}(T)]^r \quad \text{for any real non-negative } r. \tag{4.30}$$

For a Markov process we have seen in (4.8) that $\mathbf{P}(T) = \exp\{\mathbf{R}T\}$ and hence (4.30) is always satisfied. We therefore conclude that if the underlying process is Markovian and if it is observed at equal intervals of time, then it is legitimate to treat it as if it were a discrete Markov chain.

To answer the first question posed above we must take the argument a stage further. For a Markov process (4.30) holds for all positive r, and in particular for $0 < r < 1$. Thus if we are able to estimate $\mathbf{P}(T)$ then a necessary condition that it shall have arisen from a Markov process is that $\{P(T)\}^r$ ($0 < r < 1$) shall also be a transition matrix—that is, that its elements are non-negative with rows adding up to one. How do we recognize whether a given transition matrix has this property? The matter has been fully investigated by Singer and Spilerman (1976a). A simple necessary and sufficient condition when $k = 2$ was given by Kingman (1962). It is

$$\text{trace } \mathbf{P}(T) = p_{11}(T) + p_{22}(T) > 1. \tag{4.31}$$

For $k > 2$ the known necessary and sufficient conditions of Kingman (1962) are not applicable in practice, but there remain a number of necessary conditions by means of which some matrices can be ruled out as having arisen from a Markov process. Perhaps the simplest, of which (4.31) is a special case, is that

$$\prod_{i=1}^{k} p_{ii}(T) \geq \det \mathbf{P}(T) > 0 \tag{4.32}$$

where 'det' means 'the determinant of'. Singer and Spilerman (1977a) give the example of

$$\mathbf{P}(T) = \begin{pmatrix} 0.15 & 0.35 & 0.50 \\ 0.37 & 0.45 & 0.18 \\ 0.20 & 0.60 & 0.20 \end{pmatrix}$$

for which $\Pi_{i=1}^{3} p_{ii}(T) = 0.0135$ and det $\mathbf{P}(T) = 0.05$. Since (4.32) is not satisfied we conclude that $\mathbf{P}(T)$ could not have arisen from a Markov process. In practice, of course, the elements of $\mathbf{P}(T)$ would be estimated and hence subject to error. In the absence of a significance test small violations of the inequalities should be treated with caution in small samples. Other necessary conditions involve the eigenvalues of $\mathbf{P}(T)$ and full details will be found in Singer and Spilerman (1977a).

In practice the simplest way of proceeding is linked with the answer to our second question about how to find $\mathbf{R}$ given $\mathbf{P}(T)$. Since the two matrices are related by

$$\mathbf{P}(T) = \exp\{\mathbf{R}T\} \qquad (4.33)$$

the question concerns whether (4.33) has a unique solution for $\mathbf{R}$ of the required form. It turns out that this equation may have none, one, or several solutions. If there are none the matrix clearly cannot have arisen from a Markov process. If there are several solutions then more than one process could have given rise to the observed $\mathbf{P}(T)$ and there is no means of distinguishing between them. Only if there is a unique solution can we proceed with confidence. When $k = 2$ we may write:

$$\mathbf{R} = \frac{1}{T} \ln \mathbf{P}(T) = \frac{\ln(p_{11}(T) + p_{22}(T) - 1)}{p_{11}(T) + p_{22}(T) - 2}\{\mathbf{P}(T) - \mathbf{I}\}. \qquad (4.34)$$

For this to represent a legitimate $\mathbf{R}$, $p_{11}(T) + p_{22}(T) - 1$ must be real and strictly positive. Since the eigenvalues of $\mathbf{P}$ are 1, and $p_{11}(T) + p_{22}(T) - 1$ the condition may also be stated in terms of the eigenvalues. This condition is also sufficient for all values of k. Thus *if the eigenvalues of $\mathbf{P}(T)$ are real and positive then* (4.33) *has a unique solution*. In practice it seems that this condition is often met. When it is, the solution can be computed by methods discussed in Singer and Spilerman (1976a). Creedy (1979) applied their method to some data on labour market flows. There were three categories: employed (E), unemployed (U), and not in the labour force (N) and the population consisted of white male youths aged 14–24 in 1966. Over the year 1966–67 the observed transition probabilities were

	E	U	N
E	0.8600	0.0294	0.1028
U	0.6062	0.1631	0.2438
N	0.3341	0.0787	0.5872

The estimate of $\mathbf{R}$ is then

$$\mathbf{R} = \begin{pmatrix} -0.017 & 0.005 & 0.011 \\ 0.117 & -0.170 & 0.056 \\ 0.031 & 0.020 & -0.052 \end{pmatrix}.$$

If a program to compute $\mathbf{R}$ is available it is simpler to check that it is legitimate rather than calculate the eigenvalues first to see whether they are positive.

Further progress is possible towards identifying the underlying process if the system is observed at three or more points in time. We can then make direct comparison, for example, of the two-step matrix with the square of the one-step matrix. An example is provided by the classic study of Blumen and co-workers (1955) (hereafter referred to as BKM) on occupational mobility in the United States of America. The data consisted of each individual's occupational category observed at quarterly intervals. No information was available on any moves

Table 4.1 Comparison of observed and predicted eight-quarter transition probabilities for males aged 20–24. The upper figure is the observed proportion and the lower that predicted by Markov theory[a]

Industry code	A	B	C	D	E	F	G	H	J	K	U
A	0.000	0.062	0.062	0.000	0.125	0.156	0.312	0.000	0.000	0.000	0.281
	0.002	0.086	0.105	0.042	0.116	0.053	0.181	0.016	0.058	0.004	0.337
B	0.003	0.449	0.039	0.020	0.048	0.035	0.079	0.014	0.023	0.006	0.284
	0.002	0.144	0.087	0.040	0.104	0.050	0.163	0.018	0.052	0.004	0.336
C	0.002	0.037	0.461	0.023	0.046	0.021	0.101	0.007	0.022	0.002	0.278
	0.002	0.077	0.176	0.039	0.103	0.046	0.163	0.106	0.050	0.004	0.324
D	0.000	0.064	0.044	0.459	0.083	0.024	0.091	0.011	0.030	0.002	0.192
	0.001	0.070	0.080	0.218	0.099	0.046	0.141	0.015	0.047	0.003	0.279
E	0.002	0.045	0.042	0.034	0.489	0.031	0.094	0.010	0.023	0.002	0.227
	0.001	0.072	0.075	0.040	0.276	0.046	0.147	0.013	0.044	0.004	0.279
F	0.003	0.056	0.033	0.022	0.054	0.440	0.090	0.020	0.026	0.010	0.245
	0.002	0.081	0.076	0.038	0.097	0.166	0.152	0.017	0.050	0.004	0.316
G	0.002	0.047	0.051	0.025	0.046	0.038	0.491	0.020	0.044	0.002	0.235
	0.002	0.080	0.084	0.039	0.098	0.049	0.261	0.017	0.053	0.004	0.314
H	0.000	0.044	0.007	0.015	0.026	0.085	0.096	0.439	0.074	0.000	0.214
	0.001	0.077	0.077	0.035	0.090	0.048	0.170	0.158	0.052	0.004	0.287
J	0.002	0.061	0.033	0.018	0.054	0.035	0.145	0.019	0.339	0.000	0.294
	0.002	0.084	0.085	0.038	0.105	0.049	0.178	0.018	0.105	0.004	0.333
K	0.000	0.113	0.097	0.032	0.121	0.048	0.137	0.032	0.024	0.048	0.347
	0.002	0.089	0.096	0.047	0.130	0.048	0.179	0.023	0.056	0.006	0.325
U	0.001	0.069	0.068	0.035	0.077	0.040	0.153	0.018	0.055	0.004	0.482
	0.002	0.090	0.095	0.042	0.112	0.052	0.179	0.019	0.058	0.004	0.346

[a] Data from Blumen and co-workers (1955).

which might have taken place within these intervals. There were 11 categories, one consisting of the unemployed and 10 others labelled A, B, . . . , K in the table above. Since data were available for eight consecutive quarters it was possible to compare, for example, the eighth power of the one-step matrix with the eight-step matrix. The result of that comparison is shown in Table 4.1 for males aged 20–24.

The most striking feature is that the diagonal elements of the observed eight-quarter matrix are much greater than those predicted by raising the one-quarter matrix to the eighth power. We have met a similar phenomenon in our analysis of the discrete time model in Chapter 2. There we showed that it could be explained either by heterogeneity (i.e different P's) or by cumulative inertia. Each of these departures from the simple model has its analogue in the continuous time model. The latter is accommodated by allowing the time intervals between changes of state to have a distribution which implies a decreasing propensity to leave as length of stay increases. Such models will be discussed in Section 4.4. There is also a third type of departure which is possible with a continuous time model. Here we can have heterogeneity in the process governing changes of state. However, we begin our investigation with a continuous time version of the mover–stayer model.

As before, we suppose that the population is divided into two parts—the movers and the stayers. Suppose that a proportion s_i of category i are stayers which means that their transition matrix is $\mathbf{I}$; the remainder are movers with transition matrix $\mathbf{P}$. We retain the assumption that changes of state occur at points which are a realization of a Poisson process with parameter λ. Thus if τ denotes the time interval of observation (a quarter in this case) then it follows that

$$\mathbf{P}(\tau) = \mathbf{SI} + \{\mathbf{I} - \mathbf{S}\} \sum_{m=0}^{\infty} \frac{(\lambda\tau)^m}{m!} e^{-\lambda\tau} \mathbf{P}^m \qquad (4.35)$$

where $\mathbf{S}$ is a diagonal matrix whose ith element is s_i.

The question which now arises is whether, for example, the eight-quarter matrix $\mathbf{P}(8\tau)$ predicted by (4.35) agrees with that which was observed. BKM fitted the model to their data, and their results for the diagonal elements for the 20–24 age group are given in Table 4.2.

It is clear that the fit is very much better than in Table 4.1 so that in spite of its

Table 4.2 *Comparison of observed and predicted values of the diagonal elements in the eight-quarter transition matrices for males aged 20–24[a]*

Industry code	A	B	C	D	E	F	G	H	J	K	U
Observed	0.000	0.449	0.461	0.459	0.489	0.440	0.491	0.439	0.339	0.048	0.482
Predicted	0.003	0.442	0.464	0.474	0.512	0.444	0.489	0.446	0.338	0.049	0.536

[a] Data from Blumen and co-workers (1955).

crudity this mover–stayer model leaves little to be desired. More elaborate forms of heterogeneity could easily be introduced, but in this case, at least, they are hardly necessary.

The foregoing model can be viewed in two ways. We supposed above that the heterogeneity resided in the different transition matrices—**I** in one case and **P** in the other. Alternatively, we could have supposed that all individuals had the same **P** but that the rates of movement differed, being λ for the movers and 0 for the stayers. Both ways of specifying the model lead to (4.35). We now take this latter interpretation a stage further. Suppose that each individual changes state according to a Markov process characterized by a Poisson process with parameter λ and a transition matrix **P**. Heterogeneity arises by allowing λ to vary, thus giving individuals different rates of moving. The transition matrix for the interval $(0, \tau)$, $\mathbf{P}(\tau)$, is then easily found as follows. For any given individual it will be $\sum_{n=0}^{\infty} P_n(\tau; \lambda)\mathbf{P}^n$, where $P_n(\tau; \lambda)$ is the Poisson probability of n moves in $(0, \tau)$. This expression gives the required probabilities conditional on λ. To find the unconditional distribution we must find the expectation with respect to λ which is

$$\mathbf{P}(\tau) = \int_0^\infty \sum_{n=0}^{\infty} P_n(\tau; \lambda)\,\mathbf{P}^n\,\mathrm{d}F(\lambda), \qquad (4.36)$$

where $F(\lambda)$ is the distribution function of λ. Reversing the summation and integration this may be written:

$$\mathbf{P}(\tau) = \sum_{n=0}^{\infty} P_n(\tau)\mathbf{P}^n, \qquad (4.37)$$

where

$$P_n(\tau) = \int_0^\infty P_n(\tau; \lambda)\,\mathrm{d}F(\lambda). \qquad (4.38)$$

Note that (4.35) arises as a special case of this model when the distribution of λ is concentrated at the two points zero and λ.

The question now arises as to whether the inflation of the diagonal elements of the r-step matrix as compared with that predicted from the rth power of the one-step matrix which we observed with the mover–stayer model is typical of this class of models. We approach the answer to this question by way of two examples, the first constructed by BKM. They assumed a three-category system in which all members shared the same transition matrix:

$$\mathbf{P} = \begin{pmatrix} 0.70 & 0.15 & 0.15 \\ 0.20 & 0.60 & 0.20 \\ 0.25 & 0.25 & 0.50 \end{pmatrix}. \qquad (4.39)$$

Changes of state occurred according to a Poisson process for which one half of the group had $\lambda = \frac{1}{10}$ and the other half $\lambda = \frac{7}{10}$. For the whole system the number of changes of state in $(0, \tau)$ would thus be

$$P_n(\tau) = \frac{1}{2n!} \left\{ \left(\frac{\tau}{10}\right)^n e^{-\tau/10} + \left(\frac{7\tau}{10}\right)^n e^{-7\tau/10} \right\} \qquad (n = 0, 1, 2, \ldots) \quad (4.40)$$

Taking $\tau = 1$ this gives

$$P_0(1) = 0.701, \qquad P_1(1) = 0.219, \qquad P_2(1) = 0.063,$$
$$P_3(1) = 0.014, \qquad P_4(1) = 0.002.$$

The mean of this distribution is 0.395 and the variance is 0.473, indicating a higher dispersion than the Poisson distribution having the same mean. Substitution of these values and the matrix (4.39) in (4.37) gives

$$\mathbf{P}(\tau = 1) = \begin{pmatrix} 0.90 & 0.05 & 0.05 \\ 0.07 & 0.87 & 0.06 \\ 0.08 & 0.08 & 0.84 \end{pmatrix}. \qquad (4.41)$$

This is the matrix which would be calculated from the data collected for two successive quarters. Treating it as the transition matrix of a Markov chain we would predict the eight-quarter matrix by raising the matrix (4.41) to the eighth power. Thus

$$\{\mathbf{P}(\tau = 1)\}^8 = \begin{pmatrix} 0.54 & 0.25 & 0.21 \\ 0.33 & 0.44 & 0.24 \\ 0.35 & 0.30 & 0.36 \end{pmatrix}.$$

However, the eight-quarter matrix which we would expect from our present model is

$$\mathbf{P}(\tau = 8) = \begin{pmatrix} 0.63 & 0.21 & 0.17 \\ 0.26 & 0.55 & 0.19 \\ 0.28 & 0.23 & 0.48 \end{pmatrix}.$$

The heterogeneity which we have introduced thus leads us to underestimate the diagonal elements. This is exactly what Blumen and co-workers (1955) found in practice.

It is not necessary to assume that there are individual differences in the population. The same result would be obtained if the decision points occurred according to (4.40) for each person. This might be the case if decision points were of two kinds, one kind occurring at rate $\lambda = \frac{1}{10}$ and the other at rate $\lambda = \frac{7}{10}$. It is an unfortunate fact that we cannot, on the data available, distinguish between these two models for the decision point process. For predictive purposes they are identical; if we require a full explanation of the process, data on individual case histories are necessary.

Spilerman (1972b) has examined the case when λ varies continuously by assuming that

$$dF(\lambda) = \frac{c^v}{\Gamma(v)} \lambda^{v-1} e^{-c\lambda} d\lambda \qquad (\lambda \geq 0;\ c,\ v > 0).$$

This is a flexible form capable of describing patterns of variability ranging from a J-shaped distribution at one extreme to a unimodal symmetric distribution at the other. We shall use it again in this connection in Chapter 7. Substituting in (4.38) we have, for the time interval τ,

$$\begin{aligned}
P_m(\tau) &= \int_0^\infty \frac{(\lambda\tau)^m}{m!} e^{-\lambda\tau} \frac{c^v}{\Gamma(v)} \lambda^{v-1} e^{-c\lambda} d\lambda \\
&= \frac{c^v \tau^m}{m!\Gamma(v)} \int_0^\infty \lambda^{m+v-1} e^{-\lambda(c+\tau)} d\lambda \\
&= \binom{m+v-1}{m} \left(\frac{c}{c+\tau}\right)^v \left(\frac{\tau}{c+\tau}\right)^m
\end{aligned} \qquad (4.42)$$

where we define $x! = \Gamma(x+1)$ for all x. Substitution of $P_m(\tau)$ in the matrix form of (4.37) now gives

$$\mathbf{P}(\tau) = \left(\frac{c}{c+\tau}\right)^v \sum_{m=0}^\infty \binom{m+v-1}{m} \left(\frac{\tau}{c+\tau}\right)^m \mathbf{P}^m. \qquad (4.43)$$

If $\mathbf{P}$ were a scalar this would be a negative binomial series with sum

$$\left(\frac{c}{c+\tau}\right)^v \left\{\mathbf{I} - \frac{\tau}{c+\tau} \mathbf{P}\right\}^{-v}, \qquad (4.44)$$

assuming convergence. We can, of course, define (4.44) as the infinite series (4.43), but this leaves us with the problem of how to compute (4.44) bearing in mind that v need not be an integer. To avoid the problem we use the fact that $\mathbf{P}$ can be written in the form

$$\mathbf{P} = \mathbf{HDH}^{-1}$$

where $\mathbf{D}$ is a matrix having the eigenvalues of $\mathbf{P}$ on its main diagonal and zeros elsewhere, and $\mathbf{H}$ is a matrix whose ith column is the eigenvector associated with the ith eigenvalue of $\mathbf{P}$. It is easily verified that

$$\mathbf{P}^m = \mathbf{HD}^m\mathbf{H}^{-1}.$$

Hence

$$\mathbf{P}(\tau) = \left(\frac{c}{c+\tau}\right)^v \mathbf{H} \left\{ \sum_{m=0}^\infty \binom{m+v-1}{m} \left(\frac{\tau}{c+\tau}\right)^m \mathbf{D}^m \right\} \mathbf{H}^{-1}. \qquad (4.45)$$

The infinite matrix sum in brackets is a diagonal matrix whose (i, i)th element is the scalar sum

$$\sum_{m=0}^{\infty} \binom{m+v-1}{m} \left(\frac{\tau}{c+\tau} \right)^m \theta_i^m = \left(1 - \frac{\tau \theta_i}{c+\tau} \right)^{-v} \qquad (4.46)$$

where θ_i is the ith eigenvalue of $\mathbf{P}$. These series converge because, for a stochastic matrix, $\theta_i \leq 1$ for all i and $\tau/(c+\tau) < 1$ for $\tau > 0$, implying that $\tau\theta_i/(c+\tau) < 1$. An equivalent way of summing the series (4.43) is to use the spectral representation of $\mathbf{P}^m$ given in Chapter 3, equation (3.6).

Spilerman (1972b) made some numerical comparisons between the actual T-step matrix for this process and $\{\mathbf{P}(\tau)\}^T$. He found that the former had larger diagonal elements, just as with the two-point distribution for λ illustrated above.

Finally we show that this kind of result will always occur with a model for which $\mathbf{P}(\tau)$ is given by (4.36) if the eigenvalues of $\mathbf{P}$ are real. In particular we shall find a condition which ensures that

$$\text{trace } \mathbf{P}(2\tau) \geq \text{trace}[\mathbf{P}(\tau)]^2 \qquad (4.47)$$

a result which is akin to what we found in Chapter 2. From (4.37)

$$\mathbf{P}(2\tau) = \sum_{n=0}^{\infty} P_n(2\tau)\mathbf{P}^n$$

and hence,

$$\text{trace } \mathbf{P}(2\tau) = \sum_{n=0}^{\infty} P_n(2\tau) \sum_{i=1}^{k} \theta_i^n = \sum_{i=1}^{k} \sum_{n=0}^{\infty} \theta_i^n P_n(2\tau)$$

$$= \sum_{i=1}^{k} E(\theta_i^n | 2\tau), \quad \text{say}, \qquad (4.48)$$

where $\theta_1, \theta_2, \ldots, \theta_k$ are the eigenvalues of $\mathbf{P}$. Next we have

$$\{\mathbf{P}(\tau)\}^2 = \left\{ \sum_{n=0}^{\infty} P_n(\tau)\mathbf{P}^n \right\}^2 = \sum_{n=0}^{\infty} C_n \mathbf{P}^n$$

where C_n is the coefficient of $\mathbf{P}^n$ in the expansion of the square.

$$\text{trace }\{\mathbf{P}(\tau)\}^2 = \sum_{n=0}^{\infty} C_n \sum_{i=1}^{k} \theta_i^n = \sum_{i=1}^{k} \sum_{n=0}^{\infty} C_n \theta_i^n$$

$$= \sum_{i=1}^{k} \left\{ \sum_{n=0}^{\infty} P_n(t)\theta_i^n \right\}^2$$

$$= \sum_{i=1}^{k} E^2(\theta_i^n | \tau). \qquad (4.49)$$

The condition for (4.47) to hold is therefore

$$\sum_{i=1}^{k} E(\theta_i^n | 2\tau) \geq \sum_{i=1}^{k} E(\theta_i^n | \tau). \qquad (4.50)$$

Once the distribution of n is known it is a straightforward matter to test whether the condition is satisfied. When n has the negative binomial distribution of (4.42) it is easy to verify that (4.50) holds for real θ_i.

In the general case when $P_n(\tau)$ is given by (4.38) we have that

$$E(s^n|\tau) = \sum_{n=0}^{\infty} s^n \int_0^{\infty} \frac{(\lambda\tau)^n}{n!} e^{-\lambda\tau} \, dF(\lambda)$$

$$= \int_0^{\infty} e^{\lambda\tau(s-1)} dF(\lambda) = E(e^{-\lambda\tau(s-1)}).$$

Hence

$$E(s^n|2\tau) - E^2(s^n|\tau) = \text{var}\,(e^{\lambda\tau(s-1)}) \geq 0$$

for all real s. Giving to s the value of each eigenvalue in turn it follows that (4.50) holds with the inequality strict unless the distribution of λ degenerates to a single point mass.

It is thus clear that inflated diagonal elements may be attributed to variation in the rate of moving, but it must be remembered that they can also be due to heterogeneity in the transition matrices. Later in this chapter we shall see that there is a third possible explanation.

Limiting behaviour

Under certain circumstances the observed transition matrix $\mathbf{P}(\tau)$ can be made to yield information about the limiting behaviour of the underlying process. Suppose that all individuals have the same transition matrix $\mathbf{P}$, but that the times at which transitions take place are determined by an arbitrary stochastic process. Hitherto we have only considered processes obtained by mixing Poisson processes, but the result which follows is more general. Let $P_n(\tau)$ denote the probability that n changes of state take place in $(0, \tau)$. (This notation is adequate for mixtures of Poisson processes, but in the more general case the probability might be a function of how the interval was chosen.) Then it is clear that

$$\mathbf{P}(\tau) = \sum_{n=0}^{\infty} P_n(\tau)\mathbf{P}^n. \tag{4.51}$$

Let $\mathbf{p}$ be the steady-state vector associated with $\mathbf{P}$ so that $\mathbf{p} = \mathbf{p}\mathbf{P} = \mathbf{p}\mathbf{P}^2 = \ldots$. If we pre-multiply both sides of (4.51) by $\mathbf{p}$ we find

$$\mathbf{p}\mathbf{P}(\tau) = \sum_{n=0}^{\infty} P_n(\tau)\mathbf{p}\mathbf{P}^n = \sum_{n=0}^{\infty} P_n(\tau)\mathbf{p} = \mathbf{p}. \tag{4.52}$$

Thus it follows that $\mathbf{p}$ is also the steady-state vector of the observed matrix $\mathbf{P}(\tau)$. This result enables us to find the limiting structure of the embedded Markov chain. In general, as we shall see below, this is not the same as the limiting structure observed at some future distant point in time because the method takes no account of the sojourn times. However, if the underlying point process is such that the expected sojourn times are the same for each category then the $\mathbf{p}$ of (4.52) will also be the limiting structure.

If there is heterogeneity in the transition matrices, rather than in the point process, very little can be learnt about the limiting behaviour from the observation of a single $\mathbf{P}(\tau)$. The situation is essentially the same as in the discrete case discussed in Chapter 2 (p. 34). Those individuals who are subject to the transition matrix $\mathbf{P}(h)$ will settle down to a steady-state structure $\mathbf{p}(h)$, say, and the overall limiting structure will then be a weighted average of these limiting vectors.

BKM took their quarterly transition matrices and found the steady-state vectors $\mathbf{p}$ from (4.52). An example is given in Table 4.3 where $\mathbf{p}$ is compared with the observed structure.

Table 4.3   Actual and predicted occupation structure using the Markov model and the quarterly transition matrix[a]

Occupational group	(C, D, E)	G	(F, H)	(A, B, J, K)	U	Total
Average percentage of workers observed	28.2	17.0	6.8	13.7	34.3	100.0
Predicted percentage using Markov model	27.0	18.0	8.0	15.0	32.0	100.0

[a] Data from Blumen and co-workers (1955).

The agreement is very good and one may ask whether our analysis of limiting behaviour enables us to interpret this fact. The result is certainly consistent with the hypothesis that the heterogeneity lies in the varying rates of transition and that the system is close to its steady state. The steady-state structure would then be interpreted as that associated with the embedded Markov chain. The result would not support the hypothesis that $\mathbf{P}$ varied from person to person except in so far as the mover–stayer model can also be regarded as a special case of the model in which λ varies.

4.4   SEMI-MARKOV MODELS

In the last section we touched upon several generalizations of the Markov model in the course of investigating the dependence between observed transition matrices and the underlying process. Here we continue in this direction by discussing a class of models in which propensity to move depends on the length of stay in the category. This enables us, for example, to include cumulative inertia in the shape of a declining propensity to move.

The definition of a semi-Markov process may be approached in two ways which correspond with those already noted for the Markov process. One of these was via the transition rates; the other started with a Poisson process for changes of state coupled with an embedded Markov chain. The two approaches turned out to be equivalent, though one or other of them would usually be the more natural in a given application. Much the same is true of semi-Markov processes. Here we shall adopt the approach that seems most natural for modelling such things as occupational mobility.

We make the generalization of the Markov process by allowing the transition rates to be functions of the time spent in the state. Thus $r_{ij}(t)\delta t$ $(i \neq j)$ is the infinitesimal probability of a move from i to j in $(t, t + \delta t)$. Cumulative inertia would then correspond to making $r_{ij}(t)$ a decreasing function of t. Semi-Markov models arise in actuarial work and in multiple decrement analysis where the $r_{ij}(t)$'s may be risks of death from different causes expressed as functions of age. In biometry the same type of model arises in the treatment of competing risks. Most of these applications require the system to be open and we shall meet some examples of that kind in Chapter 5.

For many purposes it is useful to adopt an alternative parameterization as follows:

$$\lambda_i(t) = \sum_{\substack{j=1 \\ j \neq i}}^{k} r_{ij}(t) = -r_{ii}(t)$$

$$M_{ij}(t) = r_{ij}(t)/\lambda_i(t), \qquad (i \neq j) \tag{4.53}$$

$\lambda_i(t)$ is thus the hazard function (see Chapter 7) associated with removal from state i; $M_{ij}(t)$ is the transition probability from i to j given that the transition takes place at t. We define this to be zero if $i = j$ and then we have $\sum_{j=1}^{k} M_{ij}(t) = 1$.

In this formulation there is no explicit provision for a transition from a given state to itself. In some applications such a move would be meaningless, but in others such transitions not only occur but can be observed. For example in BKM's study of occupational mobility the states were based on industrial groupings. If we are only able to observe transitions when they cross group boundaries then the above formulation is adequate. On the other hand, if we observe changes of job some of these will involve moves within the same industrial group. It would then be an advantage to let $\lambda_i(t)$ refer to the duration of a particular job and $M_{ij}(t)$ to transitions between jobs. In this case we would not necessarily want to have $M_{ij}(t) = 0$. A process defined in this way is equivalent to a semi-Markov process, but the relationship with the $r_{ij}(t)$'s is not then given by (4.53).

As a first step in the analysis of the model we shall express several probability functions, which can be readily estimated, in terms of the transition rates. The density of sojourn time in state i, whose hazard function is $\lambda_i(t)$ is

$$f_i(t) = \lambda_i(t) \exp - \int_0^t \lambda_i(x)dx. \tag{4.54}$$

In the special case of the Markov process, $\lambda_i(t)$ is constant and so the sojourn time distribution is exponential. In order to see whether a semi-Markov model is likely to be appropriate we may therefore test the actual sojourn times for exponentiality. Kao (1974) did this when modelling the movement of coronary patients and found that the sojourn times were not exponential. He also found that the time spent in a state depended on the destination state. If we denote by $f_{ij}(t)$ the density function of the sojourn time in i for those who leave for j we can also express this in terms of the transition rates. The probability that a person remains in i until t and then moves to j in $(t, t + \delta t)$ is

$$G_i(t) r_{ij}(t) \delta t,$$

where $G_i(t) = \int_t^\infty f_i(x) \mathrm{d}x$. The integral of this expression between 0 and ∞ is the probability that an individual in i will go to j at the next step. This is an important parameter in its own right denoted by

$$m_{ij} = \int_0^\infty r_{ij}(t) G_i(t) \mathrm{d}t, \qquad (i \neq j). \tag{4.55}$$

Since

$$f_{ij}(t)\delta t = Pr\{\text{person in } i \text{ moves to } j \text{ in } (t, t + \delta t) | \text{next move is to } j\} \qquad (i \neq j)$$

we have

$$f_{ij}(t) = r_{ij}(t) G_i(t) / m_{ij}. \tag{4.56}$$

If the process can be observed continuously over a long enough period, $f_{ij}(t), f_i(t)$, and m_{ij} can be estimated directly. Once these are known, $r_{ij}(t)$ follows from (4.56).

An important special case arises when $f_{ij}(t)$ does not depend on j as Yang and Hursch (1973) found in their application to sleep patterns. This is also known as the *proportional hazards* model because it implies that $r_{ij}(t)/m_{ij}$ is a function of i and t but not of j. That is we may write

$$r_{ij}(t) = m_{ij} r_i(t), \qquad (i \neq j)$$

and hence

$$\lambda_i(t) = r_i(t) \sum_{\substack{j=1 \\ j \neq i}}^k m_{ij} = r_i(t).$$

This means that the *form* of the dependence of the rate on time is the same regardless of destination, but the *level* of the rate does vary between destinations.

In this case $M_{ij}(t) = m_{ij}$ showing that the transition probability for a person who moves at time t does depend on when the move takes place.

There are many features of the semi-Markov model which have practical interest. Kao (1974), for example, was especially interested in the paths of patients through a hospital and, in particular, with the means and variances of the sojourn times. Here we shall concentrate on just two aspects. The first is the transition probability matrix for a fixed interval of time. As we saw in the last section, when a process can only be observed discretely, we need to know how much can be inferred about the underlying process. The second aspect concerns the steady-state distribution.

Transition probabilities over a fixed interval

For a semi-Markov process these probabilities will depend on the state of the system at the time when the interval of observation begins. To begin with we consider an individual who has just entered state i. Let $p_{ij}(\tau)$ be the probability that he is in state j after an interval of length τ. An expression for this probability may be constructed as follows. If $j \neq i$ the individual must make at least one move in $(0, \tau)$. Suppose the first move is to state h and that it takes place in $(t, t + \delta t)$, then between t and τ he must move from h to j. By summing the probability of this event over h and t we obtain

$$p_{ij}(\tau) = \sum_{\substack{h=1 \\ h \neq i}}^{k} \int_0^\tau r_{ih}(t) G_i(t) p_{hj}(\tau - t) \mathrm{d}t, \qquad (i \neq j). \qquad (4.57)$$

If $j = i$ there is the further possibility that the individual remains in i throughout the interval and this has probability $G_i(\tau)$. Recalling the definition of $f_{ij}(t)$ the full result can be written:

$$p_{ij}(\tau) = \delta_{ij} G_i(\tau) + \int_0^\tau \sum_{h=1}^{k} m_{ih} f_{ih}(t) p_{hj}(\tau - t) \mathrm{d}t \qquad (i, j = 1, 2, \ldots, k), \qquad (4.58)$$

where $m_{ii} = 0$ for all i and $\delta_{ij} = 1$ if $i = j$ and otherwise is zero.

In principle these equations suffice to calculate the required transition probabilities. By treating time as a discrete variable the integral in (4.58) can be replaced by a sum and the probabilities computed recursively as in Valliant and Milkovich (1977). Such calculations do not readily give insight into the qualitative behaviour of the system so we shall pursue the mathematical analysis of (4.58) by introducing the Laplace transform. Even then, however, we shall find that the results are meagre.

The Laplace transform of a function $\phi(x)$ is defined by

$$\phi^*(s) = \int_0^\infty \phi(x) e^{-sx} \mathrm{d}x.$$

We shall also need the result that the transform of

$$\int_0^t \phi_1(x) \phi_2(t - x) \mathrm{d}x$$

is $\phi_1^*(s) \phi_2^*(s)$. Then taking the Laplace transform of both sides of (4.58) we have

$$p_{ij}^*(s) = \delta_{ij}G_i^*(s) + \sum_{h=1}^{k} m_{ih}f_{ih}^*(s)p_{hj}^*(s).\tag{4.59}$$

The sum on the right-hand side will be recognized as the (i, j)th element in the product of two matrices, one with typical element $m_{ij}f_{ij}^*(s)$ and the other with $p_{ij}^*(s)$. We may therefore rewrite (4.59) as

$$\mathbf{P}^*(s) = \mathbf{G}^*(s) + \mathbf{f}_m^*(s)\mathbf{P}^*(s)\tag{4.60}$$

where $\mathbf{G}^*(s)$ is a diagonal matrix with elements $G_i^*(s)$ and $\mathbf{f}_m^*(s)$ is the matrix with elements $\{m_{ij}f_{ij}^*(s)\}$. The Laplace transform of $\mathbf{P}^*(s)$ is thus obtained at once as

$$\mathbf{P}^*(s) = (\mathbf{I} - \mathbf{f}_m^*(s))^{-1}\mathbf{G}^*(s).\tag{4.61}$$

In spite of the apparent simplicity of this expression it is not easy to obtain explicit formulae from it or to deduce useful qualitative information about the process, but some simple special cases will be investigated below. First, however, it is instructive to see how the known results for the Markov process arise as a special case.

For the Markov process,

$$m_{ij}(t) = r_{ij}e^{-\lambda_i t}\left(\lambda_i = \sum_{j \neq i} r_{ij}\right) \quad \text{and} \quad G_i(t) = e^{-\lambda_i t}.$$

The two Laplace transforms required for (4.61) are then

$$m_{ij}f_i^*(s) = r_{ij}/(\lambda_i + s) \quad \text{and} \quad G_i^*(s) = 1/(\lambda_i + s).$$

Substitution in (4.61) gives

$$\mathbf{P}^*(s) = \{\mathbf{I} - (\mathbf{R} + \mathbf{\Lambda}_0)\mathbf{\Lambda}^*(s)\}^{-1}\mathbf{\Lambda}^*(s)\tag{4.62}$$

where $\mathbf{\Lambda}_0$ and $\mathbf{\Lambda}^*(s)$ are diagonal matrices with λ_i and $(\lambda_i + s)^{-1}$ respectively, in the ith position. Noting that $\{\mathbf{\Lambda}^*(s)\}^{-1} = \mathbf{\Lambda}_0 + s\mathbf{I}$ we have

$$\mathbf{P}^*(s) = (\mathbf{\Lambda}_0 + s\mathbf{I} - \mathbf{R} - \mathbf{\Lambda}_0)^{-1} = (s\mathbf{I} - \mathbf{R})^{-1} = \sum_{r=0}^{\infty} \frac{\mathbf{R}^r}{s^{r+1}}$$

which is the Laplace transform of $\mathbf{P}(\tau) = e^{\mathbf{R}\tau}$.

Some progress can be made in the general case if we place further restrictions on the model or consider small values of k. In the simplest case of all when $k = 2$ (4.61) becomes

$$\mathbf{P}^*(s) = \begin{bmatrix} 1 & -m_{12}f_{12}^*(s) \\ -m_{21}f_{21}^*(s) & 1 \end{bmatrix}^{-1} \begin{bmatrix} G_1^*(s) & 0 \\ 0 & G_2^*(s) \end{bmatrix}$$

$$= (1 - m_{12}m_{21}f_{12}^*(s)f_{21}^*(s))^{-1} \begin{bmatrix} G_1^*(s) & m_{12}f_{12}^*(s)G_2^*(s) \\ m_{21}f_{21}^*(s)G_1^*(s) & G_2^*(s) \end{bmatrix}.\tag{4.63}$$

Whether or not the inversion of the transforms is possible depends on the form of the $r_{ij}(t)$'s. The limiting form of $\mathbf{P}(t)$ as $t \to \infty$ can easily be found using the fact that, for any transform,

$$\lim_{t \to \infty} \phi(t) = \lim_{s \to \infty} s\phi(s).$$

In the case of (4.63) this yields

$$\mathbf{P}(\infty) = \frac{1}{\mu_1 + \mu_2} \begin{pmatrix} \mu_1 & \mu_2 \\ \mu_1 & \mu_2 \end{pmatrix} \tag{4.64}$$

where μ_1 and μ_2 are the mean sojourn times in the two states.

In the proportional hazards case the expression for the Laplace transform simplifies somewhat since then

$$\mathbf{f}_m^*(s) = \mathbf{f}^*(s)\mathbf{M}$$

where $\mathbf{f}^*(s)$ is the diagonal matrix with elements $\{f_i^*(s)\}$. Perhaps the simplest example of the proportional hazards models arises when $r_{ij}(t) = m_{ij}at^b$ $(b > -1)$ but even in this case there are no explicit expressions for the Laplace transforms required.

Considerable progress can be made if we make the additional assumption that $f_i^*(s) = f^*(s)$ for all i. This implies that the transition rates are the same for all states. In that case

$$\mathbf{P}^*(s) = \{\mathbf{I} - f^*(s)\mathbf{M}\}^{-1}\mathbf{G}^*(s) = \{\mathbf{I} - f^*(s)\mathbf{M}\}^{-1}\{1 - f^*(s)\}/s. \tag{4.65}$$

The last step follows from the fact that the Laplace transform of $\int_0^t \phi(x)\,dx$ is $\phi^*(s)/s$. Expanding the inverse as an infinite series

$$\mathbf{P}^*(s) = \left\{ \frac{1 - f^*(s)}{s} \right\} \sum_{n=0}^{\infty} \{f^*(s)\}^n \mathbf{M}^n$$

$$= \sum_{n=0}^{\infty} \left(\frac{\{f^*(s)\}^n - \{f^*(s)\}^{n+1}}{s} \right) \mathbf{M}^n. \tag{4.66}$$

Now $\{f^*(s)\}^n$ is the transform of the sum of n independent random variables with density function $f(t)$ and, by the result quoted above, $\{f^*(s)\}^n/s$ is the transform of its distribution function. Inverting both sides of (4.66) we therefore have

$$\mathbf{P}(\tau) = \sum_{n=0}^{\infty} \{F_n(\tau) - F_{n+1}(\tau)\}\mathbf{M}^n. \tag{4.67}$$

The difference between the two probabilities $F_n(\tau)$ and $F_{n+1}(\tau)$ is simply the probability that n changes of state occur in $(0, \tau)$. Finally, therefore, we have

$$\mathbf{P}(\tau) = \sum_{n=0}^{\infty} P_n(\tau)\mathbf{M}^n \tag{4.68}$$

which could clearly have been obtained by the direct argument which led to (4.51).

Here, it must be remembered that the interval $(0, \tau)$ begins with the entry of an individual to the state in question.

Illustrations

The problem in applying (4.68) lies in determining $P_n(\tau)$ from the assumed transition rates. Singer and Spilerman (1979) have considered the case where

$$r_{ij}(t) = \lambda^2 t/(1 + \lambda t), \quad \text{(all } i \text{ and } j\text{)}. \tag{4.69}$$

This represents a risk of moving which increases with duration of stay with the rate of increase tending to zero as t becomes large. These authors quote work on the stage models of developmental psychologists as providing an example where such a model is reasonable. In sociological applications the term 'cumulative stress' is used to describe an increasing removal rate in contrast to cumulative inertia where the rate decreases. For the model of (4.69) Singer and Spilerman (1979) were able to show that

$$\mathbf{P}(\tau) = e^{-\lambda t}[\cosh \lambda \tau \mathbf{M}^{\frac{1}{2}} + \mathbf{M}^{-\frac{1}{2}} \sinh \lambda \tau \mathbf{M}^{\frac{1}{2}}] \tag{4.70}$$

in which the cosh and sinh functions of matrix argument are defined in the usual way by their power series expansions. In order to investigate how such a process would differ from a Markov process we may compare $\mathbf{P}(2\tau)$ with $\{\mathbf{P}(\tau)\}^2$. Squaring both sides of (4.70) we find

$$\mathbf{P}^2(\tau) = \mathbf{P}(2\tau) + \tfrac{1}{2}e^{-2\lambda\tau}(\mathbf{I} - \mathbf{M}^{-1})(\mathbf{I} - \cosh 2\lambda\tau\mathbf{M}^{\frac{1}{2}})$$

$$= \mathbf{P}(2\tau) + \tfrac{1}{2}e^{-2\lambda\tau} \sum_{n=1}^{\infty} \frac{(2\lambda\tau)^{2n}}{(2n)!} (\mathbf{M}^{n-1} - \mathbf{M}^n). \tag{4.71}$$

It is clear that $\mathbf{P}^2(\tau)$ and $\mathbf{P}(2\tau)$ will differ, but to discover the nature of the difference we look at the traces of the two matrices as in Chapter 2. The relevant part is

$$\text{trace}(\mathbf{M}^{n-1} - \mathbf{M}^n) = \sum_{i=1}^{k} \theta_i^{n-1}(1 - \theta_i)$$

where the θ's are the eigenvalues of $\mathbf{M}$. This expression will certainly be positive if the θ's are real and positive (we know that $|\theta_i| < 1$). Hence the diagonal elements of the two-step matrix will be smaller, on average, than the Markov model would predict. A stronger result can be obtained if we consider only small values of τ when

$$\mathbf{P}^2(\tau) \doteqdot \mathbf{P}(2\tau) + \tfrac{1}{4}(1 - e^{-2\lambda\tau})(\lambda\tau)(\mathbf{I} - \mathbf{M})$$

from which it is clear that every diagonal element of $\mathbf{P}(2\tau)$ is smaller than the corresponding element of $\mathbf{P}^2(\tau)$.

It is not easy to find a decreasing transition rate for which an analysis like that just given can be carried through. We therefore consider a special case with $k = 2$ for which (4.63) can be inverted. Let us suppose that

$$f(t) = p\lambda_1 e^{-\lambda_1 t} + (1 - p)\lambda_2 e^{-\lambda_2 t}$$

We shall meet this distribution again in Chapter 7. The associated transition rate is a decreasing function of t. In this case we may write:

$$\mathbf{P}(\tau) = \begin{pmatrix} m_{11}(\tau) & 1 - m_{11}(\tau) \\ 1 - m_{11}(\tau) & m_{11}(\tau) \end{pmatrix},$$

where

$$m_{11}^*(s) = G^*(s)/[1 - \{f^*(s)\}]^2$$

$$= \left[\frac{p}{\lambda_1 + s} + \frac{p}{\lambda_2 + s} \right] \bigg/ \left[1 - \left\{ \frac{p\lambda_1}{\mu_1 + s} + \frac{(1-p)\lambda_2}{\mu_2 + s} \right\} \right]^2.$$

This is a rational algebraic fraction which means that it can be expressed in the form

$$m_{11}^*(s) = \frac{A}{s+a} + \frac{B}{s+b} + \frac{C}{s+c} + \frac{D}{s+d}$$

and hence inverted to give

$$m_{11}(\tau) = A e^{-a\tau} + B e^{-b\tau} + C e^{-c\tau} + D e^{-d\tau}.$$

The condition that the diagonal elements of $\mathbf{P}(2\tau)$ should be greater than those of $\{\mathbf{P}(\tau)\}^2$ is that

$$m_{11}(2\tau) > m_{11}^2(\tau) + \{1 - m_{11}(\tau)\}^2. \tag{4.72}$$

If we take $p = \frac{1}{2}$, $\lambda_1 = \frac{5}{2}$, $\lambda_2 = \frac{5}{8}$ it turns out that

$$m_{11}(\tau) = 0.5 + 0.1231 e^{-0.8049\tau} + 0.3769 e^{-3.8826\tau}. \tag{4.73}$$

Calculation shows that $m_{11}(\frac{1}{2}) = 0.6364$, $m_{11}(1) = 0.5628$, $m_{11}(2) = 0.5248$ from which it is easily verified that (4.46) holds for $\tau = \frac{1}{2}$ and 1.

We conjecture that cumulative inertia will always lead to an increase in the diagonal elements of $\mathbf{P}(2\tau)$ over that predicted from $\mathbf{P}^2(\tau)$ and that cumulative stress will lead to a decrease.

To summarize our conclusions from this and the previous section about the behaviour of a continuous time process observed at fixed intervals of time; we have identified three different circumstances in which departures from the Markov model can lead to over-prediction of diagonal elements, viz. heterogeneity in

1. the transition matrices $\mathbf{P}$;
2. the rate parameter λ;
3. cumulative inertia expressed as a declining transition rate.

To resolve the question of which of these influences are at work requires detailed data. In particular, (3) would be identified by looking at the distribution of intervals between moves.

Observation over an interval with arbitrary starting point

The foregoing analysis only covers time intervals which start with the entry of an individual to a given state. If we begin to observe the process at some arbitrary point the transition probabilities over that interval will depend on how long the individual has been in his present state. We can, therefore, only calculate these probabilities if we have the times which individuals have already spent in their current state. If this information is available the necessary modification is easily made. The only term in (4.58) which is affected is the first one concerning the probability of the present incumbent remaining in state i throughout the interval. For an individual who has been in the state for time x when observation begins this probability is

$$\exp - \int_x^{x+\tau} \lambda_i(y)\,\mathrm{d}y = G_i(\tau + x)/G_i(\tau).$$

The probability for a randomly chosen member of the population is then obtained by multiplying by the density function of x and integrating.

If the system is in equilibrium we can deduce the density function of x as follows. In these circumstances the period of observation is equally likely to begin at any point during the sojourn time. Thus we may write:

$$x = yt$$

where t has density function $f_i(t)$ and y is an independent uniformly distributed random variable on the interval $(0, 1)$. Hence the conditional density function of x is uniform on $(0, t)$ and this must be averaged with respect to the distribution of t. The density of the latter is not $f_i(t)$ since if we observe the system at a random point in time we are more likely to strike a large interval than a short one. The probability that the interval lies in $(t, t + \delta t)$ will thus be proportional to both $f_i(t)\delta t$ and t. Thus we have

$$f_i(x) = \int_x^\infty \frac{1}{t}\left\{\frac{tf_i(t)}{\int_0^\infty uf_i(u)\,\mathrm{d}u}\right\}\mathrm{d}t = G_i(x)/\mu_i. \qquad (4.74)$$

Hence the first term of (4.58) becomes

$$\delta_{ij} \int_0^\infty \frac{G_i(\tau + x)}{G_i(x)}\frac{G_i(x)}{\mu_i}\,\mathrm{d}x = \frac{\delta_{ij}}{\mu_i}\int_\tau^\infty G_i(x)\,\mathrm{d}x \qquad (4.75)$$

where $\mu_i = \int_0^\infty tf_i(t)\mathrm{d}t$. Noting that the Laplace transform of $\int_\tau^\infty G_i(x)\mathrm{d}x$ is $\{1 - G_i^*(s)\}/s$, (4.61) becomes

$$\mathbf{P}^*(s) = (\mathbf{I} - \mathbf{f}_m^*(s))^{-1}\mathbf{D}\{\mathbf{I} - \mathbf{G}^*(s)\} \qquad (4.76)$$

where $\mathbf{D}$ is the diagonal matrix with elements $\{1/\mu_i s\}$.

Limiting behaviour

In contrast to the transient behaviour of a semi-Markov process, the limiting behaviour is easy to obtain. It can be deduced heuristically as follows. Consider a long interval of time after the system has reached its steady state and suppose that K changes of state take place in that interval. The expected number of times that the system will visit state i will then be Km_i, where $\mathbf{m} = (m_1, m_2, \ldots, m_k)$ is the steady-state vector associated with the transition matrix $\mathbf{M}$ with elements given by (4.55). On each such visit the expected time spent in i will be μ_i and hence the total expected time for that state is $Km_i\mu_i$. The proportion of time spent in i is therefore $m_i\mu_i/\sum_{j=1}^{k} m_i\mu_i$ and this may be interpreted as the probability that an individual chosen at random will be found in i. Thus for a system of size N we write:

$$E\bar{n}_i(\infty) = Nm_i\mu_i, \qquad (i = 1, 2, \ldots, k). \tag{4.77}$$

If $f_i(t) = f(t)$ for all i the mean sojourn times will be equal and the limiting structure will be the same as that of the embedded Markov chain with transition matrix $\mathbf{M}$. This is essentially the result of (4.52) in another guise. Other special cases of (4.77) appeared in (4.29) and (4.64).

Models based on more general point processes

It is perfectly possible to envisage further generalizations obtained by introducing other point processes. The two particular assumptions of the semi-Markov process that one might wish to relax are its time homogeneity and the independence of successive sojourn times. Since all such generalizations retain the feature of the embedded Markov chain, all properties of the system which derive from this feature will remain intact. For example, the result about the stationary structure given in (4.52) holds quite generally, even though $P_n(\tau)$ may depend on aspects of the stochastic process other than the length of the interval of observation. Some extensions are almost immediate. Suppose, for example, that the point process is a time-dependent Poisson process with rate $\lambda(T)$. Then if we transform the time scale by

$$Y(T) = \int_0^T \lambda(x)\,dx$$

then, on the Y-scale, changes of state occur according to a time homogeneous process and the theory of Sections 4.2 and 4.3 applies. Results expressed on this so-called operational time scale can be converted back to 'real' time as necessary. The great practical obstacle in the way of such generalizations is the large amount of detailed data which the implementation of such models requires.

CHAPTER 5

Continuous Time Models for Open Social Systems

5.1 THEORY OF THE FIXED INPUT MARKOV MODEL

The theory of open systems is a straightforward extension of that for the closed systems treated in Chapter 4. In this chapter, therefore we shall concentrate on the new features which arise and on applications. As in Chapter 3, we regard a system as open if losses occur and we introduce two main classes of model which differ in the assumption made about the inflow. The case of a cohort arises by using a model with zero input. In terms of the theory of stochastic processes we are here moving into the realm of continuous time Markov processes with absorbing states.

The basic equations of the fixed input model can be derived from those for the discrete time model by the limiting operation carried out at the beginning of Chapter 4. Denoting the small time interval between changes of state by δT (3.1b) becomes

$$\bar{\mathbf{n}}(T + \delta T) = \bar{\mathbf{n}}(T)\mathbf{P}(\delta T) + R(T + \delta T)\mathbf{p}_0. \tag{5.1}$$

Substituting $\mathbf{P}(\delta T) = \mathbf{I} + \delta T \mathbf{R}$ and letting $T \to \infty$ we have

$$\frac{d\bar{\mathbf{n}}(T)}{dT} = \bar{\mathbf{n}}(T)\mathbf{R} + R(T)\mathbf{p}_0 \tag{5.2}$$

where now, instead of supposing that the recruits all join at $T + \delta T$, we redefine $R(T)$ to be the instantaneous rate of inflow. It is important to be clear about the properties of the matrix $\mathbf{R}$ which derive, in this case, from the corresponding matrix $\mathbf{P}(\delta T)$. The diagonal elements of $\mathbf{R}$ are given by

$$\delta T r_{ii} = p_{ii}(\delta T) - 1 = \left\{ 1 - \sum_{j=1}^{k+1} p_{ij}(\delta T) - 1 \right\} = -\delta T \sum_{\substack{j=1 \\ j \neq i}}^{k+1} r_{ij} \quad (i = 1, 2, \ldots k)$$

In the last sum, $r_{i,k+1}$ is the rate of loss to the system and it does not appear in $\mathbf{R}$ except as part of the diagonal element. The sum of the ith row of $\mathbf{R}$ is thus no longer zero but $-r_{i,k+1}$.

The system of linear differential equations in (5.2) can be integrated by standard methods to give

$$\bar{\mathbf{n}}(T) = \mathbf{n}(0)\mathbf{P}(T) + \mathbf{p}_0 \int_0^T R(x)\mathbf{P}(T - x)dx. \tag{5.3}$$

114

This equation is the continuous analogue of (3.2). The same result can be derived by a direct probability argument as follows. The first term on the right-hand side of (5.3) gives the expected number in each of the states arising from the initial stock—as in the closed model. In the second term $\mathbf{p}_0 R(x)\delta x$ is the 'initial' vector of those recruited in $(x, x + \delta x)$. After $T - x$ their distribution across the states is given by $\mathbf{p}_0 R(x)\mathbf{P}(T-x)\delta x$. Integrating x from 0 to T then provides the aggregated distribution. The same argument yields an expression for the expected number who have left the system. Using the subscript $k + 1$ to denote the set of leavers,

$$\bar{n}_{k+1}(T) = \sum_{i=1}^{k} p_{0i} \int_0^T R(x)p_{ij}(T-x)\mathrm{d}x + n_i(0)p_{i,k+1}(T). \tag{5.4}$$

The recruitment flow does not have to be continuous. It would be easy to consider discrete inflows, in which case the integrals in (5.3) and (5.4) would be replaced by sums.

As in the case of discrete time we can combine (5.3) and (5.4) into a single equation by introducing augmented versions of the vectors and matrices involved. Letting $k + 1$ denote the state comprising those who have left the system, the equation is

$$\bar{\mathbf{n}}_a(T) = \mathbf{n}_a(0)\mathbf{P}_a(T) + \mathbf{p}_{0a} \int_0^T R(x)\mathbf{P}_a(T - x)\,\mathrm{d}x. \tag{5.5}$$

The augmented version of (5.2) has $\mathbf{R}$ replaced by

$$\mathbf{R}_a = \begin{pmatrix} r_{11} & r_{12} & \cdots & r_{1k} & r_{1,k+1} \\ r_{21} & r_{22} & \cdots & r_{2k} & r_{2,k+1} \\ \vdots & \vdots & & \vdots & \vdots \\ r_{k1} & r_{k2} & \cdots & r_{kk} & r_{k,k+1} \\ 0 & 0 & & 0 & 0 \end{pmatrix}. \tag{5.6}$$

In this form the row sums are zero. There is no difficulty in regarding more than one state as absorbing as we shall do in our first example. The matrix $\mathbf{R}_a$ is formed by adding a row of zeros to $\mathbf{R}$ for each absorbing state. In order to determine the expected state vector $\bar{\mathbf{n}}(T)$ we shall use (5.5). The first step is to find $\mathbf{P}_a(T)$ by the method used in Chapter 4. The argument leading up to (4.8) holds for an open system if the augmented versions of $\mathbf{P}(T)$ and $\mathbf{R}$ are used. The stock vector then follows at once on substitution into (5.5).

Limiting behaviour

We are interested in the limiting behaviour of the vector $\bar{\mathbf{n}}(T)$ and this depends, as in the discrete case, on the nature of the input rate $R(T)$. The required result can be deduced directly from the transient solutions, but a general result of Mehlmann (1977b) throws a good deal of light on the behaviour of the system. His result concerns the class of input functions for which $R(T) \sim Re^{\alpha T}$. This includes the important case of constant input when $\alpha = 0$ which we investigate first. From (5.2) the structure required will satisfy

$$\bar{\mathbf{n}}(\infty)\mathbf{R} + R\mathbf{p}_0 = 0$$

and hence

$$\bar{\mathbf{n}}(\infty) = R\mathbf{p}_0(-\mathbf{R})^{-1}. \tag{5.7}$$

Next introduce the vector $\mathbf{v}(T) = \bar{\mathbf{n}}(T)e^{-\alpha \tau}$ from which we have

$$\frac{d\mathbf{v}(T)}{dT} = e^{-\alpha T}\frac{d\bar{\mathbf{n}}(T)}{dT} - \mathbf{v}(T).$$

Substitution into (5.2) then gives

$$\frac{d\mathbf{v}(T)}{dT} = \mathbf{v}(T)(\mathbf{R} - \alpha\mathbf{I}) + R(T)e^{-\alpha T}\mathbf{p}_0. \tag{5.8}$$

This has the same form as (5.2) except that $\mathbf{R}$ is replaced by $\mathbf{R} - \alpha\mathbf{I}$ and $R(T)$ by $R(T)e^{-\alpha T}$. Since by hypothesis,

$$\lim_{T \to \infty} R(T)e^{-\alpha T} = R$$

the limiting vector $\mathbf{v}(\infty)$ can be obtained by comparison with (5.7) as follows

$$\mathbf{v}(\infty) = R\mathbf{p}_0(\alpha\mathbf{I} - \mathbf{R})^{-1} \tag{5.9}$$

which means that

$$\bar{\mathbf{n}}(\infty) \sim e^{\alpha T}R\mathbf{p}_0(\alpha\mathbf{I} - \mathbf{R})^{-1}. \tag{5.10}$$

This result is valid only if the inverse matrix exists and this is not always so. It may be shown that the condition for (5.10) to hold is that α is larger that the greatest real parts of the eigenvalues of $\mathbf{R}$. Since these real parts are zero or negative (5.10) is certainly valid whenever the asymptotic rate of increase in the inflow is positive. It ceases to hold if the rate of contraction is too great.

Equation (5.10) allows us to say something useful about the relative sizes of the expected stocks even if the system size is increasing without limit or decreasing to vanishing point. Writing $\mathbf{q}(\infty)$ for the vector of relative expected numbers we have

$$\mathbf{q}(\infty) \propto R\mathbf{p}_0(\alpha\mathbf{I} - \mathbf{R})^{-1}. \tag{5.11}$$

The foregoing analysis is, of course, the continuous analogue of Feichtinger's (1976) work on the discrete model described in Chapter 3.

The distribution of stock numbers

Our analysis so far has been wholly concerned with the expected values of the grade sizes. It is also desirable to have some knowledge about the distribution of the grade sizes and, in particular, their variances and covariances. There is as yet no continuous version of Pollard's method which we used in the discrete time case, but Staff and Vagholkar's method extends quite easily. It only yields a tractable result if the input is a Poisson process, but this enables us to establish a continuous analogue of the earlier result about the conditions under which the grade sizes have Poisson distributions.

We require the joint distribution of the grade sizes at time T, given that the input to the system is a Poisson process with rate R. As before, we shall find the joint distribution of these numbers excluding the survivors from the initial stock. This will then enable us to deduce the asymptotic distribution by observing that all of the initial stock will ultimately be lost to the system.

Suppose that up to time Tm individuals enter the system at times $T_1, T_2, \ldots, T_m$. Then we shall first find the generating function of the numbers in the grades conditional upon $T_1, T_2, \ldots, T_m$ and m. The unconditional generating function will then be found by averaging the conditional generating function with respect to the joint distribution of the T's and m. We shall do this in two stages using the fact that, given m and T, $T_1, T_2, \ldots, T_m$ may be regarded as a random sample from the rectangular distribution with density

$$f(x) = \frac{1}{T} \qquad (0 \le x \le T)$$

and that m has a Poisson distribution with mean RT.

Let $X_{hj} = 1$ if the individual recruited at T_h is in grade j at time T and be zero otherwise. We have, in terms of the transition probabilities, that

$$Pr\{X_{hj} = 1\} = \sum_{i=1}^{k} p_{0i} p_{ij}(T - T_h). \qquad (5.12)$$

Hence the probability generating function of $X_{h1}, X_{h2}, \ldots, X_{h,k+1}$, given T_h and m, is

$$g(z_{h1}, z_{h2}, \ldots, z_{h,k+1} | T_h, m) = \sum_{j=1}^{k+1} z_{hj} \sum_{i=1}^{k} p_{0i} p_{ij}(T - T_h) \qquad (5.13)$$

and the joint probability generating function of $\{X_{hj} : h = 1, 2, \ldots, m; j = 1, 2, \ldots, k+1\}$ is

$$g(z_{11}, z_{12}, \ldots, z_{h,k+1} | \mathbf{T}, m) = \prod_{h=1}^{m} \sum_{j=1}^{k+1} z_{hj} \sum_{i=1}^{k} p_{0i} p_{ij}(T - T_h). \qquad (5.14)$$

We require the (conditional) distribution of the sums $\sum_{h=1}^{m} X_{hj}$, and their generating functions are obtained by putting $z_{hj} = z_j$, say, for $h = 1, 2, \ldots, m$, giving

$$g(z_1, z_2, \ldots, z_{k+1} | \mathbf{T}, m) = \prod_{h=1}^{m} \sum_{j=1}^{k+1} z_j \sum_{i=1}^{k} p_{0i} p_{ij}(T - T_h). \qquad (5.15)$$

Averaging with respect to $\mathbf{T}$,

$$g(z_1, \ldots, z_{k+1} | m) = \frac{1}{T^n} \int_0^T \cdots \int_0^T g(z_1, \ldots, z_{k+1} | \mathbf{T}, m) dT_1 \ldots dT_m$$

$$= \prod_{h=1}^{m} \sum_{j=1}^{k+1} z_j \sum_{i=1}^{k} p_{0i} \frac{1}{T} \int_0^T p_{ij}(T - T_h) dT_h$$

$$= \left[\frac{1}{T} \sum_{j=1}^{k+1} z_j \sum_{i=1}^{k} p_{0i} \int_0^T p_{ij}(T - x) dx \right]^m \qquad (5.16)$$

Finally,

$$g(z_1, z_2, \ldots, z_{k+1}) = \sum_{m=0}^{\infty} \frac{(RT)^m}{m!} e^{-RT} g(z_1, \ldots, z_{k+1} | m)$$

$$= \exp \left\{ -RT + R \left[\sum_{j=1}^{k+1} z_j \sum_{i=1}^{k} p_{0i} \int_0^T p_{ij}(T - x) dx \right] \right\}$$

$$= \exp \left\{ R \sum_{j=1}^{k+1} \sum_{i=1}^{k} p_{0i} \int_0^T p_{ij}(T - x) dx (z_j - 1) \right\}. \qquad (5.17)$$

This is the probability generating function of a set of $(k + 1)$ independent Poisson variates. Hence, excluding the survivors from the original stock, the numbers in the grades at time T will be independently distributed in the Poisson form with means

$$R \sum_{i=1}^{k} p_{0i} \int_0^T p_{ij}(T - x) dx \quad (j = 1, 2, \ldots, k+1), \qquad (5.18)$$

which checks with (5.3).

The result may be extended to the case when the input is a time-dependent Poisson process. In that case the input times $T_1, T_2, \ldots, T_m$ will be treated like a random sample from the distribution with density

$$f(x) = \frac{R(x)}{\int_0^T R(x) dx} \quad (0 \le x \le T)$$

and m will have a Poisson distribution with mean $\int_0^T R(x) dx$.

We have therefore shown that, after a sufficiently long time, the grade sizes will have Poisson distributions. The Poisson approximation should be good as soon as the original stock is fairly small. In a hierarchical system this will usually occur more quickly in the lower grades. If the input is not a Poisson process the argument breaks down at the point where we treat the T's as independent observations from a known distribution and so turn the product $\prod_{h=1}^{m}$ into a power. Pollard's calculations for the discrete case suggest that this step may not be critical and that the Poisson distribution may remain a good approximation, but further research on this point is required.

5.2   APPLICATIONS OF THE FIXED INPUT MODEL

Survival after treatment for cancer: a cohort study

Our first example is of a system with no input and in which there are two absorbing (or terminal), states. It has been chosen to illustrate how a well-chosen stochastic model offers advantages over the somewhat *ad hoc* procedures which have sometimes been used for this kind of problem. It is a fairly simple example of a competing risk situation concerning survival after treatment for cancer. A patient who has been treated for a disease may, at any subsequent time, be in one of a number of states. These states might be, for example, 'health', 'relapsed', and 'death'; the precise classification used will obviously depend on the objects of the enquiry and the kind of records available. A stochastic model for the post-treatment history of patients treated for cancer was developed by Fix and Neyman (1951) and discussed in more general terms by Zahl (1955). Fix and Neyman used the model to provide measures of the effectiveness of a treatment and we shall describe how this can be done below. The basic situation is of wide occurrence and there are many other possible applications.

In Fix and Neyman's model there were four states. The description of the states and the allowable transitions are indicated in Figure 5.1. The authors emphasized the difficulty of defining 'recovery' and also pointed out that it might be desirable to subdivide some of the states. For example, S_4 might be divided into those who

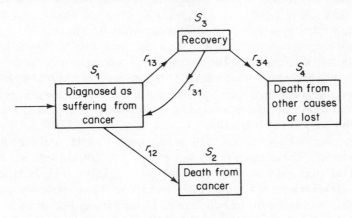

Figures 5.1

died from natural causes and those who were lost from observation. It might also be thought desirable to allow transitions between S_1 and S_4. We shall not digress to discuss these technical points since we are using the example primarily to illustrate the application of the theory of Markov processes in a social context.

The first objective in the cancer application was to estimate the transition rates. These were then used to provide measures of survival which did not suffer from the disadvantages associated with the commonly used measures. One such measure is the 'T-year survival rate'. This is the proportion of those receiving treatment who survive for at least T years. This measure would be satisfactory if cancer was the only cause of death and if all persons were under observation for the full T years. In practice this is not the case, and the T-year survival rate may be misleading. To see this we merely note that the rate would be higher if loss or death from other causes were eliminated, since then more people would survive to ultimately die of cancer. The observed value of the rate thus depends not only upon the risk of cancer death but also on other risks which have nothing to do with cancer. If a 'treatment' and 'control' group were compared by means of the crude rate the comparison would be valueless if the two groups were subject to different risks from other causes. To overcome this difficulty it is customary to calculate net rates which make allowances for such differences. Our purpose in introducing this example is to show that the stochastic model provides a more satisfactory basis for estimating net rates than the 'actuarial' method.

In their model, Fix and Neyman treated the transition rates between the states as constants. However, it is well known that the force of natural mortality for human populations is not constant but, after infancy, increases with age. It does not increase very rapidly over the middle years of life and if T is short relative to the normal life span the assumption of constancy may be quite adequate. In any event we shall show that data can be collected in such a way that the validity of the assumption can be tested. The force of mortality from various kinds of cancer has been extensively studied. The survival time following treatment appears to be highly skew; Boag (1949), for example, has suggested that it can often be adequately fitted by a skew lognormal distribution. In this case the lognormal distribution is not easy to distinguish from the exponential which would arise if the death rate were constant. The assumption of a constant rate for death from cancer is thus probably not unrealistic. Direct evidence on the true nature of the transition rates between S_1 and S_3 (recovery) and between S_3 and S_1 is lacking, but it does seem plausible to assume a constant loss rate, at least in the case of those who are lost to observation.

In the model we assume that there are N people in state S_1 at time zero and none elsewhere. The numbers in the four groups at any subsequent time T will be random variables which we denote by $n_j(T)$ ($j = 1, 2, 3, 4$); $\bar{n}_j(T)$ is the expectation of $n_j(T)$. By observing these random variables at one or more·points in time it will be possible to estimate the transition rates. Using these estimates it will then be possible to predict the numbers in the various states at future times. More important, it will be possible to estimate what these numbers would have been if death from cancer had been the only risk.

Application of the theory

The augmented matrix $\mathbf{R}_a$ in the cancer case has the form

$$\mathbf{R}_a = \begin{pmatrix} r_{11} & r_{12} & r_{13} & 0 \\ 0 & 0 & 0 & 0 \\ r_{31} & 0 & r_{33} & r_{34} \\ 0 & 0 & 0 & 0 \end{pmatrix},$$

where $r_{11} = -(r_{12} + r_{13})$ and $r_{33} = -(r_{31} + r_{34})$. The equation for the eigenvalues is $|\mathbf{I}\theta - \mathbf{R}_a| = 0$ or

$$\theta^2 (r_{11} - \theta)(r_{33} - \theta) - \theta^2 r_{13} r_{31} = 0. \tag{5.19}$$

This equation clearly has a double root equal to zero; the two remaining roots, which we label θ_3 and θ_4, are

$$\theta_3, \theta_4 = \tfrac{1}{2}\{r_{11} + r_{33} \pm \sqrt{(r_{11} - r_{33})^2 + 4r_{13}r_{31}}\}, \tag{5.20}$$

taking the positive sign for θ_3 and the negative for θ_4. It then follows from (4.24) that

$$\bar{n}_i(T) = d_{i1} + c_{i3}e^{\theta_3 T} + c_{i4}e^{\theta_4 T} \qquad (i = 1, 2, 3, 4). \tag{5.21}$$

The next step is to set down and solve the simultaneous equations for the coefficients. We first put $i = 1$ and let h take the values 2, 3, and 4, thus obtaining

$$\left. \begin{array}{l} r_{11}d_{11} + r_{31}d_{31} = 0 \\ r_{11}c_{13} + r_{31}c_{33} = \theta_3 c_{13} \\ r_{11}c_{14} + r_{31}c_{34} = \theta_4 c_{14} \end{array} \right\}. \tag{5.22}$$

Three further sets of equations are obtained for $i = 2$, 3, and 4 as follows:

$i = 2$
$$\left. \begin{array}{l} r_{12}d_{11} = 0 \\ r_{12}c_{13} = \theta_3 c_{23} \\ r_{12}c_{14} = \theta_4 c_{24} \end{array} \right\} \tag{5.23}$$

$i = 3$
$$\left. \begin{array}{l} r_{13}d_{11} + r_{33}d_{31} = 0 \\ r_{13}c_{13} + r_{33}c_{33} = \theta_3 c_{33} \\ r_{13}c_{14} + r_{33}c_{34} = \theta_4 c_{34} \end{array} \right\} \tag{5.24}$$

$i = 4$
$$\left. \begin{array}{l} r_{34}d_{31} = 0 \\ r_{34}c_{33} = \theta_3 c_{43} \\ r_{34}c_{34} = \theta_4 c_{44} \end{array} \right\}. \tag{5.25}$$

It follows at once that $d_{11} = d_{31} = 0$ and hence the first equation in each group can be ignored in what follows. The initial conditions require that all members of the system are in S_1 at time zero. Let us suppose, therefore, that $n_1(0) = 1$ and $n_i(0) = 0$, $i > 1$. If $n_1(0) = N$ the appropriate values of $\bar{n}_i(T)$ can be obtained simply by multiplying by N those obtained assuming $n_1(0) = 1$. In addition to the equations listed above we now have

$$\left.\begin{array}{l} d_{11} + c_{13} + c_{14} = 1 \\ d_{21} + c_{23} + c_{24} = 0 \\ d_{31} + c_{33} + c_{34} = 0 \\ d_{41} + c_{43} + c_{44} = 0 \end{array}\right\}. \tag{5.26}$$

To solve the equations we proceed as follows. Adding both sides of (5.22) and using the initial conditions we obtain

$$\theta_3 c_{13} + \theta_4 c_{14} = r_{11}. \tag{5.27}$$

A similar operation on (5.23) yields

$$\theta_3 c_{23} + \theta_4 c_{24} = r_{12}. \tag{5.28}$$

But this equation can be expressed in terms of c_{13} and c_{14} from (5.23) to give

$$c_{13} + c_{14} = 1. \tag{5.29}$$

The pair of simultaneous equations (5.27) and (5.28) may then be solved, giving

$$c_{13} = \frac{\theta_4 - r_{11}}{(\theta_4 - \theta_3)}, \qquad c_{14} = \frac{r_{11} - \theta_3}{(\theta_4 - \theta_3)},$$

and hence

$$c_{23} = \frac{r_{12}(\theta_4 - r_{11})}{\theta_3(\theta_4 - \theta_3)}, \qquad c_{24} = \frac{r_{12}(r_{11} - \theta_3)}{\theta_4(\theta_4 - \theta_3)}.$$

If this whole procedure is repeated on (5.24) and (5.25) we obtain

$$\left.\begin{array}{l} c_{33} = \dfrac{-r_{13}}{(\theta_4 - \theta_3)}, \qquad c_{34} = \dfrac{r_{13}}{(\theta_4 - \theta_3)} \\[2ex] c_{43} = \dfrac{-r_{13}r_{34}}{\theta_3(\theta_4 - \theta_3)}, \qquad c_{44} = \dfrac{r_{13}r_{34}}{\theta_4(\theta_4 - \theta_3)} \end{array}\right\}.$$

Only two constants remain to be determined; these are d_{21} and d_{41}. Using the initial conditions we find

$$d_{21} = -c_{23} - c_{24} = -r_{12}r_{33}/\theta_3\theta_4, \tag{5.30}$$
$$d_{41} = -c_{42} - c_{43} = r_{13}r_{34}/\theta_3\theta_4. \tag{5.31}$$

We now turn to consider how these results may be used to make valid comparisons of survival rates. When $N = 1$, $\bar{n}_i(T)$ may be regarded as the probability of being in S_i at time T. Thus $\bar{n}_2(T)$ and $\bar{n}_4(T)$ may be interpreted as the crude risks of death from cancer and natural causes respectively. However, $\bar{n}_4(T)$ also depends on the force of natural mortality, and, as we pointed out above, this reduces its value as a measure of risk. We really require a net measure of risk from which the effect of natural mortality is eliminated. The actuarial approach to the problem defines a net rate of mortality due to cancer by the formula

$$_A\bar{n}_2(T) = \bar{n}_2(T)/\{1 - \tfrac{1}{2}\bar{n}_4(T)\}. \tag{5.32}$$

This measure purports to give the expected number of cancer deaths that would occur in $(0, T)$ if natural mortality did not exist. The derivation of (5.32) will be clearer if we write it in the form

$$\bar{n}_2(T) = {}_A\bar{n}_2(T) - \tfrac{1}{2}\bar{n}_4(T){}_A\bar{n}_2(T). \tag{5.33}$$

The second term on the right-hand side of (5.33) is an estimate of the number of people who would have died from cancer in the period had they not in fact died from natural causes. It is obtained by assuming that the probability of death from cancer being preceded by death from natural causes is one-half. Our model provides an alternative method of estimating net rates. We can eliminate the effect of natural mortality by putting $r_{34} = 0$ in the model. The net risk may then be written:

$$\bar{n}_2^0(T) = \frac{r_{12}r_{31}}{\theta_3^0\theta_4^0} + \frac{r_{12}(\theta_4^0 - r_{11})}{\theta_3^0(\theta_4^0 - \theta_3^0)} e^{\theta_3^0 T} + \frac{r_{12}(r_{11} - \theta_3^0)}{\theta_4^0(\theta_4^0 - \theta_3^0)} e^{\theta_4^0 T} \tag{5.34}$$

where the superfix on $\bar{n}_2(T)$, θ_3 and θ_4 denotes that r_{34} has been set equal to zero.

The use of these results may be illustrated by two numerical examples. We assume the following transition intensities:

	r_{12}	r_{13}	r_{31}	r_{34}
Example 1	1.0	2.0	0.5	0.2
Example 2	0.5	0.5	0.5	0.5

Substituting these values in (5.20) we find, for Example 1,

$$\left.\begin{aligned}\bar{n}_2(T) &= 0.6364 - 0.3764e^{-0.3260T} - 0.2600e^{-3.3740T} \\ \bar{n}_2^0(T) &= 1 - 0.7344e^{-0.1492T} - 0.2657e^{-3.3508T}\end{aligned}\right\} \tag{5.35}$$

and, for Example 2,

$$\left.\begin{aligned}\bar{n}_2(T) &= \tfrac{2}{3} - \tfrac{1}{2}e^{-\frac{1}{2}T} - \tfrac{1}{6}e^{-(3/2)T} \\ \bar{n}_2^0(T) &= 1 - 0.7236e^{-0.1910T} - 0.2764e^{-1.3090T}\end{aligned}\right\}. \tag{5.36}$$

One unsatisfactory feature of the actuarial risk is seen from its limiting behaviour as $T \to \infty$. Instead of approaching one, as we would expect a reasonable measure to do, it tends to a limit less than one in both cases. Inspection of (5.32) shows that

Table 5.1   *A comparison of net risks of cancer death calculated by (a) the actuarial method and (b) using the stochastic model*

	T	0.5	1	2	5	∞
Example 1	$\bar{n}_2(T)$	0.269	0.356	0.440	0.563	0.636
	$\bar{n}_2^0(T)$	0.269	0.358	0.455	0.652	1.000
	$_A\bar{n}_2(T)$	0.272	0.370	0.477	0.656	0.778
Example 2	$\bar{n}_2(T\cdot)$	0.199	0.326	0.474	0.626	0.667
	$\bar{n}_2^0(T)$	0.199	0.328	0.486	0.721	1.000
	$_A\bar{n}_2(T)$	0.201	0.338	0.515	0.733	0.800

this result always holds. It also appears to be generally true that $_A\bar{n}_2(T) < \bar{n}_2^0(T)$ if T is sufficiently large. Some numerical values are given in Table 5.1.

This example provides a good illustration of the use of a stochastic model for measuring a social phenomenon. It also shows that the use of 'common-sense' corrections to crude measures may seriously underestimate the quantity being measured. These arguments presuppose that the model used provides an adequate description of the phenomenon. If, in fact, the transition intensities are not constant the simpler actuarial estimate may be preferable because it is 'distribution-free'. As we show below, rough methods are available for testing the adequacy of the model.

We have conducted the foregoing discussion as if the transition rates were known. In practice they will not be known and must therefore be estimated from the data. General methods for doing this were mentioned in Chapter 4, but for our present purposes the simpler method of Fix and Neyman will suffice. At time T we can observe the numbers of original patients in each of the four states. These numbers may be treated as estimates of the $\bar{n}_i(T)$'s, which in turn are functions of the unknown parameters. In the present case this method would yield four equations for estimating the four unknown parameters. Unfortunately the equations are not linearly independent because

$$\sum_{i=1}^{4} \bar{n}_i(T) = N,$$

the total number observed. The situation would be even worse if there were other non-zero intensities in **R**. The difficulty can be overcome if the state of the system can be observed at several points in time. An alternative method is to observe some additional feature of the system by, for example, adopting the proposal of Fix and Neyman to count the number of returns to S_1 in $(0, T)$. If sufficient observational material is available it will not only be possible to estimate all the parameters but also to test the fit of the model.

The limiting structure, $\bar{\mathbf{n}}(\infty)$, can be derived directly without recourse to the full treatment just described, but (5.21) yields the result immediately.

From equations (5.30) and (5.31) we have

$$\left. \begin{array}{l} \bar{n}_2(\infty) = d_{21} = r_{12}(r_{31} + r_{34})/\theta_3\theta_4 \\ \bar{n}_4(\infty) = d_{41} = r_{13}r_{34}/\theta_3\theta_4 \end{array} \right\}. \tag{5.37}$$

The remaining limiting expectations are zero. The relative values of $\bar{n}_2(\infty)$ and $\bar{n}_4(\infty)$ thus depend in a simple way on the transition rates. The form of this dependence can be most clearly seen by writing the ratio as follows:

$$\frac{n_2(\infty)}{n_4(\infty)} = \frac{r_{12}}{r_{13}}\left(1 + \frac{r_{31}}{r_{34}}\right) \tag{5.38}$$

in which r_{12}/r_{13} is the ratio of the intensities out of the state 'diagnosed as suffering from cancer' and r_{31}/r_{34} is the ratio of intensities out of the state 'recovery'. A high recovery rate r_{13} tends to increase the proportion of patients who die of 'other causes', but this effect will be counteracted to some extent if there is also a high rate of relapse, r_{31}.

We have already pointed out that the model was originally developed to provide a basis for measuring the effect of a treatment. One such measure is provided by $\bar{n}_2^0(T)$, the net proportion who would die of cancer if death from other causes was eliminated. Fix and Neyman argue that $\bar{n}_2^0(T)$ is not the only nor necessarily the most appropriate measure of survival. A discussion of this point would be outside the scope of the book, but we mention it in order to observe that the quantities $\bar{n}_2(T)$ or $\bar{n}_2^0(T)$ are likely to be useful in constructing alternative measures. For example, Fix and Neyman suggest the use of the expected normal life in a period $(0, T)$ if cancer were the only risk of death. Since $\bar{n}_2^0(T)$ is the distribution function of 'normal' life in the absence of other risks the expectation may be written

$$e_2 = T\{1 - \bar{n}_2^0(T)\} + \int_0^T x \frac{d\bar{n}_2^0(x)}{dx} dx$$

or

$$e_2 = \int_0^T \{1 - \bar{n}_2^0(x)\} dx. \tag{5.39}$$

A hierarchical manpower system

Continuous time models for hierarchical systems were first proposed by Seal (1945) and Vajda (1948). Their models were non-Markovian, but both authors discussed some special cases which coincide with those derived from our general theory. We consider a system which can be represented diagrammatically, as shown in Figure 5.2. This system has one terminal state which we have labelled S_{k+1}. Promotion takes place only from one grade to the one above it and new entrants all go into grade 1. The augmented matrix of transition intensities for the

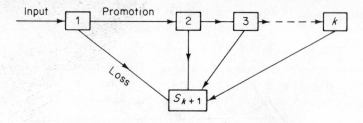

Figure 5.2

system we have described has the form

$$\mathbf{R}_a = \begin{pmatrix} r_{11} & r_{12} & 0 & \cdots & 0 & r_{1,k+1} \\ 0 & r_{22} & r_{23} & \cdots & 0 & r_{2,k+1} \\ 0 & 0 & r_{33} & \cdots & 0 & r_{3,k+1} \\ \cdot & \cdot & \cdot & \cdot & & \cdot \\ \cdot & \cdot & \cdot & \cdot & & \cdot \\ \cdot & \cdot & \cdot & \cdot & & \cdot \\ 0 & 0 & 0 & \cdot & \cdot & r_{k+1,k+1} \end{pmatrix}$$

where

$$r_{ii} = -(r_{i,i+1} + r_{i,k+1}), \qquad i < k$$
$$r_{kk} = -r_{k,k+1}, \qquad r_{k+1,k+1} = 0.$$

The simple triangular structure of $\mathbf{R}_a$ enables us to obtain explicit formulae for the eigenvalues and the coefficients $\{c_{ij}\}$ which appear in the expressions for the transition probabilities $\{p_{ij}(T)\}$. We find at once that $\theta_i = r_{ii}(i = 1, 2, \ldots, k + 1)$. The equations which determine the c's obtained from (4.19) are

$$r_{i-1,i}c_{i-1,h} + r_{ii}c_{ih} = r_{hh}c_{ih} \qquad (i, h = 1, 2, \ldots, k+1)$$

and

$$\left.\begin{array}{l} \\ \\ \sum_{h=1}^{k+1} c_{ih} = 1 \quad \text{if } i = 1 \\ \qquad\qquad = 0 \quad \text{if } i > 1 \end{array}\right\}. \qquad (5.40)$$

The initial conditions represented by this last pair of equations follow from the fact that all new entrants begin their careers in grade 1. The set of equations (5.41) may be solved to give

$$\left.\begin{array}{l} c_{ih} = \prod_{j=1}^{i-1} r_{j,j+1} \left/ \prod_{\substack{j=1 \\ j \neq h}}^{i} (r_{hh} - r_{jj}) \right. \qquad (i = 2, 3, \ldots, k+1; h = 1, 2, \ldots, i) \\ c_{11} = 1 \\ c_{ih} = 0 \text{ otherwise} \end{array}\right\} \cdot (5.41)$$

We shall be interested only in $\bar{n}_j(T)$ for $j \leq k$, in which case, from (5.3),

$$\bar{n}_j(T) = \int_0^T R(x)p_{1j}(T-x)\mathrm{d}x + \sum_{i=0}^{j} n_i(0)p_{ij}(T) \qquad (j = 1, 2, \ldots, k). \quad (5.42)$$

The coefficients obtained from (5.40) give

$$p_{1j}(T) = \sum_{h=1}^{k} c_{jh}e^{r_{hh}T} \qquad (5.43)$$

and this may be substituted in (5.42). A similar expression can be found for $p_{ij}(T), i > 1$ and all j, using the appropriate initial conditions, but they can easily be deduced from those for $p_{1j}(T)$ when we have a simple hierarchy. An entrant who starts his career in grade i of a k-state system is in the same position as one who starts in grade 1 in a $(k-(i-1))$-stage system. By replacing k by $k-i+1$ and relabelling the transition intensities the required expressions are obtained. We give an example below. It is obvious that $p_{ij}(T) = 0$ if $i > j$, which is why the upper limit of summation in the last term of (5.42) is j.

The model which we have described is slightly more general than the Markov version of Vajda's (1948) model. He assumed a constant rate of input to the system and a constant loss rate; his results may thus be obtained from ours by putting $R(x) = R$ and $r_{i,k+1} = r_{k+1}$, say, for $i \leq k$. We have also given the expected grade sizes for all T, whereas Vajda discussed only the limiting case.

As we have pointed out on several occasions, the theory which we have used above requires that the quantities $r_{ii}(i = 1, 2, \ldots, k+1)$ are all distinct. In the case we are discussing $r_{ii} = -(r_{i,i+1}+r_{i,k+1})$ for $i \leq k$, so that equalities between the r_{ii} would occur if the total loss rates for certain grades were equal. A case of particular interest in which this happens occurs if $r_{i,i+1} = r$ and if $r_{i,k+1} = r_{k+1}$ for $i < k$. This corresponds to a situation in which promotion rates and loss rates are the same for all grades except the last. The appropriate modifications to the theory may be obtained by allowing the eigenvalues $r_{ii}(i < k)$ to approach one another in (5.43). The resulting expression for $p_{1j}(T)$ is then

$$p_{1j}(T) = \frac{r^{j-1}}{(j-1)!}T^{j-1}e^{-(r+r_{k+1})T} \qquad (j = 1, 2, \ldots, k-1). \quad (5.44)$$

If $j = k$ the expression is slightly more complicated but we shall see below that it is not needed. For $p_{ij}(T)$ we use the device mentioned above and find that

$$p_{ij}(T) = \frac{r^{j-i}}{(j-i)!}T^{j-i}e^{-(r+r_{k+1})T} \qquad (i \leq j < k). \quad (5.45)$$

Using (5.44) and (5.45) we may determine $\bar{n}_j(T)$ for $j < k$ by substitution in (5.42); $\bar{n}_k(T)$ is then obtained from

$$\bar{n}_k(T) = \sum_{j=1}^{k} \bar{n}_j(T) - \sum_{j=1}^{k-1} \bar{n}_j(T). \quad (5.46)$$

The total size at time T is obtained by summing both sides of (5.42) over j and using the fact that

$$\sum_{j=1}^{k} p_{ij}(x) = 1 - p_{i,k+1}(x) = 1 - e^{-r_{k+1}x}.$$

This then gives

$$\sum_{j=1}^{k} \bar{n}_j(T) = \int_0^T R(x) e^{-r_{k+1}(T-x)} dx + N(0) e^{-r_{k+1}T}, \qquad (5.47)$$

where

$$N(0) = \sum_{j=1}^{k} n_j(0).$$

We illustrate the theory by taking $k = 3$,

$$r_{i,i+1} = 1 \quad (i = 1, 2), \qquad r_{i,4} = 2 \quad (i = 1, 2, 3)$$

for which

$$p_{ij}(T) = \frac{1}{(j-i)!} T^{j-i} e^{-3T} \quad (i = 1, 2; \; i \le j < 3). \qquad (5.48)$$

Let the input be an exponential function of time with $R(x) = 100 e^{\alpha x}$ and let $n_1(0) = 100$, $n_2(0) = 50$, $n_3(0) = 20$. We leave the parameter α unspecified in order to examine the effect which it has on the solution.

Equation (5.42) yields

$$\left. \begin{aligned} \bar{n}_1(T) &= \frac{100}{3+\alpha} e^{\alpha T} + 100 e^{-3T} \left\{ \frac{2+\alpha}{3+\alpha} \right\} \\ \bar{n}_2(T) &= \frac{100}{(3+\alpha)^2} e^{\alpha T} + 100 e^{-3T} \left\{ T + \frac{1}{2} - \frac{T}{3+\alpha} - \frac{1}{(3+\alpha)^2} \right\} \end{aligned} \right\}. \qquad (5.49)$$

The expectation $\bar{n}_3(T)$ is now obtained by subtracting $\bar{n}_1(T) + \bar{n}_2(T)$ from

$$\sum_{j=1}^{3} \bar{n}_j(T) = \frac{100 e^{\alpha T}}{(2+\alpha)} + 100 e^{-2T} \left\{ 1.7 - \frac{1}{2+\alpha} \right\}. \qquad (5.50)$$

The crucial dependence of $\bar{n}_j(T)$ on α is evident from these formulas. If $\alpha > 0$ the input increases exponentially with time as also do the expected sizes of the grades. If $\alpha < 0$ the expected grade sizes tend to zero with increasing time. The reader should compare these results with those obtained in Chapter 3 for the discrete time model with geometric input. In the case of constant input, obtained by putting $\alpha = 0$, the $\bar{n}_j(T)$'s approach finite limits as $T \to \infty$. These limits are

$$\bar{n}_1(\infty) = 33.3, \qquad \bar{n}_2(\infty) = 11.1, \qquad \bar{n}_3(\infty) = 5.6,$$

the overall size being 50. This last result could have been found directly from the fact that the average time spent in the system is $r_{i4}^{-1} = \frac{1}{2}(i = 1, 2, 3)$ and that the input is 100 per unit time. The limiting size is, of course, independent of the initial size, and the parameter values we have chosen bring about a reduction in overall size from 170 to 50. Inspection of (5.49) shows that the lowest grade reaches its limit more rapidly than the middle and highest grades. We observed the same phenomenon with the discrete time model.

It is interesting to observe that the *relative* sizes of the $\bar{n}_j(T)$'s for large T depend upon α in a simple fashion. If $\alpha > -2$ then

$$\left. \begin{array}{ll} \bar{n}_1(T) \sim \dfrac{100}{3+\alpha} \, e^{\alpha T} & \bar{n}_2(T) \sim \dfrac{100}{(3+\alpha)^2} \, e^{\alpha T} \\[3mm] \bar{n}_3(T) \sim \dfrac{100}{(2+\alpha)(3+\alpha)^2} \, e^{\alpha T} & \end{array} \right\}. \tag{5.51}$$

The effect of increasing the rate of input is to increase the sizes of the lower two grades relative to the highest. Conversely, a decreasing rate of input leads to a concentration at the upper end of the hierarchy. Had we repeated the calculation with large values of k we should have found that, for large T, the grade sizes were in geometric progression except for the kth grade. In particular, this result holds for the case of constant input obtained by putting $\alpha = 0$. In this case we find

$$\left. \begin{array}{ll} \bar{n}_j(\infty) = \dfrac{R}{r}\left(\dfrac{r}{r+r_{k+1}}\right)^j & (j = 1, 2, \ldots, k-1) \\[4mm] \bar{n}_k(\infty) = \dfrac{R}{r_{k+1}} - \displaystyle\sum_{j=1}^{k-1} \bar{n}_j(\infty) = \dfrac{R}{r_{k+1}}\left(\dfrac{r}{r+r_{k+1}}\right)^{k-1} \end{array} \right\}, \tag{5.52}$$

where R is the rate of input to the system. It thus follows that a hierarchical system with constant promotion rates throughout, a constant loss rate and a constant input rate will tend to a geometric structure. The exception to this general rule is that the kth grade will be larger than the term in the geometric series corresponding to $j = k$. A system of the kind that we have discussed will thus have a tendency to become 'top-heavy'. If $R(x)$ has the exponential form considered earlier the same result holds for the relative expected grade sizes, as we have already noted in our example. Later in this chapter we shall consider how far this general conclusion remains true when the rather special assumptions of the present model are relaxed.

5.3   THEORY OF THE FIXED SIZE MODEL

If the total size of the system is a known function of T the expected stock numbers can be found by a method which closely parallels that for the discrete time model. We develop the theory on the assumption that the total size, $N(T)$, is a continuous function of T and we define $M(T) = dN(T)/dT$. The equations we require are obtained from (5.2) or (5.3) by substituting the appropriate expression for $R(T)$. As before this is composed of two parts, one to replace losses and the other to fill now positions. The expected number of new entrants in $(T, T + \delta T)$ is thus

$$R(T)\delta T = \sum_{i=1}^{k} \bar{n}(T) r'_{i,k+1} \, \delta T + M(T)\delta T. \tag{5.53}$$

On substitution into (5.2) we have the continuous version of (3.54) as

$$\frac{d\bar{n}(T)}{dT} = \bar{n}(T)\mathbf{R} + \mathbf{n}(T)\mathbf{r}'_{k+1}\,\mathbf{p}_0 + M(T)\mathbf{p}_0$$

$$= \bar{n}(T)\{\mathbf{R} + \mathbf{r}'_{k+1}\mathbf{p}_0\} + M(T)\mathbf{p}_0. \tag{5.54}$$

The matrix in brackets is the continuous equivalent of $\mathbf{Q}$ in Chapter 3 and has the property that its row sums are zero. The equation has the same form as (5.2) and so by making the appropriate changes in (5.3) we deduce that

$$\bar{\mathbf{n}}(T) = \mathbf{n}(0)\mathbf{P}^+(T) = \mathbf{p}_0 \int_0^T M(\mathbf{x})\mathbf{P}^+(T-x)dx, \tag{5.55}$$

where $\mathbf{P}^+(T)$ is the transition matrix calculated from the intensity matrix $\mathbf{R} + \mathbf{r}'_{k+1}\mathbf{p}_0$ by the methods of Chapter 4. In the case of both (5.3) and (5.5) it may be convenient to work with the Laplace transform which for (5.3) is

$$\mathbf{n}^*(s) = \{\mathbf{n}(0) + \mathbf{p}_0 R^*(s)\}\mathbf{P}^*(s). \tag{5.56}$$

For (5.55) we merely replace $R^*(s)$ by $M^*(s)$ and $\mathbf{P}^*(s)$ becomes the transform of $\mathbf{P}^+(T)$ which is $(s\mathbf{I} - \mathbf{R} - \mathbf{r}'_{k+1}\mathbf{p}_0)^{-1}$.

Limiting behaviour

This may be investigated by a similar approach to that used by Feichtinger (1976) for the discrete time case. From (5.53) we have

$$R(T) = \bar{\mathbf{n}}(T)\mathbf{r}'_{k+1} + M(T).$$

Let $\mathbf{l} = (1, 1, \ldots 1)$, then we have that

$$\mathbf{R}\mathbf{l}' = -\mathbf{r}'_{k+1} \quad \text{and} \quad \bar{\mathbf{n}}(T)\mathbf{l}' M(T)/N(T) = M(T).$$

Substituting for $\mathbf{r}_{k+1}$ and $M(T)$ in (5.54) we find an alternative version of the equation:

$$\frac{d\bar{\mathbf{n}}(T)}{dT} = \bar{\mathbf{n}}(T)\{\mathbf{R} + (\mathbf{I}\, M(T)/N(T) - \mathbf{R})\mathbf{l}'\mathbf{p}_0\}. \tag{5.57}$$

This will only have a steady-state solution if $N(T)$ tends to a limit so we consider the relative values of the expected stocks by introducing

$$\mathbf{q}(T) = \bar{\mathbf{n}}(T)/\bar{\mathbf{n}}(T)\mathbf{l}'$$

which yields

$$\frac{d\mathbf{q}(T)}{dT} = \mathbf{q}(T)\left(\mathbf{R} - \frac{M(T)}{N(T)}\mathbf{I}\right)\left(\mathbf{I} - \mathbf{l}'\mathbf{p}_0\right). \tag{5.58}$$

This equation may be compared with the corresponding equation for a time-dependent Markov process. If such a process has transition intensity matrix $\mathbf{R}(T)$ then the state probabilities satisfy

$$\frac{d\mathbf{q}(T)}{dT} = \mathbf{q}(T)\mathbf{R}(T).$$

If the matrix on the right-hand side of (5.58) has the same form as an intensity matrix we can use the theory of time-dependent processes to deduce the behaviour of our process. The condition we require is that the matrix has negative elements on the diagonal, positive elements elsewhere and row sums equal to zero. Mehlmann (1977b) shows that this requirement is met if

$$M(T)/N(T) > \max_i(-r_{i,k+1}) \quad \text{for all } T. \tag{5.59}$$

This implies that there is a limit on the rate at which the system can contract.

For a limit to exist we require that $\lim_{T\to\infty} \mathbf{R}(T)$ exists and this condition holds for (5.58) if

$$\lim_{T\to\infty} M(T)/N(T) = c.$$

In that case the steady-state structure satisfies

$$\mathbf{q}(\infty)(\mathbf{R} - c\mathbf{I})(\mathbf{I} - \mathbf{l}'\mathbf{p}_0) = \mathbf{0}$$

or

$$\mathbf{q}(\infty) \propto \mathbf{p}_0(c\mathbf{I} - \mathbf{R})^{-1}. \tag{5.61}$$

The constant of proportionality is chosen so that the elements of $\mathbf{q}(\infty)$ add up to one.

An important special case arises when $c = 0$, meaning that the rate of growth tends to zero. When this happens the limiting relative structure is the same regardless of the pattern of growth. In particular it is the same as when there is no growth at all. Recalling the form of the Laplace transform of $\mathbf{P}(T)$ we may also write:

$$\mathbf{q}(\infty) \propto \mathbf{p}_0 \int_0^\infty \mathbf{P}(T)e^{-cT}\,dT. \tag{5.62}$$

An illustration

Consider a manpower system with $k = 3$, $p_{01} = 1$, $p_{02} = p_{03} = 0$, $r_{12} = r_{23} = r$ and $r_{14} = r_{24} = r_{34} = r_4$, say. Assume also that $n_1(0) = N$, $n_2(0) = n_3(0) = 0$. We shall find the expected stock numbers starting with the Laplace transform of (5.55). First we require the transform of $\mathbf{P}^+(T)$ which involves the matrix

$$\mathbf{R} + \mathbf{r}'_{k+1}\mathbf{p}_0 = \begin{pmatrix} -r & r & 0 \\ r_4 & -r-r_4 & r \\ r_4 & 0 & -r_4 \end{pmatrix}.$$

The transform we require is then

$$(s\mathbf{I} - \mathbf{R} - \mathbf{r}'_{k+1}\mathbf{p}_0)^{-1} = \frac{1}{s\{s+(r+r_4)\}^2}$$

$$\begin{pmatrix} (s+r+r_4)(s+r_4) & r(s+r_4) & r^2 \\ r_4(s+r_4)+rr_4 & (s+r)(s+r_4) & r(s+r) \\ r_4(s+r+r_4) & rr_4 & (s+r)(s+r_4)-rr_4 \end{pmatrix}. \quad (5.63)$$

Suppose we consider the case of exponential growth setting $M(T) = Me^{\alpha T}$ $(M > 0)$ for which $M^*(s) = M/(s-\alpha)$, $(s > \alpha)$. The stock vectors are now obtained from

$$\bar{\mathbf{n}}^*(s) = \{(N, 0, 0) + (1, 0, 0)M/(s-\alpha)\}(s\mathbf{I} - \mathbf{R} - \mathbf{r}'_{k+1}\mathbf{p}_0)^{-1}. \quad (5.64)$$

For example,

$$\bar{n}_1^*(s) = (s+r_4)(N(s-\alpha)+M)/s(s-\alpha)(s+r+r_4). \quad (5.65)$$

Resolution into partial fractions gives

$$\bar{n}_1^*(s) = \frac{A_1}{s} + \frac{B_1}{s-\alpha} + \frac{C_1}{s+r+r_4}, \quad (5.66)$$

provided that $\alpha \neq 0$. The coefficients in the expansion are

$$A_1 = \left(\frac{N\alpha - M}{\alpha}\right)\left(\frac{r_4}{r+r_4}\right)$$

$$B_1 = \frac{M}{\alpha}\frac{(r_4+\alpha)}{(r+r_4+\alpha)}$$

$$C_1 = \frac{r\{N(r+r_4)+N\alpha-M\}}{(r+r_4)(r+r_4+\alpha)}.$$

We may note that $A_1 + B_1 + C_1 = N$. If $\alpha = 0$ the partial fraction representation is

$$\bar{n}_1^*(S) = \frac{A_1'}{s} + \frac{B_1'}{s^2} + \frac{C_1'}{(s+r+r_4)}, \quad (5.67)$$

where

$$A'_1 = \frac{Mr}{(r+r_4)^2} + \frac{Nr_4}{(r+r_4)}$$

$$B'_1 = \frac{Mr_4}{(r+r_4)}$$

$$C'_1 = \frac{r\{N(r+r_4)-M\}}{(r+r_4)^2},$$

Inverting the leading expressions in (5.66) and (5.67) we find

$$\begin{aligned}\bar{n}_1^*(T) &= A_1 + B_1 e^{\alpha T} + C_1 e^{-(r+r_4)T} \quad (\alpha \neq 0)\\ &= A'_1 + B'_1 T + C_1 e^{-(r+r_4)T} \quad (\alpha = 0)\end{aligned}\Bigg\}. \qquad (5.68)$$

The last term in each part of the right-hand side of (5.68) tends to zero as T increases. For large values of T the behaviour of $\bar{n}_1(T)$ depends critically on the value of α. Three cases must be distinguished as follows:

$$\begin{aligned}\bar{n}_1(T) &= B_1 e^{\alpha T} \quad \text{if } \alpha > 0\\ &\sim B'_1 T \quad \text{if } \alpha = 0\\ &\sim A_1 \quad \text{if } \alpha < 0.\end{aligned}$$

The determination of $\bar{n}_2(T)$ follows similar lines to that of $\bar{n}_1(T)$ and yields a solution of the form

$$\bar{n}_2(T) = A_2 + B_2 e^{\alpha T} + C_2 e^{-(r+r_4)T} + D_2 T e^{-(r+r_4)T} \qquad (\alpha \neq 0). \qquad (5.69)$$

If $\alpha = 0$ the second term in this equation is replaced by one which is linear in T and the coefficients have different values. Their determination is left to the reader. The approach to the limit is somewhat slower for $\bar{n}_2(T)$ than for $\bar{n}_1(T)$ because of the factor T which is present in the last term. The general form of $\bar{n}_3(T)$ is identical with (5.69) and a similar remark about the rate of approach to the limit applies.

The relative grade sizes in the limit can be deduced from the expression for $\mathbf{n}(T)$ or directly from (5.61) or (5.62).

Thus

$$\mathbf{q}(\infty) \propto (1,0,0) \begin{pmatrix} \alpha+r+r_4 & -r & 0 \\ 0 & \alpha+r+r_4 & -r \\ 0 & 0 & \alpha+r_4 \end{pmatrix}^{-1},$$

hence

$$\mathbf{q}(\infty) = \left\{ \frac{(r_4+\alpha)}{(r+r_4+\alpha)}, \frac{r(r_4+\alpha)}{(r+r_4+\alpha)^2}, \frac{r^2}{(r+r_4+\alpha)^2} \right\}. \qquad (5.70)$$

Inspection of this result shows that the effect of increasing the exponential rate of growth is to increase the relative size of the lowest grade and decrease that of the highest. In fact, if α becomes very large the lowest grade dominates the rest, as we observed with the discrete time model. In the limit, as $\alpha \to \infty$, $\mathbf{q}(\infty) \to \mathbf{p}_0$, which, in the present example is $(1, 0, 0)$.

5.4   SYSTEMS WITH FIXED INPUT AND LOSS RATE DEPENDING ON LENGTH OF STAY

Theory

Up to this point we have assumed the transition intensities to be constant, but the argument leading to the basic equation (4.4) holds even if $\mathbf{R}$ is a function of T. The assumption of constant rates appears to have been adequate for the application to cancer survival, but it would almost certainly not be so for application to flows in a manpower system. Promotion flows usually depend on seniority within the grade and, as we shall see in Chapter 7, propensity to leave the system depends strongly on T, the length of service. In this section we shall investigate the consequences of relaxing the assumption of constant loss rate. Finally, in Section 5.5, we shall allow the other rates to depend on length of stay in the grade.

General methods are available for the solution of the basic equations of the Markov process in the time-dependent case. However, in the particular case of interest to us it is possible to obtain an explicit solution very easily. To do this we restrict attention to hierarchical systems in which there are no demotions. We set out the transition rates in an augmented matrix in which the $(k+1)$th category corresponds to the outside world. Thus

$$\mathbf{R}_a(T) = \begin{pmatrix} r_{11}(T) & r_{12} & \cdots & r_{1k} & r_{k+1}(T) \\ 0 & r_{22}(T) & \cdots & r_{2k} & r_{k+1}(T) \\ \vdots & \vdots & & \vdots & \vdots \\ 0 & 0 & \cdots & r_{kk}(T) & r_{k+1}(T) \\ 0 & 0 & \cdots & 0 & 0 \end{pmatrix}.$$

The differential equation for $\mathbf{P}(T)$ may be broken down into two parts to give

$$\frac{d p_{ij}(T)}{dT} = \sum_{h=1}^{j-1} r_{hj} p_{ih}(T) + r_{jj}(T) p_{ij}(T) \qquad (i = 1, 2, \ldots, k; j = 1, 2, \ldots, k)$$

$$(5.71)$$

$$\frac{d p_{i,\,k+1}(T)}{dT} = \sum_{h=1}^{k} r_{k+1}(T) p_{ih}(T)$$

$$= r_{k+1}(T)\{1 - p_{i,\,k+1}(T)\} \qquad (i = 1, 2, \ldots, k).$$

$$(5.72)$$

The set of equations given by (5.72) can be solved immediately, using the initial condition $p_{i,k+1}(0) = 0$, to give

$$p_{i,k+1}(T) = 1 - \exp\left\{-\int_0^T r_{k+1}(x)dx\right\} \qquad (i = 1, 2, \ldots, k). \qquad (5.73)$$

This probability is, of course, the distribution function of completed length of service and it could have been found directly. In fact, the appropriate choice of $r_{k+1}(T)$ for a given problem would usually be made on the basis of the observed distribution $p_{i,k+1}(T)$.

The system of equations (5.71) cannot be solved in the manner of Chapter 4 because some of its coefficients are not constants. It can, however, be transformed into a set which has constant coefficients as follows. Let

$$\left.\begin{array}{ll} r'_{ij} = r_{ij} & (i \neq j) \\ r'_{jj} = r_{jj}(T) + r_{k+1}(T) & (j = 1, 2, \ldots, k) \end{array}\right\}. \qquad (5.74)$$

Note that r'_{jj} is a constant for all j in consequence of the definition of $r_{jj}(T)$. We introduce a new set of probabilities, denoted by $\{p'_{ij}(T)\}$, as follows:

$$p_{ij}(T) = p'_{ij}(T)\{1 - p_{i,k+1}(T)\} \qquad (i, j = 1, 2, \ldots, k). \qquad (5.75)$$

The probability $p'_{ij}(T)$ defined in this equation is the conditional probability of the transition from grade i to grade j in $(0, T)$, given that no loss occurs in the same interval. Substituting from (5.75) into (5.71) we obtain

$$\frac{dp'_{ij}(T)}{dT} = \sum_{h=1}^{j} r'_{hj} p'_{ih}(T) \qquad (i, j = 1, 2, \ldots, k). \qquad (5.76)$$

This system of equations can be solved by the methods of Section 4.1. By first finding $p'_{ij}(T)$ from (5.76) we are able to determine $p_{ij}(T)$ from (5.75).

If the probabilities $\{p_{ij}(T)\}$ have been computed for the system on the assumption of a constant loss rate it is a simple matter to obtain the probabilities for a time-dependent loss rate. To illustrate the procedure let $i = 1$. Then we may write:

$$p'_{1j}(T) = \sum_{h=1}^{k} c'_{jh} e^{r'_{hh}T} \qquad (j = 1, 2, \ldots, k). \qquad (5.77)$$

Reference to (5.41) shows that the coefficients $\{c'_{jh}\}$ are the same as those for the system with constant loss rate because $r_{jj} - r_{hh} = r'_{jj} - r'_{hh}$ and $r_{j,j+1} = r'_{j,j+1}$. Using (5.75) we then have

$$p_{1j}(T) = \sum_{h=1}^{k} c_{jh} \exp\left\{r'_{hh}T - \int_0^T r_{k+1}(x)dx\right\} \qquad (5.78)$$

where the coefficients $\{c_{jh}\}$ are those previously computed on the basis of a constant loss rate. The only alteration required by our generalization is in the form of the exponent where $r_{hh}T$ is replaced by

$$r'_{hh}T - \int_0^T r_{k+1}(x)\mathrm{d}x.$$

Illustration of the theory

We are now in a position to assess the likely consequences of erroneously assuming the loss rate to be constant. We shall illustrate the procedure using the second example of Section 5.2. In that case we took $k = 3, r_{12} = r_{23} = r = 1$ and $r_{14} = r_{24} = r_{34} = r_4 = 2$. The discussion will be confined to the case of a constant rate of input with $R = 100$ and will be concerned solely with the limiting grade structure $\bar{\mathbf{n}}(\infty)$. We first note that the expected total size of the system depends only on the mean of the distribution of completed length of service. We shall therefore choose $r_4(T)$ to be such that the mean length of stay in the system is $\frac{1}{2}$ as before. In the case of a constant loss rate the length of stay has an exponential distribution. In our examples below we shall consider two extreme departures from this form.

Consider first a mixed exponential distribution with density function

$$f(T) = \tfrac{1}{2}\{\lambda_1 e^{-\lambda_1 T} + \lambda_2 e^{-\lambda_2 T}\}. \tag{5.79}$$

In order that this shall give the same mean length of service as the earlier example we must choose λ_1 and λ_2 to satisfy

$$\lambda_1^{-1} + \lambda_2^{-1} = 1. \tag{5.80}$$

To obtain the transition probabilities from those already found for a constant loss rate we require the result that, for the mixed exponential distribution,

$$\exp\left\{-\int_0^T r_{k+1}(x)\mathrm{d}x\right\} = \tfrac{1}{2}\{e^{-\lambda_1 T} + e^{-\lambda_2 T}\}.$$

The case $\lambda_1 = \lambda_2 = 2$ yields the formulas for the exponential distribution of length of stay. Making the necessary substitutions in (5.78) we find

$$
\left.
\begin{aligned}
\bar{n}_1(\infty) &= 50\left\{\frac{1}{\lambda_1 + 1} + \frac{\lambda_1 - 1}{2\lambda_1 - 1}\right\} \\
\bar{n}_2(\infty) &= 50\left\{\frac{1}{(\lambda_1 + 1)^2} + \frac{(\lambda_1 - 1)^2}{(2\lambda_1 - 1)^2}\right\} \\
\bar{n}_3(\infty) &= 50 - \bar{n}_1(\infty) - \bar{n}_2(\infty)
\end{aligned}
\right\}
\tag{5.81}
$$

where λ_2 has been eliminated using (5.80) and $\lambda_1 \geq 1$. As λ_1 varies between one and infinity, $\bar{\mathbf{n}}(\infty)$ takes on all its possible values. It may readily be shown that the expected grade sizes given by (5.81) have extreme values at $\lambda_1 = 2$, which is the exponential case, at $\lambda_1 = 1$ and at $\lambda_1 = \infty$. The last two cases are equivalent because $f(T)$ is symmetrical in λ_1 and λ_2. The extreme structures attainable are given in the first two columns of Table 5.2. We shall discuss the results below.

At the opposite extreme to that considered in the last paragraph we may suppose that the length of stay is a constant. That is, we take

$$
\begin{aligned}
r_4(T) &= 0 \qquad (T \le r_4^{-1}) \\
&= \infty \qquad (T > r_4^{-1})
\end{aligned} \Bigg\}. \tag{5.82}
$$

In this case, with $r_4 = 2$, the limiting grade sizes will be given by

$$
\bar{n}_j(\infty) = 100 \int_0^{\frac{1}{2}} p_{1j}^0(x)\,\mathrm{d}x \qquad (j = 1, 2, 3) \tag{5.83}
$$

where $p_{1j}^0(x)$ is obtained from (5.44) by setting $r_{k+1} = 0$. The range of integration is restricted because

$$
\exp\left\{-\int_{\frac{1}{2}}^T r_4(x)\,\mathrm{d}x\right\} = 0 \quad \text{for } T > \tfrac{1}{2}.
$$

Substituting the numerical values for r we obtain

$$
\begin{aligned}
\bar{n}_1(\infty) &= 100(1 - e^{-\frac{1}{2}}) = 39.4 \\
\bar{n}_2(\infty) &= 100(1 - \tfrac{3}{2}e^{-\frac{1}{2}}) = 9.0 \\
\bar{n}_3(\infty) &= 50 - \bar{n}_1(\infty) - \bar{n}_2(\infty) = 1.6
\end{aligned} \Bigg\}. \tag{5.84}
$$

The foregoing results are brought together in Table 5.2.

The figures in Table 5.2 represent, in a certain sense, the greatest variation in structure that can occur. They relate only to one particular example, but calculations for other cases suggest that the general pattern revealed here is typical. It can be seen from the table that the greater the variability of the length of service distribution the greater the size of the highest grade. In absolute terms, the change in the highest grade is roughly balanced by that in the lowest, but, in relative terms, the highest grade depends most critically on the assumption of constant loss rate. In practice, length of service distributions have been found to be highly skew and they have been successfully graduated by mixed exponential distributions. With constant promotion rates we would therefore expect the

Table 5.2 *Grade structure for the example under various extreme assumptions about the loss intensity*

	$\lambda_1 \to 1$ or ∞	$\lambda_1 = 2$	Fixed length of stay
$\bar{n}_1(\infty)$	25.0	33.3	39.4
$\bar{n}_2(\infty)$	12.5	11.1	9.0
$\bar{n}_3(\infty)$	12.5	5.6	1.6
$N(\infty)$	50.0	50.0	50.0

limiting structure to be more like that in the first column of the table with a relatively large number of members in the highest grade. Any factor which tends to increase the variability of length of service with the mean held constant is therefore likely to increase the size of the higher grades at the expense of the lower.

In the preceding discussion we have considered a constant rate of input. We would have reached similar conclusions had we dealt with expanding organizations with a rate of growth satisfying $\lim_{T \to \infty} M(T)/N(T) = 0$ because the limiting structures are the same in both cases. A fuller generalization allowing the promotion rates to depend on length of stay would be of great interest.

5.5 SEMI-MARKOV OPEN MODELS

As Gilbert (1973) has pointed out the extension of the semi-Markov theory of Chapter 4 to open systems is almost immediate. The basic formula (5.3) still holds. To use it for a semi-Markov model requires the calculation of $\mathbf{P}(T)$ according to the prescription of Section 4.4. As with closed systems the calculation may be fairly complicated. We shall therefore demonstrate two approaches, applicable in the special case of a simple hierarchy, which are sufficient to yield worthwhile results. The first is a direct approach to the solution of the basic integral equations and the second involves a device involving the introduction of hypothetical grades.

A direct approach

Although empirical evidence about how promotion rates depend on seniority is less abundant than that relating to leaving, it is common for firms to take seniority into account when making promotions. In this section, therefore, we shall try to throw some light on the effect of seniority-dependent promotion rates on our earlier conclusions.

In the discussion of discrete time models in Chapter 2 we encountered a similar problem in trying to accommodate the principle of cumulative inertia. In that case we expanded the state space so that seniority within a class became part of the specification of the class. This was possible because seniority was measured in the same units as the basic time scale. Ginsberg (1971) extended this idea by making use of the theory of semi-Markov processes to construct continuous time models. In this approach the transitions are governed by a probability transition matrix and the times at which transitions take place by probability distributions—one for each possible transition. Such a process can be specified in terms of transition rates expressed as functions of the time spent in a given state; this is the approach we shall adopt here to emphasize the continuity with what has gone before.

We consider initially a closed system of hierarchical form in which promotion is into the next highest grade and recruitment is into the lowest. The only transition probabilities that we require are then $p_{1j}(T)$ $(j = 1, 2, \ldots, k)$, where k is the number of grades. When these have been found we can go back to (5.3) and obtain the expected grade sizes for the open system in which we are interested. Let $r_{j,j+1}(\tau)$ $(j = 1, 2, \ldots, k-1)$ denote the promotion intensity from grade j to grade $j+1$ for a person who entered grade j a time τ ago. As before, $r_{j,k+1}(T)$ is the intensity of loss for an individual with present length of service T. The equations for the probabilities $p_{ij}(T)$ are obtained by what is essentially a special case of the argument leading to (4.58), but they have a much simpler form. This feature arises from the hierarchical structure which we have assumed. Thus if we consider the transition from state 1 to state j there is only one possible sequence of transitions, namely, $1 \to 2 \to 3 \to \ldots \to j$. We thus have

$$p_{1j}(T) = \int_0^T p_{11}(\tau)r_{12}(\tau)p_{2j}(T - \tau)\mathrm{d}\tau \qquad (j = 2, 3, \ldots, k). \qquad (5.85)$$

The probabilities $\{p_{1j}(T)\}$ may be obtained recursively since $p_{2j}(T - \tau)$ is the same as $p_{1,j-1}(T - \tau)$ for the $(k - 1)$-grade system which results when grade 1 is deleted. In practice, it is not easy to obtain explicit solutions, especially for large values of j. Since we shall present an alternative method which is sufficiently general for most purposes a detailed discussion of these integral equations will not be necessary. The chief value of the direct approach is that it easily yields a general solution for $j = 1$ which, in turn, leads to $\bar{n}_1(T)$. We shall now use it for that purpose.

Suppose that there is a maximum time that may be spent in the lowest grade. If promotion or loss has not occurred by that time then promotion follows automatically. Such an assumption requires that the promotion intensity $r_{12}(\tau)$ becomes infinite at $\tau = b$, where b is the maximum length of service in the grade. A simple, increasing function of length of service having this property was suggested by Vajda (1947) who took

$$r_{12}(\tau) = c/(b - \tau) \qquad (0 \le \tau < b; c > 0).$$

We shall assume that the loss intensity has a similar form with

$$r_{1,k+1}(\tau) = u/(v - \tau) \qquad (0 \le \tau < v; u > 0) \qquad (5.86)$$

where v is the maximum time that can be spent in the organization. (In grade 1 total length of service and seniority within the grade are synonymous. Hence we may use either T or τ to denote it.) If (5.86) obtains, the length of completed service distribution associated with this loss rate has the density function

$$f(\tau) = \frac{u}{v}\left(1 - \frac{\tau}{v}\right)^{u-1} \qquad (0 \le \tau < v), \qquad (5.87)$$

and the mean length of service is

$$\mu = v/(u + 1). \qquad (5.88)$$

Under the foregoing assumptions we easily find that

$$p_{11}(T) = \left(\frac{b-T}{b}\right)^c \left(\frac{v-T}{v}\right)^u \qquad (0 \leq T < \min(b, v)). \qquad (5.89)$$

The expected size of the lowest grade may now be found by substituting from the last equation into (5.3). Assuming a constant rate of input we have, for the kind of hierarchy being considered in this section,

$$\left.\begin{aligned} \bar{n}_1(T) &= R \int_0^T \left(1 - \frac{x}{b}\right)^c \left(1 - \frac{x}{v}\right)^u dx + n_1(0)\left(1 - \frac{T}{b}\right)^c \left(1 - \frac{T}{v}\right)^u \\ &\qquad\qquad\qquad\qquad\qquad\qquad (0 \leq T < \min(b, v)) \\ &= R \int_0^{\min(b,v)} \left(1 - \frac{x}{b}\right)^c \left(1 - \frac{x}{v}\right)^u dx \qquad (\min(b, v) \leq T < \infty) \end{aligned}\right\} \qquad (5.90)$$

It is obvious from this equation that the limiting value of $\bar{n}_1(T)$ is attained at $T = \min(b, v)$ and is given by the second of the two integrals.

We may now compare these results with those obtained when we assumed constant promotion and loss rates. To do this meaningfully we must arrange that the Markov system has the same average length of stay as the one considered above. This is achieved by taking

$$r_{1,k+1} = (u+1)/v, \qquad r_{12} = (c+1)/b. \qquad (5.91)$$

If we restrict the comparison to the limiting case we have shown that, for the Markov system, $\bar{n}_1(\infty) = R/(r_{12} + r_{1,k+1})$. This has to be compared with the second expression in (5.90), which may now be written:

$$\bar{n}_1(\infty) = \int_0^{\min(b,v)} \left(1 - \frac{x}{b}\right)^{r_{12}b-1} \left(1 - \frac{x}{v}\right)^{r_{1,k+1}v-1} dx. \qquad (5.92)$$

If both v and b tend to infinity

$$\bar{n}_1(\infty) \to R \int_0^\infty e^{-x(r_{12}+r_{1,k+1})} dx = R/(r_{12} + r_{1,k+1})$$

which agrees with the known result for constant transition rates. The greatest divergence between the two assumptions will occur when b and v are both small. The extreme case occurs when

$$b = r_{12}^{-1} \quad \text{and} \quad v = r_{1,k+1}^{-1}$$

(smaller values of b or v would make u or c negative). In this case

$$\bar{n}_1(\infty) = R \min(r_{12}^{-1}, r_{1,k+1}^{-1}). \qquad (5.93)$$

It thus follows that, for any system with transition rates given by (5.85) and (5.86),

$$R/(r_{12} + r_{1,k+1}) \leq \bar{n}_1(\infty) \leq R \min(r_{12}^{-1}, r_{1,k+1}^{-1}). \qquad (5.94)$$

The effect of making promotion chances increase with length of service is thus to increase the expected size of the lowest grade. Since the total expected size does not depend on the form of the promotion intensity we further conclude that the *relative* expected size of the lowest grade will also be increased.

As b and v increase, the approach to the lower bound of the inequalities (5.94) is quite rapid, as may be seen from the fact that when $b = v$

$$\bar{n}_1(\infty) = R/(r_{12} + r_{1,k+1} - v^{-1}).\tag{5.95}$$

We may therefore conclude that the assumption of a constant promotion intensity may not be crucial, at least for the lowest grade. This investigation could be pursued for the second lowest grade, but we shall use an alternative method.

A method involving hypothetical grades

This method has been widely adopted in the theory of queues for dealing with non-Markovian systems. In essence it involves the replacement of the actual system by a Markov system of greater complexity. It depends on the following argument. Suppose that we have two grades in series with constant promotion rates as shown in Figure 5.3 and that there are no losses. The lengths of stay in stages I and II will now be exponential with means $1/r'_{12}$ and $1/r'_{23}$ respectively. Suppose now that the transitions between I and II are not observable and that the members of the two grades cannot be distinguished. The distribution of length of stay for the combined grade is thus the sum of two independent exponential variates. There is little loss in generality if we assume that $r'_{12} = r'_{23} = r$, say, in which case the length of service in the combined grade has density function

$$f(T) = r^2 T e^{-rT} \qquad (T \geq 0).\tag{5.96}$$

Promotions from the combined grade will now appear *as if* the promotion rate was

$$r_{12}(T) = f(T) \Big/ \int_T^\infty f(x)\,\mathrm{d}x = r\left(\frac{rT}{rT+1}\right)$$

$$= 2r_{12}\left(\frac{r_{12}T}{r_{12}T + \frac{1}{2}}\right)\tag{5.97}$$

where r_{12} is the constant rate which would lead to the same mean length of stay.

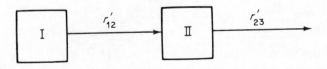

Figure 5.3

We have thus produced a Markov system which behaves, when viewed in a particular way, as a non-Markov system with $r_{12}(T)$ given by this equation.

The argument just given may be generalized. Instead of two stages we may consider g stages for which we find

$$f(T) = \frac{r^g}{(g-1)!} T^{g-1} e^{-rT} \qquad (T \geq 0). \qquad (5.98)$$

Under these circumstances $r_{12}(T)$ is always an increasing function of T with rate of increase depending on g, and it always tends to a limit as $T \to \infty$.

The method we adopt is to replace each grade of the actual system by an appropriate number, g, of what will be termed 'subgrades.' The loss rates must be the same for each subgrade within a given grade. The expected sizes of the subgrades are determined by the standard theory for Markov processes. Those for the actual grades are then simply obtained by summation over the relevant subgrades. We illustrate the method on the kind of simple hierarchy discussed in Section 5.2 for the case of constant transition rates. Again we restrict attention to the limiting behaviour. Let us assume that the loss rate is the same for all grades and denoted by r_{k+1} and that there is a constant rate of input R. Although not necessary, it is convenient to assume that the same value of g is appropriate for each grade. Our system is to be replaced by one of gk grades with constant and equal promotion intensities. To facilitate comparison with the results given in (5.52) we denote these intensities by gr. Let $\bar{z}_j(\infty)$ denote the limiting expected size of the jth subgrade; it may be obtained from (5.52) by replacing r by rg and k by gk. Thus

$$\bar{z}_j(\infty) = \frac{R}{gr}\left(\frac{gr}{gr+r_{k+1}}\right)^j \qquad (j = 1, 2, \ldots, gk-1)$$

and

$$\bar{z}_{gk}(\infty) = \frac{R}{r_{k+1}}\left(\frac{gr}{gr+rk}\right)^{gk-1} \qquad\qquad (5.99)$$

The expected number in the jth grade of the original organization is then

$$\bar{n}_j(\infty) = \frac{R}{gr} \sum_{i=g(j-1)+1}^{gj} \left(\frac{gr}{gr+r_{k+1}}\right)^i$$

$$= \frac{R}{r_{k+1}}\left(\frac{gr}{gr+r_{k+1}}\right)^{g(j-1)}\left\{1-\left(\frac{gr}{gr+r_{k+1}}\right)^g\right\} \qquad (5.100)$$

$$(j = 1, 2, \ldots, k-1)$$

If $g = 1$ this reduces to the expression given in (5.52) for constant promotion rates. By increasing g we increase the dependence of promotion on length of service in the grade. In the extreme case as $g \to \infty$ all promotions take place after a fixed length of service r^{-1}, when we find

$$
\left.
\begin{aligned}
&\bar{n}_j(\infty) = \frac{R}{r_{k+1}} \exp\left\{ -\frac{r_{k+1}}{r}(j-1) \right\} \left[1 - \exp\left\{ -\frac{r_{k+1}}{r} \right\} \right] \\
&\hspace{4cm} (j = 1, 2, \ldots, k-1)
\end{aligned}
\right\} . \quad (5.101)
$$

and

$$
\bar{n}_k(\infty) = R/r_{k+1} - \sum_{j=1}^{k-1} \bar{n}_j(\infty)
$$

An interesting feature revealed by these results is that the relative expected grade sizes, except the last, form a geometric progression, whatever g. The constant factor of the progression varies between $(1 + r_{k+1}/r)^{-1}$ for the constant promotion intensity and $\exp(-r_{k+1}/r)$ for promotion after a fixed length of service. Some numerical values are given in Table 5.3.

Table 5.3 *Comparison of the constant factors appropriate for the two extreme promotion rules*

r_{k+1}/r	2	1	$\frac{1}{2}$	$\frac{1}{5}$	0
$\left(1 + \dfrac{r_{k+1}}{r}\right)^{-1}$	0.333	0.500	0.667	0.833	1.000
$\exp\left\{ -\dfrac{r_{k+1}}{r} \right\}$	0.135	0.368	0.607	0.819	1.000

The importance of the assumption about the promotion rates thus depends upon the ratio r_{k+1}/r. If the ratio is small, meaning that promotion is much more likely than loss, the assumption is not critical. On the other hand, if r_{k+1}/r is large the kind of assumption we make will be much more important. This point is illustrated by Table 5.4 where we have compared grade structures for high and low values of the ratio.

The effect of making the promotion intensities increasing functions of seniority is to increase the relative sizes of the lower groups at the expense of the higher. A similar result was obtained when we allowed the loss rate to depend on total length of service. Another similarity between the two cases is that the size of the higher grades depends more critically on the assumptions. In general, therefore, we may expect the Markov model to overestimate the sizes of the higher grades and underestimate the lower.

Although it seems more realistic to suppose that promotion rates are increasing functions of seniority the method of substages can be adapted for use when they are decreasing functions. One way of doing this is to set up Markov systems with

Table 5.4   *The relative grade structures $q(\infty)$ for promotion after a fixed or random length of stay when $k = 4$*

	Grade	1	2	3	4
$\dfrac{r_{k+1}}{r} = 2$	Random	0.667	0.222	0.074	0.037
	Fixed	0.865	0.117	0.016	0.002
$\dfrac{r_{k+1}}{r} = \tfrac{1}{2}$	Random	0.333	0.222	0.148	0.296
	Fixed	0.393	0.239	0.145	0.223

grades in parallel. The idea can be demonstrated on the system shown in Figure 5.4 with two stages in parallel and constant transition intensities. The length of stay of those members passing through I will be exponential with parameter r'_{13}. Those who pass through II will have an exponential length of stay distribution with parameter r'_{23}. If we are unable to distinguish the two stages the apparent

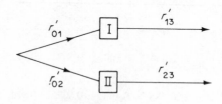

Figure 5.4

density function of length of stay will be a mixture of the two exponential densities. More precisely, the density function will be

$$f(T) = \left(\frac{r'_{01}}{r'_{01} + r'_{02}}\right) r'_{13}\, e^{-r'_{13}T} + \left(\frac{r'_{02}}{r'_{01} + r'_{02}}\right) r'_{23}\, e^{-r'_{23}T} \qquad (T \geq 0). \quad (5.102)$$

It can easily be shown that the apparent intensity of loss for the combined grades,

$$f(T) \Big/ \int_{T}^{\infty} f(x)\,dx,$$

is a strictly decreasing function of T for this distribution.

    We now show how this result may be used to study a three-grade hierarchy with promotion intensities of the kind leading to the density function of (5.102). Consider the Markov system shown in Figure 5.5 with the possible routes of transfer indicated by arrows. Each pair of grades within the dotted rectangles is assumed to have the same loss rate. If the rectangles represent the grades of the actual system, this Markov system will have decreasing promotion intensities

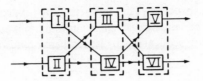

Figure 5.5

when its subgrades are combined. The transition rates can be chosen to give the required average throughputs both for the individual grades and for the system as a whole. It should be noticed that the matrix of transition intensities arising from the system described above is triangular so that much of the simplicity associated with simple hierarchical systems is retained. The general effect of introducing this kind of dependence is to make the higher grades relatively larger than they would have been with constant promotion intensities.

5.6   COMPLEMENTS TO CHAPTERS 4 AND 5

Theory

The theory of continuous time Markov processes is often treated rather briefly in the elementary texts and even then the discussion is commonly limited to special cases such as the birth and death process. Cox and Miller (1965) is a useful starting point for further reading and Isaacson and Madsen (1976) contains an introductory chapter. Bartlett (1955) provides a concise treatment of the essentials of the theory, but for a full account one must turn to a book such as Chung (1967). The more advanced treatments tend to be much preoccupied with the many subtle features which arise in the passage to continuous time such as the possibility of infinitely many changes of state in a finite interval of time. Such refinements are of little consequence in applied work where, at best, the model is no more than a crude approximation to reality. It is chiefly for this reason that we have linked our treatment closely to the discrete time theory where the scope for pathological behaviour is more limited.

The theory of semi-Markov processes also receives scant attention in the books on stochastic processes though here, again, Cox and Miller (1965) set out the essentials.

The pioneering papers by Pyke (1961a, b) are often quoted, but the survey paper by Çinlar (1975) is perhaps, the most useful source of information and it provides a gateway to the substantial literature on the subject. For the reader interested in social applications Ginsberg (1971) is the most accessible account. The titles of Pyke's and Çinlar's papers refer to Markov renewal theory rather than semi-Markov processes. Markov renewal processes and semi-Markov processes are equivalent and some writers seem to treat the two terms as synonyms. The distinction lies in two different ways of constructing what is essentially the same process. The semi-Markov approach, which we have adopted, focuses on the sequence of states and the times at which changes occur. The Markov renewal process arises by generalizing renewal processes which we shall meet in Chapter 8.

Applications

For a variety of reasons the application of continuous time models has lagged behind that of their discrete time counterparts. Fix and Neyman's (1951) work on cancer survival is an early example of a Markov model as is that of Herbst (1963) which occurs in Chapter 7. Recently there have been signs that the usefulness and tractability of the Markov model are becoming more widely recognized. In addition to Creedy (1979) we may mention unpublished work by Plewis (1980) to model teacher's ratings of classroom behaviour as a dynamic process. This was a closed system with $k = 3$. Plewis was able to show that, although the Markov model did not fit particularly well when applied to all 300 pupils in the study, it was much better when the data were disaggregated into a 'deprived' and 'control' group. Closed Markov process models have been proposed for the dynamics of group structure and social networks by Sørenson and Hallinan (1977) and Holland and Leinhardt (1977). In the latter case a relationship such as 'liking' is defined between each pair of individuals. The state of the system at any time is described by an incidence matrix showing the pairs between which the relationship exists. Transitions between states are then supposed to occur according to a Markov process with specified transition rates. The practical usefulness of such models is somewhat limited by the very large state spaces which are typically required. A somewhat different kind of model of friendship choice, based on the birth and death process, was proposed by Yamaguchi (1980). This generalizes an earlier model of Coleman (1964b) by allowing the birth and death rates for friendship to depend on whether the tie is direct or indirect.

An open system modelled by Slivka and Cannavale (1973) concerns the passage of defendants through a court system. Their data consisted of a random sample of 1067 adult arrests drawn from the Philadelphia Police Department files. The states formed a single hierarchy as follows: Arrest; Preliminary arraignment (1), Preliminary hearing (2), Indictment (3), Arraignment (4), Trial (5). Arrest constitutes the inflow and so the states were as numbered, with (6) denoting release which could occur from states 2, 3, and 5. The form of the transition intensity matrix with estimated values was

$$
\mathbf{R}_a = \begin{pmatrix}
-1.680 & 0.0585 & 0 & 0 & 0 & 0.1095 \\
0 & -0.0310 & 0.0308 & 0 & 0 & 0.0002 \\
0 & 0 & -0.0110 & 0.0110 & 0 & 0 \\
0 & 0 & 0 & -0.0190 & 0.0190 & 0 \\
0 & 0 & 0 & 0 & -0.9600 & 0.9600 \\
0 & 0 & 0 & 0 & 0 & 0
\end{pmatrix}
$$

The relative steady-state frequencies for a constant inflow to the system, which the reader may verify as an exercise, are

$$n_1 = R(r_{12} + r_{16})^{-1}, \qquad n_2 = Rr_{12}/(r_{12} + r_{16})(r_{23} + r_{26});$$
$$n_3 = Rr_{12}r_{23}/(r_{12} + r_{16})(r_{23} + r_{26})r_{34},$$
$$n_4 = Rr_{12}r_{23}/(r_{12} + r_{16})(r_{23} + r_{26})r_{45};$$
$$n_5 = Rr_{12}r_{23}/(r_{12} + r_{16})(r_{23} + r_{26})r_{56}.$$

Applications of semi-Markov models outside the traditional fields of competing risk and multiple decrement analysis are becoming more common. We have already mentioned Yang and Hursch (1973) on sleep patterns. These authors used a discrete time version of the model, but the theory follows that which we have given for continuous time very closely. Kao's (1974) work on the movement of coronary patients is particularly concerned with the lengths of time spent in various states which he graduates by Weibull distributions. His paper also includes estimates of the embedded transition matrices. A mainly theoretical treatment of an application to clinical trials is provided by Weiss and Zelen (1965). McClean (1980) has proposed a semi-Markov model for multi-grade populations with Poisson input. Valliant and Milkovich (1977) attempt to compare the semi-Markov and Markov models in a personnel forecasting exercise. Since the Markov model is a special case of the semi-Markov model one can say in advance that the latter must provide a fit which is at least as good as the former. The authors come to the opposite conclusion largely, it would appear, because of their estimation techniques. As in many applications the system was observed for a limited period so that many of the sojourn times were incomplete. The problems which this poses were recognized, but not overcome. Thompson (1980) has shown how this situation should be handled in a study of residential mobility in Ontario. Relevant material is also to be found in Chiang (1968), Singer and Spilerman (1976a), Bartholomew (1977a), and Tuma and coworkers (1979).

A major application of semi-Markov models to labour mobility in Scandinavia has been reported by Ginsberg (1978a, 1978b, 1978c, 1979a, 1979b, 1979c) in a series of publications. These are largely concerned with the statistical analysis of moves and the estimation of the relevant parameters.

Variations and extensions

There are many variations in the way the theory could be presented. For example, we could have introduced Laplace transforms at an earlier stage. Using the fact that the transform of $\exp(\mathbf{R}T)$ is $(s\mathbf{I} - \mathbf{R})^{-1}$; the matrix $\mathbf{P}(T)$ and vector $\bar{\mathbf{n}}(T)$ would then have been found by inverting the relevant transforms. We have scarcely done justice to the extensive work by Singer and Spilerman on the embedding problem which is concerned with finding solutions to the equation

$$\frac{1}{T} \ln \mathbf{P}(T) = \mathbf{Q}.$$

From a practical point of view the situation is greatly eased by the fact that most observed matrices turn out to have real positive eigenvalues in which case there are no serious difficulties.

In the previous editions we gave a different treatment of the fixed size model of Section 5.3. There, it was shown that the Laplace transform of the expected stock vector was given by

$$\bar{\mathbf{n}}^*(s) = \{ s\mathbf{n}(0)\mathbf{p}_{k+1}^*(s) + M^*(s) \} \, \mathbf{p}_0 \mathbf{P}^*(s) \, (1 - s \, \mathbf{p}_0 \mathbf{p}_{k+1}^*(s))^{-1} + \mathbf{n}(0) \, \mathbf{P}^*(s).$$

This is equivalent to our present result and we have used the same example to illustrate its application. Our treatment here follows Mehlmann (1977b) and aims to bring out more clearly the close parallel with the discrete time theory.

CHAPTER 6

Control Theory for Markov Models

6.1 THE PROBLEM

In the preceding chapters we have formulated various versions of the Markov chain model for social systems. For the most part we have been concerned with the effect of constant transition probabilities on the changing class structure. This enabled us to see how the aggregate behaviour of the system depended on the propensities to move of its members. We saw in Chapter 3, for example, how a manpower system with reasonable-looking promotion rates showed a tendency for the higher grades to grow at the expense of the lower. This feature is often observed in practice where it commonly occurs at the end of a period of expansion. The inertia inherent in human systems ensures that promotion rates which have become inflated during the expansion tend to be maintained after growth stops. The result is that the structure becomes top-heavy. This phenomenon was experienced by many universities after the expansion of the 1960s.

In some fields of application it is possible to exert some control over the system. Indeed, manpower and education planning are largely concerned with how to construct and operate a system so as to meet specified needs. Such an exercise will usually begin with a forecast of what will happen 'if present trends continue' in the manner described in earlier chapters, but this is only a first step. The projected structures will rarely coincide with what is desired and so the question arises as to what can be done to alter things. It is at this juncture that the need for a control theory arises. In developing such a theory for Markov models we proceed in the opposite direction to that of the earlier chapters. There, we were given the transition probabilities and wished to know what future stock numbers would be. Now, the desired stocks will be given and the problem is to find the flow numbers required to achieve them.

As stated the problem is trivial. Any desired structure can be attained at once if there are no restraints on when and where people can be moved. In practice such arbitrary action would be highly undesirable if not impossible. Measures designed to control a system affect the prospects of individuals whose aspirations must be taken seriously if the system is to function harmoniously and efficiently. We must therefore investigate what can be achieved under the constraints which the requirements of good management impose. In practice this means that only some flows can be controlled and even then there may be limits to the degree of control which can be exercised.

The control variables

The theory of control is most fully developed for discrete time models. Apart from a brief digression at the end of Section 6.2 we shall be concerned entirely with the open discrete time model discussed in Chapter 3. The objects of control may be various. For example, we may wish to operate the system within a fixed budget or subject to certain restrictions on grade sizes. Many such objectives can be expressed in terms of the expected stock vector $\mathbf{n}(T)$ for some future values of T. A number of possible formulations are mentioned in the Complements section. Our treatment here will suppose that there is some desired structure, or sequence of structures, which we wish to attain or maintain. Such goal structures will be denoted by an asterisk. In practice it will not usually be the case that such a goal can be specified precisely. It will normally be sufficient if the actual structure is reasonably close to the ideal. In any event, random variation will ensure minor fluctuations even if the ideal is attained. It might, therefore, seem more reasonable to specify our goals in the form of a set of acceptable structures rather than as a single $\mathbf{n}^*$. This would complicate the analysis but without commensurate gains in understanding. Our primary aim is to gain insight into the operation of systems and this can be achieved using the simpler way of specifying goals.

As to the means of control, this must be exercised through the flows of people which it is convenient to classify into three categories:

(a) Wastage (the vector $\mathbf{w}$ which is identical to the loss vector $\mathbf{p}_{k+1}$ of Chapter 3).
(b) Promotion, taken here to include transfers and demotion (the matrix $\mathbf{P}$);
(c) Recruitment. This has two parts, the total number $\{R(T)\}$ and their allocation to the grades $\mathbf{r} \equiv \mathbf{p}_0$. It is with the latter that we shall be primarily concerned in this chapter.

In the model, wastage and promotion are intimately connected. Because

$$\sum_{j=1}^{k} p_{ij} + w_i = 1 \qquad (i = 1, 2, \ldots, k)$$

any change in either w_i or a p_{ij} requires compensating changes in one or more of the other parameters to maintain the equality. This will not cause any particular problems of interpretation in the applications we shall discuss and we prefer to keep the distinction.

The wastage flow can be controlled to some extent. It can be increased by dismissing people or by offering them financial or other inducements to leave, and decreased by improved conditions or inducements to stay. These methods of control are, however, somewhat uncertain in their operation and, in the case of dismissal certainly undesirable.

More precise control can be exercised over the promotion flows in that these result from direct management decisions. Even here, however, there are often compelling reasons for varying the promotion rates as little as possible. For example, an increase in the promotion rate may involve the promotion of inadequately qualified people and make it difficult to return to the original standards should this prove necessary later. Equally, a decrease in the promotion rates is likely to create problems among people who see what they regarded as their expectations for advancement being eroded.

The recruitment flows offer the most attractive means of control since decisions to recruit more or fewer people at a given level can be taken without the same immediate impact on those already serving. There may be practical difficulties in finding enough recruits at the required levels and existing members may express concern about the effects, for example, of high recruitment near the top on their own promotion prospects. Nevertheless, of the three main methods of control the adjustment of the recruitment vector seems to offer the least painful means of control. In some applications this is the only control variable. Thus if the classes refer to age (or length of service) 'promotion' is the process of moving from one age group to the next and this clearly cannot be controlled.

Control has two aspects which we call *attainability* and *maintainability*. Attainability is concerned with whether or not a goal can be reached and, if so, by what means. Maintainability has to do with remaining at the goal structure once it has been attained. Although the question of attainability is prior to that of maintainability in many applications, it is convenient to treat them in the reverse order. This is because it turns out that the solution to the maintainability problem partly solves the attainability problem also.

In this chapter an important distinction is made between what is called a deterministic and a stochastic environment. This is essentially the same distinction as that already made between deterministic and stochastic models, but here it refers to whether or not the *uncontrolled* flows are random. When the Markov model was used in a forecasting mode we were able to move easily between the deterministic and stochastic interpretations and, for many purposes, it was hardly necessary to distinguish between them. In the control situation more care must be taken as both the meaning of the problem and the interpretation of the solution depend on whether a deterministic or stochastic environment is assumed. We deal first with control in a deterministic environment in Sections 6.2 and 6.3. Real environments are, of course, usually stochastic, but the deterministic analysis provides both an approximation to the more realistic stochastic case and a foundation on which the fuller stochastic treatment can be based.

6.2   MAINTAINABILITY IN A DETERMINISTIC ENVIRONMENT

If $\mathbf{n}^*$ is the structure to be maintained then there must exist values of $\mathbf{P}$, $\mathbf{w}$, and $\mathbf{r}$ such that

$$\mathbf{n}^* = \mathbf{n}^*\mathbf{P} + \mathbf{n}^*\mathbf{w}'\mathbf{r}. \qquad (6.1)$$

Note that in this situation $R(T) = \mathbf{n}^*\mathbf{w}'$ since if the structure is to be maintained the total size is necessarily fixed. If recruitment is the only flow subject to control then $\mathbf{P}$ and $\mathbf{w}$ are fixed and $\mathbf{r}$ is to be determined. We shall refer to this as recruitment control. Promotion control arises when $\mathbf{w}$ and $\mathbf{r}$ are given and $\mathbf{P}$ is at choice.

Control by recruitment

There are many reasons why we may wish to maintain a given structure over a period of time. For example, the question may arise as part of a planning exercise when a grade structure has been decided upon and it is required to know what promotion and recruitment policies are compatible with this structure. Alternatively an existing system may have a structure which is becoming too top-heavy and then the question is how to prevent matters becoming any worse. Not every structure can be maintained and an important part of the investigation will be to delineate those structures which can be maintained from those which cannot.

As our prime interest is in the relative sizes of the grades it will be convenient throughout this section to work in terms of the variables

$$q_i(T) = \bar{n}_i(T)/N \qquad (i = 1, 2, \ldots, k),$$

where N is the total size of the system. The basic difference equation for the Markov model is then

$$\mathbf{q}(T+1) = \mathbf{q}(T)\mathbf{P} + \mathbf{q}(T)\mathbf{w}'\mathbf{r}. \tag{6.2}$$

A structure $\mathbf{q}(T)$ can be maintained if we can find values of the control parameters such that $\mathbf{q}(T+1) = \mathbf{q}(T)$. In other words, we are interested in values of the parameters which satisfy

$$\mathbf{q} = \mathbf{q}\mathbf{P} + \mathbf{q}\mathbf{w}'\mathbf{r}. \tag{6.3}$$

Since the recruitment vector $\mathbf{r}$ is the only set of parameters amenable to control we have to find an $\mathbf{r}$ satisfying (6.3). The possibility of doing this is easily investigated by solving (6.3) for $\mathbf{r}$ to give

$$\mathbf{r} = \mathbf{q}(\mathbf{I} - \mathbf{P})/\mathbf{q}\mathbf{w}'. \tag{6.4}$$

It is easy to check that the elements of $\mathbf{r}$ so obtained add up to one, but they may not all be positive. If they are, (6.4) gives the unique policy meeting the requirements; if not, the structure is not maintainable.

It would be useful to be able to characterize the set of structures which can be maintained with a given $\mathbf{P}$ and $\mathbf{w}$. A simple characterization of the maintainable region which follows directly from (6.4) is that it is the set of $\mathbf{q}$'s for which

$$\mathbf{q} \geq \mathbf{q}\mathbf{P}.$$

To obtain a more illuminating characterization we view the problem geometrically. Any structure $\mathbf{q}$ can be represented by a point in k-dimensional

Euclidean space with the elements of $\mathbf{q}$ as its coordinates. Since all the $\mathbf{q}$'s must be non-negative and add up to one, all allowable structures lie on the hyper-plane $\sum_{i=1}^{k} q_i = 1$ in the positive orthant. When $k = 2$ this is the hypotenuse of the right-angled triangle with vertices $(0, 0)$, $(0, 1)$, $(1, 0)$. When $k = 3$ it is the equilateral triangle illustrated in Figure 6.1. In higher dimensions the position is difficult to visualize but the geometrical terminology is still useful. The set of possible structures will be denoted by $\mathcal{X}$. The maintainable region $\mathcal{M}$ will be a subset of $\mathcal{X}$ whose boundaries we wish to determine.

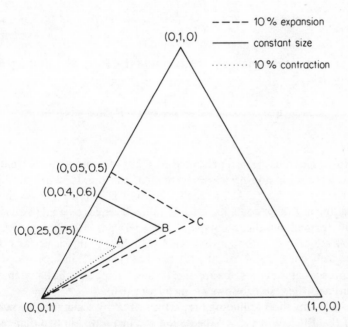

A = (0.158, 0.210, 0.632), B = (0.286, 0.286, 0.428), C = (0.386, 0.307, 0.307)

Figure 6.1   The maintainable region for recruitment control with $k = 3$ and

$$\mathbf{P} = \begin{pmatrix} 0.5 & 0.4 & 0 \\ 0 & 0.6 & 0.3 \\ 0 & 0 & 0.8 \end{pmatrix}$$

The boundary of $\mathcal{M}$ may be found as follows. From (6.3)

$$\mathbf{q}(\mathbf{I} - \mathbf{P}) = \mathbf{q}\mathbf{w}'\mathbf{r}.$$

Hence

$$\mathbf{q} = \mathbf{q}\mathbf{w}'\mathbf{r}(\mathbf{I} - \mathbf{P})^{-1}; \qquad (6.5)$$

the inverse always exists, as we noted earlier in (3.4). The vector $\mathbf{r}$ may be written $\mathbf{r} = \Sigma_{i=1}^{k} r_i \mathbf{e}_i$, where $\mathbf{e}_i$ is the vector with a one in the ith position and zeros elsewhere. Substituting in (6.5) then gives

$$\mathbf{q} = \mathbf{q}\mathbf{w}' \sum_{i=1}^{k} r_i \{\mathbf{e}_i (\mathbf{I} - \mathbf{P})^{-1}\}. \qquad (6.6)$$

Post-multiplying both sides of (6.6) by a column vector of ones gives

$$1 = \mathbf{q}\mathbf{w}' \sum_{i=1}^{k} r_i d_i,$$

where d_i is the sum of the elements in the ith row of $(\mathbf{I} - \mathbf{P})^{-1}$. Substituting for $\mathbf{q}\mathbf{w}'$ in (6.6) then yields

$$\mathbf{q} = \sum_{i=1}^{k} \left(\frac{r_i d_i}{\sum_{j=1}^{k} r_j d_j} \right) \frac{1}{d_i} \mathbf{e}_i (\mathbf{I} - \mathbf{P})^{-1}. \qquad (6.7)$$

The vector $\mathbf{q}$ has thus been represented as a convex combination (that is, a weighted average with non-negative weights) of the points with coordinates $d_i^{-1} \mathbf{e}_i (\mathbf{I} - \mathbf{P})^{-1}$ $(i = 1, 2, \ldots, k)$. For each $\mathbf{r}$ there is exactly one set of weights and hence one $\mathbf{q}$, and as $\mathbf{r}$ ranges over its possible values $\mathbf{q}$ ranges over all points in the convex hull formed by the points $d_i^{-1} \mathbf{e}_i (\mathbf{I} - \mathbf{P})^{-1}$ $(i = 1, 2, \ldots, k)$. The maintainable region is thus the convex region with these points as vertices. These vertices are easily computed by taking the rows of $(\mathbf{I} - \mathbf{P})^{-1}$ in turn and scaling their elements so that the row sums are one. If the w's happen to be equal then $\mathbf{q}\mathbf{w}'$ is a constant and the result follows at once from (6.6).

We now have a method of finding the vertices of $\mathscr{M}$ by scaling the elements in the rows of $(\mathbf{I} - \mathbf{P})^{-1}$. When $k = 3$ the region can be plotted in two dimensions; this has been done for a system with matrix

$$\mathbf{P} = \begin{pmatrix} 0.5 & 0.4 & 0 \\ 0 & 0.6 & 0.3 \\ 0 & 0 & 0.8 \end{pmatrix}$$

in Figure 6.1. The region is the one with vertices $(0, 0, 1)$, $(0, 0.4, 0.6)$, $(0.286, 0.286, 0.428)$. We note that all of the structures which can be maintained by recruitment are top-heavy. In this example the least top-heavy structure is represented by the structure at the point B on the diagram, which has a large top grade. In higher dimensions it is less easy to represent the region diagrammatically, but inspection of the vertices gives a good idea of the structures which are maintainable.

The numerical example is a special case of an important class of transition matrices, which we shall describe as super-diagonal, having the form

$$\mathbf{P} = \begin{pmatrix} p_{11} & p_{12} & 0 & 0 & 0 \\ 0 & p_{22} & p_{23} & 0 & 0 \\ \cdot & \cdot & \cdot & \cdot & \cdot \\ \cdot & \cdot & \cdot & \cdot & p_{k-1,k} \\ 0 & \cdot & \cdot & \cdot & p_{kk} \end{pmatrix}$$

This means that promotion is possible only into the next higher grade; it also arises if the grades are defined by age groups. The matrix $(\mathbf{I} - \mathbf{P})$ can easily be inverted; its ith row is

$$\mathbf{e}_i(\mathbf{I} - \mathbf{P})^{-1} = \left(0, \ldots, 0, \frac{1}{1 - p_{ii}}, \frac{p_{i,i+1}}{(1 - p_{ii})(1 - p_{i+1,i+1})}, \ldots, \right.$$

$$\left. \frac{1}{1 - p_{ii}} \prod_{r=i}^{k-1} \frac{p_{r,r+1}}{(1 - p_{r+1,r+1})} \right) \qquad (i = 1, 2, \ldots, k - 1) \qquad (6.8)$$

and

$$\mathbf{e}_k(\mathbf{I} - \mathbf{P})^{-1} = \left(0, 0, \ldots, 0, \frac{1}{1 - p_{kk}} \right).$$

There is no simple expression for the row sums of $(\mathbf{I} - \mathbf{P})^{-1}$ but the vertices of $\mathcal{M}$ are easily calculated numerically. We shall see later that the structures corresponding to the vertices are maintained by recruitment into a single grade. The first row of $(\mathbf{I} - \mathbf{P})^{-1}$ is of particular interest because it gives the structure which can be maintained by recruitment into the bottom grade of a hierarchy. Under these circumstances the jth grade has size proportional to

$$\frac{1}{(1 - p_{jj})} \prod_{r=1}^{j-1} \frac{p_{r,r+1}}{(1 - p_{r+1,r+1})} \qquad (6.9)$$

where the product is defined to be one if $j = 1$. This should be compared with (3.23) which gives the structure which will arise in equilibrium with a constant rate of recruitment at the lowest level. The two situations are essentially the same and we note again the crucial importance of $(1 - p_{jj})$ in determining what size of grade can be maintained. Other maintainable structures are weighted averages of the rows, suitably scaled, of $(\mathbf{I} - \mathbf{P})^{-1}$. Since the first row is the only one with a non-zero entry in the lowest grade it is obvious that no structure requiring a bottom grade larger than $1/d_1(1 - p_{11})$ can be maintained.

These expressions simplify even further if the diagonal elements in $\mathbf{P}$ are also zero. This occurs if the k grades are based on age (or seniority) groups of the same length as the discrete time interval between events. Only the numerators in (6.8) remain and the non-zero elements in each row then form a decreasing sequence as intuition suggests they must. It does not follow that any maintainable age structure must be a decreasing function of age because the scaling and weighting of the rows may produce 'humps'. For example, a heavy recruitment in the middle age range will obviously inflate the size of that and, to a lesser extent, the following age groups. If 'age' refers to length of service in the system then recruits must, by definition, enter at the lowest level; only the first row applies and any maintainable length of service structure must be monotonic non-increasing. This way of defining the grades provides an important link between Markov chain models and the renewal theory models to be discussed in later chapters.

Control by recruitment under growth and contraction

The set of maintainable structures turns out to be surprisingly small for plausible-looking $\mathbf{P}$ matrices. In particular, if $\mathbf{P}$ is upper-triangular, $\mathcal{M}$ usually excludes the typical 'bottom-heavy' structure of the traditional staff 'pyramid'. In this section we shall examine how the position is affected if the system is growing or contracting. This investigation may be regarded from two points of view. Expansion or contraction may be looked upon as a further control parameter by means of which the set of maintainable structures can be varied or as something imposed from outside over which we have no control. In one sense we cannot maintain a structure if the system is changing size since some grade sizes must inevitably change. We can, however, consider what Forbes (1970) has termed a quasi-stationary state in which the relative sizes of the grades are to be maintained. This is likely to be the most relevant case in practice if the relative grade sizes are determined by the *kind* of work the organization does and the total size by the *amount* of work it does.

Let $N(T)$ be the total size of the system at time T and let $N(T+1) = (1 + \alpha)N(T)$. The basic equation for an organization of fixed size (3.54a) is then

$$\bar{\mathbf{n}}(T + 1) = \bar{\mathbf{n}}(T)\mathbf{P} + \bar{\mathbf{n}}(T)\mathbf{w'r} + \alpha N(T)\mathbf{r}. \tag{6.10}$$

Introducing $\mathbf{q}(T) = \bar{\mathbf{n}}(T)/N(T)$ the equation becomes

$$(1 + \alpha)\mathbf{q}(T + 1) = \mathbf{q}(T)\mathbf{P} + (\mathbf{q}(T)\mathbf{w'} + \alpha)\mathbf{r}. \tag{6.11}$$

The condition for a quasi-stationary structure $\mathbf{q}$ to be maintainable is thus that

$$(1 + \alpha)\mathbf{q} = \mathbf{q}\mathbf{P} + (\mathbf{q}\mathbf{w'} + \alpha)\mathbf{r} \tag{6.12}$$

for some $\mathbf{r} \geq \mathbf{0}$ with $\Sigma_{i=1}^{k} r_i = 1$. If such an $\mathbf{r}$ exists it is given by

$$\mathbf{r} = \{\mathbf{q}(\mathbf{I} - \mathbf{P}) + \alpha\mathbf{q}\}/(\mathbf{q}\mathbf{w'} + \alpha). \tag{6.13}$$

This should be compared with (6.4). As before, the elements of $\mathbf{r}$ always sum to one, but we are interested in those $\mathbf{q}$'s for which they are all non-negative. It is clear that any $\mathbf{q}$ for which $\mathbf{r}$ is given by (6.4) is non-negative will also yield a non-negative $\mathbf{r}$ in (6.13) if $\alpha > 0$. Hence, if the system is expanding the maintainable region will be larger than when it is constant in size. The converse holds if $\alpha < 0$. Thus expansion increases the range of maintainable structures and contraction decreases it.

The argument leading to the determination of the vertices of $\mathcal{M}$ goes through in the present case, with obvious modifications to give the vertices of $\mathcal{M}$ with coordinates proportional to

$$\mathbf{e}_i(\mathbf{I}(1+\alpha) - \mathbf{P})^{-1} \qquad (i = 1, 2, \ldots, k). \tag{6.14}$$

The only change required in (6.8) is the replacement of $1 - p_{ii}$ by $1 + \alpha - p_{ii}$ for all i in the denominators. If α is positive this clearly increases the amount by which the vertex structures taper towards the top. If α is negative a complication arises, as we see from the fact that $1 + \alpha - p_{ii}$ will be negative if $|\alpha| > 1 - p_{ii}$. Suppose that the present size of the ith grade is $n_i(T)$; then at $T + 1$ the size of that grade will have to be $(1 + \alpha)n_i(T)$ if the structure is to be maintained. However, the expected number of losses is $w_i n_i(T)$ so that if $|\alpha| < w_i$ the structure certainly cannot be maintained. The region $\mathcal{M}$ will thus only be non-empty if $|\alpha| \geq w_i$ for all i, and the maximum rate of contraction for which at least one maintainable structure exists is $\alpha = -\min_i w_i$.

To illustrate these results we return to the numerical example with $k = 3$, shown in Figure 6.1. The new vertices of the region for 10 per cent expansion and 10 per cent contraction are as follows:

10 *per cent expansion*	(0, 0, 1), (0, 0.5, 0.5), (0.386, 0.307, 0.307)
10 *per cent contraction*	(0, 0, 1), (0, 0.25, 0.75), (0.158, 0.210, 0.632).

These regions are also plotted on the figure, from which it is clear that a change in the total size of this amount has a large effect on the size of the region, especially around the vertex represented by the last points listed above.

The foregoing calculations can be given a different interpretation. Suppose that we had set out to investigate the effect of changing the wastage rate on the size of the maintainable region. If we had done this by adding an amount α on to each wastage rate and subtracting the same amount from each p_{ii} the recruitment vector required to maintain $\mathbf{q}$ would be

$$\mathbf{r} = \mathbf{q}\{\mathbf{I} - (\mathbf{P} - \alpha\mathbf{I})\}/(\mathbf{q}\mathbf{w}' + \alpha). \tag{6.15}$$

Equation (6.15) is identical with (6.13) and the effects of a change in size and of a fixed change in the wastage rate are equivalent. This interpretation can only be made when $\alpha > 0$ if $\alpha < \min_i p_{ii}$.

Control by promotion

There are circumstances in which the recruitment vector is not amenable to control, as, for example, when it is linked to the supply at different levels or dictated by management policy. In such a case we wish to know what degree of control can be exercised by the promotion (or demotion) rates. For a fixed size organization the problem is that of finding a matrix $\mathbf{P}$ satisfying

$$\mathbf{q} = \mathbf{qP} + \mathbf{q}\mathbf{w}'\mathbf{r}. \tag{6.16}$$

To delineate the maintainable region we must determine the set of $\mathbf{q}$'s for which there exists an admissible $\mathbf{P}$. Such a $\mathbf{P}$ must have non-negative elements with ith row summing to $1 - w_i (i = 1, 2, \dots, k)$. We cannot proceed to solve (6.16) for $\mathbf{P}$ in the way we solved it for $\mathbf{r}$ to see whether the solution has the desired property because there is not, in general, a unique solution. However, it is easy to determine the maintainable region if no other restrictions are placed upon $\mathbf{P}$. Writing (6.16) in the form

$$\mathbf{qP} = \mathbf{q} - \mathbf{q}\mathbf{w}'\mathbf{r} \tag{6.17}$$

we see that the condition

$$\mathbf{q} \geq \mathbf{q}\mathbf{w}'\mathbf{r} \tag{6.18}$$

is necessary for $\mathbf{q}$ to be maintainable since all the elements of the vector $\mathbf{qP}$ must be non-negative. It is also sufficient because it is obviously possible to choose the p_{ij}'s so as to distribute those who do not leave in any way whatsoever.

The case of a general $\mathbf{P}$ is of rather limited practical interest because it is rare for all types of transition to be possible. In a hierarchical organization, $\mathbf{P}$ is usually upper-triangular or super-diagonal (that is non-zero elements in the main diagonal and the diagonal above it) in form. The latter case also lends itself to simple analysis, so we shall concentrate on this case in the remainder of this section.

When $\mathbf{P}$ has the super-diagonal form the equation (6.17) admits a unique solution which may then be tested to see whether its elements are non-negative. It is more convenient to abandon the matrix notation temporarily and to write (6.17) as

$$\left.\begin{aligned}
q_1 p_{11} &= q_1 - r_1 \sum_{i=1}^{k} q_i w_i \\[2mm]
q_1 p_{12} + q_2 p_{22} &= q_2 - r_2 \sum_{i=1}^{k} q_i w_i \\[2mm]
&\dotfill \\[2mm]
q_{k-1} p_{k-1,k} + q_k p_{kk} &= q_k - r_k \sum_{i=1}^{k} q_i w_i
\end{aligned}\right\} \tag{6.19}$$

Eliminating p_{ii}, using $p_{ii} = 1 - w_i - p_{i,i+1}$ $(i = 1, 2, \dots, k-1)$, and solving for $p_{i,i+1}$ we obtain

$$p_{i,i+1} = \sum_{j=1}^{i} r_j \left\{ \sum_{j=1}^{k} \frac{q_j w_j}{q_i} \right\} - \sum_{j=1}^{i} \frac{q_j w_j}{q_i} \qquad (i = 1, 2, \dots, k-1), \tag{6.20}$$

where the last term is defined as zero if $i = 1$. A structure $\mathbf{q}$ is thus maintainable if

$$0 \le p_{i,\, i+1} \le 1 - w_i \qquad (i = 1, 2, \ldots, k-1)$$

and so the maintainable region is that set of $\mathbf{q}$'s such that

$$0 \le \sum_{j=1}^{i} r_j \left\{ \sum_{j=1}^{k} q_j w_j \right\} - \sum_{j=1}^{i} q_j w_j \le q_i (1 - w_i) \qquad (i = 1, 2, \ldots, k-1).$$

$$(6.21)$$

Equation (6.20) has a simple intuitive derivation most easily seen when $r_1 = 1$, $r_i = 0$ $(i > 1)$, in which case

$$p_{i,\, i+1} = \sum_{j=i+1}^{k} q_j w_j / q_i.$$

In words, this says that the proportions requiring to be promoted from grade i must be equal to the number leaving from grades $i + 1$ to k divided by the size of grade i. This result is, in effect, a discrete time version of a result we shall meet later in the discussion of renewal models in Chapter 8.

The shape of the maintainable region given by (6.21) is interesting, especially when it is compared with the corresponding regions for control by recruitment. Figure 6.2 shows regions plotted for a three-grade system having the same wastage rates as the system illustrated in Figure 6.1 (viz. 0.1, 0.1, 0.2) and two recruitment vectors $(1, 0, 0)$ and $(0.5, 0.5, 0)$. In contrast to the situation in Figure 6.1, the very top-heavy structures are now excluded from $\mathcal{M}$ and many of the bottom-heavy structures are included. The precise situation depends, of course, on $\mathbf{r}$; $\mathbf{r} = (1, 0, 0)$ is more favourable to the maintainance of bottom-heavy structures than any other $\mathbf{r}$. If the problem is to arrest a tendency to grow at the top it is clear that promotion control is likely to be more effective than recruitment control.

If both $\mathbf{P}$ and $\mathbf{r}$ are subject to control it is easy to see that any structure is maintainable. One way of maintaining a given structure is to make all the promotion rates zero and to recruit to each grade the same number of people as leave. There will be other, more reasonable, ways.

There is no difficulty about extending the argument of the last section to cover the case of an expanding or contracting system. We shall state the main results and the reader should have no difficulty in justifying them.

The maintainable region is defined by

$$\mathbf{q} \ge (\mathbf{q}\mathbf{w}' + \alpha)\mathbf{r} / (1 + \alpha) \qquad (6.22)$$

for general $\mathbf{P}$. For super-diagonal $\mathbf{P}$ the elements of $\mathbf{P}$ must satisfy

$$q_1 p_{11} = q_1 (1 + \alpha) - r_1 \left\{ \sum_{i=1}^{k} q_i w_i + \alpha \right\}$$

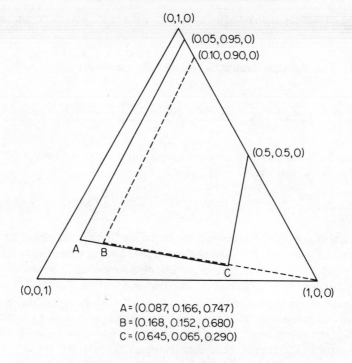

A = (0.087, 0.166, 0.747)
B = (0.168, 0.152, 0.680)
C = (0.645, 0.065, 0.290)

Figure 6.2   The maintainable regions for promotion control with $k = 3$, wastage vector $(0.1, 0.1, 0.2)$ and recruitment vectors (a) $(1, 0, 0)$ and (b) $(0.5, 0.5, 0)$. The dashed boundary gives the region using (a) and the solid region using (b).

and

$$q_i p_{i,i+1} + q_{i+1} p_{i+1,i+1} = q_{i+1}(1+\alpha) - r_{i+1}\left\{ \sum_{i=1}^{k} q_i w_i + \alpha \right\}$$

$$(i = 1, 2, \ldots, k-1). \qquad (6.23)$$

Solving, as before, for $p_{i,i+1}$ and making use of the inequalities $0 \le p_{i,i+1} \le 1 - w_i$ we arrive at the following specification of the maintainable region:

$$0 \le \sum_{j=1}^{i} r_j \sum_{j=1}^{k} q_j(w_j + \alpha) - \sum_{j=1}^{i} q_j(w_j + \alpha) \le q_i(1 - w_i) \qquad (i = 1, 2, \ldots, k-1).$$

$$(6.24)$$

Maintainability in continuous time

Many of the results on maintainability carry over to continuous time with scarcely any modification. This is because the maintainable region, for example, is unaffected by the limiting operation by which we arrived at the continuous time Markov process in Chapter 4. Thus the condition $\mathbf{q} \geq \mathbf{qP}$ for a structure $\mathbf{q}$ to be maintainable by recruitment becomes

$$\mathbf{q} > \mathbf{q}(\mathbf{I} + \delta T \mathbf{R})$$

or

$$\mathbf{qR} > 0 \tag{6.25}$$

where $\mathbf{R}$ is the matrix of transition rates. The vertices of the maintainable region are obtained by normalizing the rows of $(-\mathbf{R})^{-1}$ which is proportional to $(\mathbf{I} - \mathbf{P})^{-1}$. The recruitment vector which maintains $\mathbf{q}$ is easily shown to be

$$\mathbf{r} = \mathbf{qR}(\mathbf{qRl'})^{-1} \tag{6.26}$$

which is simply a normalization of $\mathbf{qR}$. If the system is expanding continuously at a rate α, that is so that $N(T) = N(0)e^{\alpha T}$, the condition for maintainability becomes

$$\mathbf{q}(\mathbf{I}\alpha - \mathbf{R}) \geq 0. \tag{6.27}$$

A similar analysis can be made for the case of promotion control.

6.3 ATTAINABILITY IN A DETERMINISTIC ENVIRONMENT

Attainability is a more complex concept than maintainability. A structure which is attainable from one starting point may not be from another. The number of steps required for the change and the route followed may be subject to variation and we may, or may not, wish to keep the total size fixed at the intermediate stages. A full analysis would have to go into all these matters, but for practical purposes a relatively limited investigation will suffice. We begin by observing that there is little point in trying to attain a structure which is not in the maintainable region, $\mathcal{M}$. If we did attain such a point we could not stay there so, at best, it would only buy time to make more fundamental changes in the system. Provided we agree to restrict discussion to goals in $\mathcal{M}$ there is available a simple result which settles the most important question. If we equate 'attain' with 'get arbitrarily close to' we can assert that '*any maintainable* $\mathbf{q}^*$ *can be attained from any starting structure*'. This result is easily proved by appeal to results on the limiting structure given in Chapter 3. We give the argument for recruitment control, that for promotion control is similar. It is sufficient to show that the result holds if the total size is held fixed. Suppose we start from $\mathbf{q}(0)$ and ask what would happen if we repeatedly used the recruitment vector $\mathbf{r}^*$ which we define as that which maintains $\mathbf{q}^*$. We know that the limiting structure in this case satisfies

$$\mathbf{q} = \mathbf{qP} + \mathbf{qwr}^*.$$

But by definition, $\mathbf{r}^*$ also satisfies

$$\mathbf{q}^* = \mathbf{q}^*\mathbf{P} + \mathbf{q}^*\mathbf{wr}^*$$

which implies that $\mathbf{q} = \mathbf{q}^*$. Such an $\mathbf{r}$ only exists, of course, when $\mathbf{q}^*$ is in $\mathcal{M}$. This argument goes further than showing the existence of such a strategy: it also provides one. It tells us to recruit in the same fixed proportions as if we were already at the goal and wished to remain there. The question which now arises is: can we do better? This can only be answered by reference to some criterion of optimality. There are many ways in which optimality might be defined—in terms of time or cost, for example. We shall describe two approaches to the problem, both using the methods of mathematical programming. The first approach, called 'free time' control, aims first to determine the smallest number of steps in which the goal can be attained and then to choose among the possible strategies on the basis of cost of some kind. The second method aims to get as close as possible to the goal in a prescribed number of steps. This leads on to the idea of applying such strategies repeatedly so as to ultimately converge on the goal.

Free time control by recruitment

In formulating our problem in programming terms it is desirable to work in terms of numbers rather than proportions. Accordingly we start with the equation

$$\mathbf{n}(T+1) = \mathbf{n}(T)\mathbf{P} + \mathbf{f}(T+1) \tag{6.28}$$

where $\mathbf{f}(T+1) = R(T+1)\mathbf{r}$ is the vector of the numbers recruited at time $T+1$. In this chapter we shall not distinguish notationally between stock numbers and their expectations. We shall suppose that the total size of the organization, $N(T)$, is given, but this will be taken care of through the restraints to be imposed below. The problem is to find a T^* and a sequence of vectors $\mathbf{f}(1), \mathbf{f}(2), \ldots, \mathbf{f}(T^*)$ such that T^* is the smallest T for which $\mathbf{n}(T) = \mathbf{n}^*$, the target structure. The restraints which the model imposes on the unknowns are

$$\mathbf{n}(T+1) = \mathbf{n}(T)\mathbf{P} + \mathbf{f}(T+1) \qquad (T = 0, 1, 2, \ldots, T^*-2) \tag{6.29a}$$

$$\mathbf{n}^* = \mathbf{n}(T^*-1)\mathbf{P} + \mathbf{f}(T^*) \tag{6.29b}$$

$$\sum_{i=1}^{k} n_i(T) = N(T) \qquad (T = 1, 2, \ldots, T^*-1) \tag{6.29c}$$

$$\mathbf{n}(T) \geq 0 \qquad (T = 1, 2, \ldots, T^*-1);$$
$$\mathbf{f}(T) \geq 0 \qquad (T = 1, 2, \ldots, T^*). \tag{6.29d}$$

If $T^* = 1$, (6.29a) and (6.29c) disappear. These equations are linear in the $n_j(T)$'s and the $f_j(T)$'s, and we may therefore use the standard methods of mathematical programming to find a feasible solution. Since T^* is unknown we must put $T^* = 1, 2, \ldots$ in turn until the first value is found for which a feasible solution exists. Notice that a solution to the restraints (6.29) will give not only the vectors $\mathbf{f}(T)$ which we set out to find but also the intermediate structures $\mathbf{n}(T)$ through which the system passes. When T has been found there will normally be many solutions and it is therefore open to us to select one set by minimizing some function of economic or social interest. One possible choice in the manpower context is to minimize the total salary bill. If the average salary cost in grade i is c_i the total salary cost associated with a strategy is

$$\sum_{T=1}^{T^*-1} \sum_{i=1}^{k} c_i n_i (T) \tag{6.30}$$

and then we have a standard linear programming problem. (The argument which follows is not affected if future costs are discounted.) A minimization algorithm would be used to minimize (6.30) as soon as a T^* had been found for which a feasible solution exists.

The test for the existence of a feasible solution is trivial when $T^* = 1$ but of considerable practical interest. In this case only (6.29b) is operative and we have to find whether there is an $\mathbf{f}(1) \geq 0$ such that

$$\mathbf{n}^* = \mathbf{n}(0)\mathbf{P} + \mathbf{f}(1). \tag{6.31}$$

If there is it will be given by

$$\mathbf{f}(1) = \mathbf{n}^* - \mathbf{n}(0)\mathbf{P}. \tag{6.32}$$

The condition for $\mathbf{f}(1)$ to be non-negative is, of course, the condition for one-step attainability. The fixed time methods in the following section take this as their starting point.

We can make some interesting deductions about the kind of recruitment strategies which minimize the cost function (6.30) by enumerating the number of unknowns and equations in (6.29). Altogether the vectors $\mathbf{n}(T)(T = 1, 2, \ldots, T^* - 1)$ and $\mathbf{f}(T)(T = 1, 2, \ldots, T^*)$ give $2kT^* - k$ variables in all. The number of equations may be enumerated as follows:

$$
\begin{array}{ll}
(6.29a) & k(T^* - 1) \\
(6.29b) & k \\
(6.29c) & \underline{T^* - 1} \\
& kT^* + T^* - 1
\end{array}
$$

A well-known result in the theory of linear programming states that the number of non-zero variables in a basic feasible solution cannot exceed $kT^* + T^* - 1$. Unless degeneracy occurs it will be equal to that number. If all the elements of $\mathbf{n}(0)$ are positive it is clear from (6.29a) that all subsequent vectors $\mathbf{n}(T)$ will have positive elements and that this accounts for $k(T^* - 1)$ of those available. The remaining $(kT^* + T^* - 1) - (kT^* - k) = T^* + k - 1$ non-zero values (at most) must therefore be distributed among the kT^* positions in the $\mathbf{f}(T)$ vectors. It therefore follows that at least $(k - 1)(T^* - 1)$ of the recruitment elements will be zero, which means that recruitment will tend to be concentrated at a few levels.

The argument may be carried further to obtain an idea of how the non-zero elements will be distributed. Suppose that we have reached time $T^* - 1$. We then have a problem which we know can be solved in one step. Hence the accounting procedure used above can be applied with $T^* = 1$ and we deduce that all of the elements in the last vector in the recruitment vector will be positive. This leaves $T^* + k - 1 - k$ non-zero elements to be distributed among the preceding $T^* - 1$ vectors. Provided that $N(T)$ does not decrease sufficiently fast for redundancy to occur *some* recruits will be needed at each T, so there will have to be at least one non-zero element in each vector. However, with only $T^* - 1$ non-zero elements available there can be no more than one per vector. Hence a typical recruitment strategy minimizing (6.30) will consist of recruiting into one grade only at each time point until the last step when recruits will be taken in at every level. This argument needs slight modification if degeneracy occurs or if the initial vector $\mathbf{n}(0)$ contains zeros (as it does in a numerical example considered later). The kind of strategy which the programming formulation dictates might be acceptable in practice if recruitment was always at the same level, but if it called for repeated changes in the recruitment level it would be very difficult to implement. This suggests that we ought to impose further restrictions designed to limit the range of feasible solutions to those which could be implemented.

If we decided to drop the $(T^* - 1)$ restrictions of (6.29c) implying that there is no need to control the intermediate size of the organization, there could be no more than kT^* non-zero elements in the optimal solution. Of these only $kT^* - k(T^* - 1) = k$ could be elements of the recruitment vectors. In general, therefore, there would be no recruitment at all until $T^* - 1$. The absurdity of this strategy in most practical situations highlights the need for some restrictions of the kind of (6.29c).

Free time control by promotion

The problem in this case is to find a sequence of matrices $\mathbf{P}(T)$ which, when coupled with a given recruitment vector $\mathbf{r}$, will take the system from $\mathbf{n}(0)$ to $\mathbf{n}^*$ in the shortest possible time. The basic equation may now be written:

$$\mathbf{n}(T+1) = \mathbf{n}(T)\mathbf{P}(T) + \mathbf{n}(T)\mathbf{w}'\mathbf{r} + M(T+1)\mathbf{r}, \qquad (6.33)$$

where $M(T+1) = N(T+1) - N(T)$. The unknowns are the vectors $\mathbf{n}(T)$ and the matrices $\mathbf{P}(T)$, but the basic equation is not linear in these variables. We must therefore recast the problem in a form which makes the restrictions linear by expressing all the internal flows as numbers rather than proportions. It is easier to count the equations and to recognize linear dependence in the equations if we temporarily abandon matrix notation. Let $n_{ij}(T) = n_i(T)p_{ij}$ be the number flowing from grade i to grade j in $(T, T+1)$. Then we have to find a T^* and values of the variables $n_{ij}(T) (i, j = 1, 2, \ldots, k)$ and $n_i(T)(i = 1, 2, \ldots, k)$ such that

$$n_j(T+1) = \sum_{i=1}^{k} n_i(T) + r_j \left\{ \sum_{i=1}^{k} n_i(T)w_i + M(T+1) \right\}$$

$$(j = 1, 2, \ldots, k; T = 0, 1, \ldots, T^* - 2) \quad (6.34a)$$

$$n_j^* = \sum_{i=1}^{k} n_{ij}(T^*-1) + r_j \left\{ \sum_{i=1}^{k} n_i(T^*-1)w_i + M(T^*) \right\}$$

$$(j = 1, 2, \ldots, k) \qquad (6.34b)$$

$$\sum_{i=1}^{k} n_{ij}(T) = n_i(T)(1-w_i)$$

$$(i = 1, 2, \ldots, k-1; \; T = 0, 1, \ldots, T^*-1) \qquad (6.34c)$$

$$\sum_{j=1}^{k} n_j(T) = N(T) \qquad (T = 1, 2, \ldots, T^*-1) \qquad (6.34d)$$

$$\mathbf{n}(T) \geq 0 \qquad (T = 1, 2, \ldots, T^*-1);$$

$$n_{ij}(T) \geq 0 \qquad (i, j = 1, 2, \ldots k; T = 0, 1, \ldots T^*-1). \qquad (6.34e)$$

The $N(T)$'s are assumed to be given as before and (6.34d) drops out if $T^* = 1$. The case $i = k$ has been excluded from (6.34c) because it is implied by (6.34a) together with (6.34c). By summing both sides of (6.34a) over j and subtracting (6.34c) summed over i we obtain the missing equation.

As in the case of recruitment control a good deal can be deduced from an enumeration of the number of variables and equations. The number of variables will depend on whether $\mathbf{P}$ has a special structure; the enumeration is made for three structures in the table below.

Variables	Full matrix	Upper-triangular form	Super-diagonal form
$n_j(T)$'s	$k(T^*-1)$	$k(T^*-1)$	$k(T^*-1)$
$n_{ij}(T)$'s	$k^2 T^*$	$\frac{1}{2}k(k+1)T^*$	$(2k-1)T^*$
Total number	$k(k+1)T^* - k$	$\frac{1}{2}k(k+3)T^* - k$	$3kT^* - T^* - k$

The number of equations is as follows.

Set	Number of equations
(6.34a)	$k(T^*-1)$
(6.34b)	k
(6.34c)	$T^*(k-1)$
(6.34d)	T^*-1
Total	$2kT^*-1$

In general, therefore, there will be $2kT^* - 1$ non-zero elements in a basic feasible solution. Again, this is small compared with the total number of elements so an optimal control strategy (in the sense of minimizing a cost function such as (6.29)) will consist of a sequence of matrices with a large number of zeros. If recruitment takes place into every grade the $n_j(T)$'s must have non-zero elements in every position. This leaves $(2kT^* - 1) - k(T^* - 1) = kT^* + k - 1$ to be distributed among the T^* P-matrices. At the penultimate stage, when $T^* = 1$, there are $2k - 1$ non-zero elements for the k^2 places in the last transition matrix. This leaves $k(T^* - 1)$ non-zero elements for the preceding $(T^* - 1)$ matrices. Every matrix must have at least one element in each row so we conclude there will have to be exactly one element in each row. This implies that at every step all of the people in a grade who do not leave will either stay where they are or be moved *en bloc* to another grade. If the matrix is to be of super-diagonal form this means that either everyone is promoted or no one is promoted. Extreme policies like this kind are clearly not practicable. However, there will be other feasible (not basic feasible) solutions which reach the goal in the same number of steps. It may be possible to find one among these which is more acceptable but it will not have minimum cost. If recruitment is not allowed into every grade then some $n_j(T)$'s may be zero and so more non-zero variables will be available for the transition matrices. This hardly improves the situation because it will be equally impracticable to empty whole grades from time to time.

These results emphasize the limitations of the linear programming approach to control in the simple form we have adopted. However, they clearly demonstrate that acceptable ways of changing the structure by promotion control are unlikely to be either the quickest or the cheapest. Perhaps the main use of the linear programming algorithm is for determining whether a goal is attainable in a reasonably small number of steps. Even this result must be interpreted with caution because of the stochastic nature of the flows. The programming algorithm will give the number of steps if the system keeps to its expected path. In reality the number of steps will be a random variable.

Fixed time control by recruitment and promotion

In practice, the target structure may not be something which has to be attained precisely but rather an indication of the direction in which we should aim. What is then required is to get somewhere near the target in a reasonably short time. Indeed there may be a fixed time available in which to do the best we can.

Let the time available be denoted by T^* so that the problem is now to get as close to the goal as possible in T^* steps. To formulate this requirement mathematically we must define closeness in terms of a measure of distance between two structures. Suppose such a measure is denoted by $D(\mathbf{n}^*, \mathbf{n}(T^*))$ for the situation at time T^*. The problem is then to minimize this distance with respect to the sequence of control parameters and subject to the appropriate restraints. Formally, then, we have the following two problems:

1. Control by recruitment

$$\text{Minimize:} \quad D(\mathbf{n}^*, \mathbf{n}(T^*)).$$

Subject to:

$$\mathbf{n}(T+1) = \mathbf{n}(T)\mathbf{P} + \mathbf{f}(T+1) \qquad (T = 0, 1, \ldots, T^*-1) \tag{6.35a}$$

$$\sum_{i=1}^{k} n_i(T) = N(T) \qquad (T = 1, 2, \ldots, T^*) \tag{6.35b}$$

$$\mathbf{n}(T) \geq \mathbf{0}\,(T = 1, 2, \ldots, T^*); \ \mathbf{f}(T+1) \geq 0 \qquad (T = 0, 1, \ldots, T^*-1). \tag{6.35c}$$

2. Control by promotion

$$\text{Minimize:} \quad D(\mathbf{n}^*, \mathbf{n}(T^*)).$$

Subject to:

$$n_j(T+1) = \sum_{i=1}^{k} n_{ij}(T) + r_j \left\{ \sum_{i=1}^{k} n_i(T)w_i + M(T+1) \right\}$$
$$(j = 1, 2, \ldots, k; T = 0, 1, \ldots, T^*-1) \tag{6.36a}$$

$$\sum_{j=1}^{k} n_{ij}(T) = n_i(T)(1 - w_i)$$
$$(i = 1, 2, \ldots, k-1; T = 0, 1, \ldots, T^*-1) \tag{6.36b}$$

$$\sum_{i=1}^{k} n_i(T) = N(T) \qquad (T = 1, 2, \ldots, T^*). \tag{6.36c}$$

The ease with which these problems can be solved depends on the form of the distance function. This function ought to reflect the penalties attached to having the various grades over or under strength. A fairly general function which will be the basis of the analysis in this section is

$$D_a = \sum_{i=1}^{k} W_i |n_i^* - n_i(T^*)|^a \qquad (a > 0) \tag{6.37}$$

where $W_1, W_2, \ldots, W_k$ are a set of non-negative weights chosen to reflect the importance attached to the correct manning of each grade. A large value of a gives greater relative weight to large discrepancies. When $a = 1$ the problems posed in (6.35) and (6.36) can be converted into linear programming form; when $a = 2$ they are quadratic programmes. In the remainder of this section we shall explore a special case of the general problem which combines practical interest with sufficient mathematical simplicity to provide explicit solutions.

We first assume that the organization is to be of constant size. This enables us to express the problem in terms of the proportions in the grades and so dispense with (6.35b) and (6.36c). Secondly, we shall assume that $T^* = 1$. At first sight this appears to be an unduly severe restriction until we recall that the stochastic nature of the flows will require us to recalculate the strategy at each step. Putting $T^* = 1$ means that, at each step, we shall move as far as possible towards the goal. In view of the uncertainties of the manpower planning environment it may be highly desirable as well as mathematically convenient to make as much progress towards the goal at each step. Our third restriction is that we shall only consider the case of control by recruitment. This has wider applications than promotion control because of its relevance to the control of age distributions. Recruitment strategies determined by progamming methods are also less apt to be impractical than those involving promotion.

The particular case we have decided to examine in detail may therefore be expressed as follows:

Minimize:

$$D_a = \sum_{i=1}^{k} W_i |q_i^* - q_i(1)|^a \text{ with respect to } \mathbf{r}(1).$$

Subject to:

$$\mathbf{q}(1) = \mathbf{q}(0)\mathbf{P} + \mathbf{q}(0)\mathbf{w}'\mathbf{r}(1) \qquad (\mathbf{r}(1) \geq \mathbf{0}). \qquad (6.38)$$

Now

$$\mathbf{q}^* - \mathbf{q}(1) = \mathbf{q}^* - \mathbf{q}(0)\{\mathbf{P} + \mathbf{w}'\mathbf{r}(1)\}$$

$$= \mathbf{q}(0)\mathbf{w}'\left[\frac{\mathbf{q}^* - \mathbf{q}(0)\mathbf{P}}{\mathbf{q}(0)\mathbf{w}'} - \mathbf{r}(1)\right]$$

$$\propto \mathbf{y} - \mathbf{r}(1)$$

where $\mathbf{y} = \{\mathbf{q}^* - \mathbf{q}(0)\mathbf{P}\}/\mathbf{q}(0)\mathbf{w}'$. Note that $\Sigma_{i=1}^{k} y_i = 1$. Thus, instead of D_a, we can minimize the same function of the differences $y_i - r_i$. Making the appropriate substitutions our problem finally becomes:

Minimize:

$$D_a = \sum_{i=1}^{k} W_i |y_i - r_i|^a \qquad (a > 0). \qquad (6.39)$$

Subject to:

$$\sum_{i=1}^{k} r_i = 1 \qquad (\mathbf{r} \geq \mathbf{0}),$$

where, for brevity, we have dropped the argument of $\mathbf{r}$. If the goal is attainable in one step the minimum of D_a will be zero and the solution is $r_i = y_i$. If at least one y_i is negative the solution is the 'nearest' $\mathbf{r}$ to $\mathbf{y}$.

We first show that $r_i = 0$ for each i such that $y_i \leq 0$. (This is also true for a much wider class of distance functions.) Let Σ^+ denote summation over those i for which $y_i > 0$ and Σ^- summation over the remaining i. Then

$$D_a = \sum{}^+ W_i(y_i - r_i)^a + \sum{}^- W_i(|y_i| + r_i)^a.$$

D_a is decreased if any r_i in the first summation is increased up to the point when $r_i = y_i$ and it is increased if any r_i in the second summation is increased. Hence we should make the r_i's in the second sum as small as possible. There is nothing to prevent them all being made zero because $\Sigma^+ y_i > 1$ and so we can increase the r_i's in the first sum without the need for $r_i > y_i$ for any i. Hence we can replace D_a in (6.39) by $\Sigma^+ W_i(y_i - r_i)^a$ and add the restrictions $r_i \leq \min(y_i, 1)$.

The case a = 1

$$D_1 = \sum{}^+ W_i y_i - \sum{}^+ W_i r_i.$$

The minimum will occur when $\Sigma^+ W_i r_i$ is a maximum subject to $0 \leq r_i \leq \min(y_i, 1)$, $\Sigma^+ r_i = 1$. The solution is to take the grade with the largest W_i and make that r_i as large as possible. If this r_i is less than one we go to the grade with the next largest W_i and make its r_i as large as the restraints allow, and so on. In the special case when $W_i = W$ for all i, D_1 is independent of the r_i's (because $\Sigma^+ r_i = 1$) and so any solution satisfying $r_i \leq y_i$ if $y_i > 0$ and $r_i = 0$ if $y_i \leq 0$ yields the same minimum value. A particular member of this class of solutions which will be investigated numerically is one which takes $r_i \propto y_i$ if $y_i > 0$. In practical terms this means that each grade which is undermanned, after leavers have been taken out, receives the same proportion of its needs. This strategy embodies a principle of 'fairness' which might commend itself on other than mathematical grounds.

The case a = 2

$$D_2 = \sum{}^+ W_i(y_i - r_i)^2.$$

This has to be minimized with respect to the r_i's subject to $\Sigma^+ r_i = 1$ and $r_i \geq 0$ for all i. Let

$$\phi = \sum{}^+ W_i(y_i - r_i)^2 + 2\alpha \sum{}^+ r_i$$

where α is an undetermined multiplier. Then, if $\mathbf{r}^*$ is the minimizing value of $\mathbf{r}$, at this point

$$\left.\begin{array}{l} \dfrac{\partial\phi}{\partial r_i} = 0 \quad \text{if } r_i^* > 0 \\[2ex] \dfrac{\partial\phi}{\partial r_i} > 0 \quad \text{if } r_i^* = 0 \end{array}\right\} . \tag{6.40}$$

For the function ϕ defined above this implies

$$\left.\begin{array}{ll} r_i^* = y_i - \dfrac{\alpha}{W_i} & \text{if } y_i > \alpha/W_i \\[2ex] = 0 & \text{if } y_i \leq \alpha/W_i \end{array}\right\} \tag{6.41}$$

where α must be chosen such that $\Sigma^+ r_i^* = 1$. If, for example, the W_i's are equal to one and

$$y_1 = 0.7, \qquad y_2 = -0.5, \qquad y_3 = 0.1, \qquad y_4 = 0.4, \qquad y_5 = 0.3$$

it may be verified that $\alpha = 0.133$ and hence that

$$r_1^* = 0.567, \qquad r_2^* = r_3^* = 0, \qquad r_4^* = 0.267, \qquad r_5^* = 0.167.$$

The same result may be reached by the following algorithm.
Compute:

$$\left. \begin{aligned} y_i' &= 0 \quad \text{if } y_i \le 0 \\ y_i' &= y_i - (\textstyle\sum^+ y_i - 1)/W_i \textstyle\sum_j^+ W_j^{-1} \end{aligned} \right\} \tag{6.42}$$

If $\mathbf{y}' \ge 0$ then $\mathbf{r} = \mathbf{y}'$; otherwise treat the y_i''s as original y_i's and repeat the process until a non-negative vector is obtained. Similar strategies can be devised for other values of a, and it may easily be verified that if the W_i's are equal they are all equivalent to that described for $a = 2$. Note that the solution given by (6.41) also minimizes D_1, but it will tend to have fewer non-zero elements in the control vectors than the strategy which takes $r_i \propto y_i$.

Numerical comparison of strategies for control by recruitment

To illustrate the calculations we take $k = 3$ and use the parameter values on which Figure 6.1 is based, namely

$$\mathbf{P} = \begin{pmatrix} 0.5 & 0.4 & 0 \\ 0 & 0.6 & 0.3 \\ 0 & 0 & 0.8 \end{pmatrix}, \qquad \mathbf{w} = (0.1, 0.1, 0.2).$$

As the goal we take $\mathbf{q}^* = (0.286, 0.286, 0.428)$ which is the vertex B in Figure 6.1; this is the least top-heavy structure in the maintainable region. For the starting structure we take $\mathbf{q}(0) = (0, 0, 1)$. All points along the path from $\mathbf{q}(0)$ to $\mathbf{q}^*$ can be regarded as new starting structures so the calculations are more comprehensive than appears at first sight. The fixed strategy which causes the structure to converge on $\mathbf{q}^*$ is $\mathbf{r}^* = (1, 0, 0)$. A free-time analysis of the problem shows that the smallest number of steps in which the goal can be attained is seven. In Tables 6.1a and b we compare the performance of the following four strategies:

(a) Fixed—using $\mathbf{r}^* = (1, 0, 0)$ repeatedly.
(b) S_1—using the adaptive strategy minimizing D_1 with $r_1 \propto y_i$ if $y_i > 0$.
(c) S_2—using the adaptive strategy minimizing D_2.
(d) Optimum—the free-time strategy with grade costs in the ratio 1:2:3.

The main difference revealed by the tables is between S_1 and S_2 on the one hand and the fixed and optimum strategies on the other. The former pair allow a

Table 6.1a   A numerical comparison of control strategies with k = 3: the grade structures

T	Strategy	$q_1(T)$	$q_2(T)$	$q_3(T)$	T	Strategy	$q_1(T)$	$q_2(T)$	$q_3(T)$
0		0	0	1					
1	Fixed	0.200	0	0.800	5	Fixed	0.300	0.253	0.447
	S_1	0.100	0.100	0.800		S_1	0.223	0.268	0.508
	S_2	0.100	0.100	0.800		S_2	0.253	0.253	0.495
	Optimum	0.200	0	0.800		Optimum	0.300	0.253	0.447
2	Fixed	0.280	0.080	0.640	6	Fixed	0.295	0.272	0.433
	S_1	0.151	0.179	0.670		S_1	0.237	0.276	0.487
	S_2	0.165	0.165	0.670		S_2	0.264	0.264	0.472
	Optimum	0.280	0.080	0.640		Optimum	0.295	0.272	0.433
3	Fixed	0.304	0.160	0.536	7	Fixed	0.291	0.281	0.428
	S_1	0.182	0.228	0.590		S_1	0.247	0.280	0.473
	S_2	0.207	0.207	0.586		S_2	0.272	0.272	0.457
	Optimum	0.304	0.160	0.536		Optimum	0.286	0.286	0.429
4	Fixed	0.306	0.218	0.477		Goal	0.286	0.286	0.428
	S_1	0.205	0.254	0.540					
	S_2	0.235	0.235	0.531					
	Optimum	0.306	0.218	0.477					

Table 6.1b   A numerical comparison of control strategies with k = 3: the recruitment vectors r'(T)

Strategy	T							Relative cost
	1	2	3	4	5	6	7	
S_1	0.500	0.559	0.641	0.719	0.782	0.831	0.868	1.06
	0.500	0.441	0.359	0.281	0.218	0.169	0.132	
	0	0	0	0	0	0	0	
S_2	0.500	0.639	0.747	0.827	0.883	0.922	0.949	1.05
	0.500	0.361	0.253	0.173	0.117	0.078	0.051	
	0	0	0	0	0	0	0	
Optimum	1	1	1	1	1	1	0.965	1.00
	0	0	0	0	0	0	0.032	
	0	0	0	0	0	0	0.003	

substantial spread of recruits over the two lower grades, whereas the latter concentrate on the lowest grade. Although the paths to the goal show corresponding differences there is very little to choose between them after seven steps. The fixed and optimum strategies are very similar in this case, but this will not always be so. For example, had we approached the same goal from $\mathbf{q}(0)$ = (1, 0, 0) the optimum and adaptive strategies would have concentrated recruitment at the upper levels. Extensive calculations for various goal structures and starting points suggest that the one-step adaptive strategies usually perform well in terms of reaching the neighbourhood of the goal. Calculations given in Bartholomew (1975) suggest that they are little inferior to free time strategies or to fixed time strategies which look further ahead. This conclusion will be reinforced when we come to consider the stochastic aspects in Sections 6.4 and 6.5.

6.4 MAINTAINABILITY IN A STOCHASTIC ENVIRONMENT

When we turn to the behaviour of a system in a stochastic environment new and subtle features arise. In the deterministic case a structure either could or could not be maintained. In a stochastic environment it will be possible to maintain any structure if sufficient people leave. Nevertheless, some structures can be maintained more easily than others. We therefore have to introduce the notion of the probability that a structure can be maintained. The value of this probability will depend on the time at which the decision about recruitment has to be taken. If the uncontrolled flows have already occurred at this stage we shall have more information and so be better placed to maintain the structure. A second difference between the deterministic and stochastic treatments arises out of the first. In the deterministic case a structure which could be maintained could be maintained indefinitely. In a stochastic environment few structures can be maintained in perpetuity and hence we have to consider how to remain as close as possible to the goal over a period of time. This implies that the problems of maintainability and attainability are bound up together and we cannot deal adequately with the former without introducing the latter.

The probability of maintaining a structure by recruitment

A structure can be maintained by recruitment if and only if, for each grade, the sum of flows into the grade (before recruitment) does not exceed the target size. Thus if $\mathbf{n}$ is the structure to be maintained we require that

$$n_j \geq \sum_{i=1}^{k} n_{ij}, \qquad (j = 1, 2, \ldots, k), \tag{6.43}$$

where we have omitted the argument of $n_{ij}(T)$ for simplicity. Before any flows take place the n_{ij}'s are random variables and we may therefore calculate the probability that it will be possible to maintain the structure as

$$P_M = Pr\{n_j \geq \sum_{i=1}^{k} n_{ij}, \ j = 1, 2, \ldots, k | \mathbf{n}\}. \tag{6.44}$$

According to the Markov model the flows out of grade i, $n_{i1}, n_{i2}, n_{ik}, n_{i,k+1}$, have a multinomial distribution with parameters $n_i, p_{i1}, p_{i2} \ldots p_{ik}, w_k$ $(i = 1, 2 \ldots k)$ and so in principle, P_M can be determined.

Before discussing how to do this in practice we consider, briefly, what the position would be if the recruitment numbers $\{f_i\}$ had to be fixed before the n_{ij}'s were known. We might then reasonably act on the assumption that they would be equal to their expectations and choose $\mathbf{f} = \mathbf{n} - \mathbf{nP}$. This policy would only maintain the structure if it turned out that the assumption was precisely true which is clearly an event of negligible probability. The consideration of P_M would thus be of no real value in that situation. We shall therefore proceed on the assumption that the n_{ij}'s are already known when $\mathbf{f}$ has to be chosen.

In certain special cases the evaluation of P_M is easy. For example, let $\mathbf{n}$ represent an age (or length of service) distribution by year of age, n_i being the number aged i. Then a year later all of these individuals will have left or be of age $i + 1$. The transition probabilities are then

$$P_{i,j} = 1 - w_i \quad \text{if } j = i + 1$$
$$= 0 \quad \text{otherwise.}$$

After wastage has taken place the vector $\mathbf{n}$ will have been reduced to

$$(0, n_1 - W_1, n_2 - W_2 \ldots n_{k-1} - W_{k-1}),$$

where W_i is the number leaving at age i. This can be restored to $\mathbf{n}$ by choosing as recruitment vector

$$\mathbf{f} = (n_1, n_2 - n_1 + W_1, n_3 - n_2 + W_2, \ldots, n_k - n_{k-1} + W_{k-1}).$$

We require the probability that all the elements of this vector are non-negative. Since the W_i's are independently and binomially distributed

$$P_M = Pr\{\mathbf{f} \geqslant \mathbf{0}\} = \prod_{i=1}^{k-1} Pr\{W_i \geq \max(0, n_i - n_{i+1})\}$$

$$= \prod_{i=1}^{k-1} \sum_{W_i = m_i}^{n_i} \binom{n_i}{W_i} w_i^{W_i} (1 - w_i)^{n_i - W_i} \tag{6.45}$$

where $m_i = \max(0, n_i - n_{i+1})$. Note that $P_M = 1$ if $n_1 \leq n_2 \leq \ldots \leq n_k$ which means that some structures, at least, can be maintained exactly in spite of the random variation.

In the case of a general $\mathbf{P}$ the probability does not factorize as in (6.45), but it can still be found if N is not too large and some values are given below in Table 6.2. Otherwise a good approximation can be found by invoking the central limit theorem. Thus let

$$X_j = n_j - \sum_{i=1}^{k} n_{ij}$$

then

$$E(X_j) = n_j - \sum_{i=1}^{k} n_i p_{ij}, \qquad \text{var}(X_j) = \sum_{i=1}^{k} n_i p_{ij}(1 - p_{ij}).$$

$$\text{cov}(X_j, X_h) = - \sum_{i=1}^{k} n_i p_{ih} p_{ij}, \qquad (j, h = 1, 2, \ldots, k; j \neq h).$$

The X_j's are then treated as having a multinormal distribution with means and variances as above. Introducing a continuity correction we then have

$$P_M = Pr\{X_j \geqslant 0; j = 1, 2, \ldots k\} =$$

$$Pr\left\{ u_j \geq \frac{- E(x_j) - \frac{1}{2}}{\left\{ \sum_{i=1}^{k} n_i p_{ij}(1 - p_{ij}) \right\}^{\frac{1}{2}}}; j = 1, 2, \ldots k \right\} \qquad (6.46)$$

where $\mathbf{u}$ has a standard normal distribution with the same correlation matrix as the X_j's. This approximation is particularly useful when N is large and k is small. Some indication of its accuracy can be obtained from Table 6.2.

The approximation is good enough for most practical purposes even when the grade sizes are quite small. The structures in the left half of the table represent the

Table 6.2   *Exact and approximate values of P_M for the system with $k = 3$ and*

$$\mathbf{P} = \begin{pmatrix} 0.7 & 0.2 & 0 \\ 0 & 0.8 & 0.1 \\ 0 & 0 & 0.9 \end{pmatrix}$$

n			Exact	Approx.	n			Exact	Approx.
8	8	8	0.369	0.358	2	5	11	0.707	0.646
18	18	18	0.304	0.304	6	15	33	0.775	0.708
36	36	36	0.270	0.271	12	30	66	0.870	0.858
72	72	72	0.247	0.248	24	60	132	0.955	0.950

least top-heavy maintainable structure and those on the right are in the middle of $\mathcal{M}$. A fuller picture of how P_M depends on the structure is given for the same $\mathbf{P}$ and $N = 90$ on Figure 6.3. The deterministic maintainable region is marked on the figure and it is evident that P_M is small outside the region and large inside. However, there is often an appreciable probability that structures inside $\mathcal{M}$ cannot be maintained in a stochastic environment. As might be expected it can be

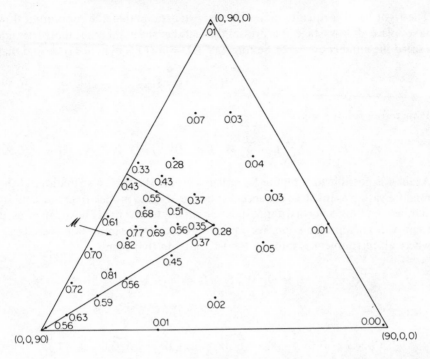

Figure 6.3 Probabilities of maintaining various structures by recruitment control with

$$P = \begin{pmatrix} 0.7 & 0.2 & 0 \\ 0 & 0.8 & 0.1 \\ 0 & 0 & 0.9 \end{pmatrix}$$

(From Bartholomew, 1977b; reproduced by permission of Applied Probability Trust.)

shown that $P_M \to 1$ inside $\mathcal{M}$ and $P_M \to 0$ outside. On the boundary the limit is a value intermediate between 0 and 1.

The probability of maintaining a structure by promotion

Here we limit the discussion to the case where the grades form a hierarchy with promotion into the next higher grade only. In this case the accounting equations are

$$n_j(T+1) = n_j(T) - n_{j, j+1} + n_{j-1, j} - W_j + f_j \qquad (j = 1, 2, \ldots k) \quad (6.47)$$

with $n_{01} = 0$, $n_{k, k+1} = 0$. The condition for the structure $\mathbf{n}$ to be maintained is that the flow numbers shall satisfy the inequalities

$$0 \le n_{j, j+1} \le n_j - W_j, \qquad (j = 1, 2, \ldots k-1). \tag{6.48}$$

The right-hand inequality arises because we assume that the promotion flow takes place after wastage; it expresses the fact that the number promoted cannot exceed the number available. Setting $n_j(T+1) = n_j(T) = n_j$ in (6.47) we find that

$$n_{j,j+1} = \sum_{i=1}^{j} f_i - \sum_{i=1}^{j} W_i \qquad (j = 1, 2, \ldots k-1). \qquad (6.49)$$

It therefore follows that

$$P_M = Pr\left\{0 \le \sum_{i=1}^{j} f_i - \sum_{i=1}^{j} W_i \le n_j - W_j; j = 1, 2, \ldots k-1)\right\}. \qquad (6.50)$$

Again it is possible to compute P_M exactly because the exact distributions of the random variables involved are specified by the Markov model. In practice this is cumbersome, but a normal approximation similar to that used above can be used to provide an approximation. We concentrate here on the important special case where all recruitment is into the lowest grade. In that event

$$f_1 = \sum_{i=1}^{k} W_i, f_2 = f_3 = \ldots f_k = 0,$$

whence

$$P_M = Pr\left\{0 \le \sum_{i=1}^{k} W_i - \sum_{i=1}^{j} W_i \le n_j - W_j; j = 1, 2, \ldots k-1\right\}$$
$$= Pr\left\{\sum_{i=j}^{k} W_i \le n_j; j = 1, 2, \ldots k-1\right\}. \qquad (6.51)$$

The last step follows from the fact that the left-hand inequality is trivially satisfied. Since the W_i's are independently and binomially distributed their partial sums will be approximately normal with

$$E(Z_j) = \sum_{i=j}^{k} n_i w_i, \qquad \text{var}(Z_j) = \sum_{i=j}^{k} n_i w_i (1 - w_i)$$

$$\text{corr}(Z_j, Z_h) = \left\{\sum_{i=h}^{k} n_i w_i (1 - w_i) \Big/ \sum_{i=j}^{k} n_i w_i (1 - w_i)\right\}^{\frac{1}{2}}, \qquad h > j,$$

where $Z_j = \sum_{i=j}^{k} W_i$. Using the normal approximation with continuity correction

$$P_M \doteq Pr\left\{u_j \ge \left(-n_j + \sum_{i=j}^{k} n_i w_i - \tfrac{1}{2}\right) \Big/ \left(\sum_{i=j}^{k} n_i w_i (1 - w_i)\right)^{\frac{1}{2}}; j = 1, 2, \ldots k-1\right\},$$

$$(6.52)$$

where the u's are standard normal variables with the same correlation coefficients as the Z's.

If $\mathbf{r} \neq (1, 0, \ldots 0)$ a similar approach can be used, but the form of the region over which the volume of the multinormal distribution has to be found is more complicated. We also have to distinguish between the allocation of recruits to grades in fixed proportions or with fixed probabilities. As yet there is no theory available in the latter case, but the former is treated in Bartholomew (1979) which contains an example showing how P_M depends on $\mathbf{n}$. As with recruitment control the deterministic boundary of $\mathcal{M}$ divides those structures for which $P_M \to 1$ from those for which $P_M \to 0$.

Strategies for maintaining a structure by recruitment

The foregoing analysis provides only a partial solution to the maintainability problem. Provided that the inequalities (6.43) are satisfied we choose

$$f_j = n_j - \sum_{i=1}^{k} n_{ij} \qquad (j = 1, 2, \ldots k-1),$$

but the question of what to do if some of the f_j's are negative remains to be answered. We describe two approaches to the problem.

In the first approach we abandon the attempt to maintain a structure exactly. Instead we ask only that the expected stock vector should be equal to $\mathbf{n}$ for any T. This is easily achieved by taking

$$\mathbf{f} = \mathbf{n}(\mathbf{I} - \mathbf{P}) \tag{6.53}$$

as may be verified by substituting $\mathbf{n}(0) = \mathbf{n}$ and $R(T - \tau)\mathbf{p}_0 = \mathbf{f}$ in (3.2). (This argument ignores the fact that $\mathbf{f}$ must consist of integers. It can be remedied by randomly rounding the elements in such a way that the expectation of $\mathbf{f}$ is given by (6.53).) Since the structure is no longer fixed it is useful to have some way of judging how much the structure is likely to deviate from $\mathbf{n}$. This is provided by the variances and covariances which can be obtained directly from results given in Chapter 3. The term $n_{0j}(T + 1)$ in (3.38) is now a constant and so all the terms arising from this in the derivation disappear. The effect of this is to delete $\mu_0(T + 1)$ from the answer given in (3.40). It is useful in the present context to extract the variances and covariances and set them out in a square matrix in the usual way. The difference equation for the variance–covariance matrix, $\mathbf{V}(T)$ then has the form

$$\mathbf{V}(T+1) = \mathbf{P}'\mathbf{V}(T)\mathbf{P} + [\bar{\mathbf{n}}(T)\mathbf{P}]_d - \mathbf{P}'[\bar{\mathbf{n}}(T)]_d\mathbf{P} \tag{6.54}$$

with $\mathbf{V}(0) = \mathbf{0}$.

This strategy has the advantage that it can be applied before the current wastage figures are available, but this also means that the total size will vary. If the wastage is known before the recruitment is decided upon an alternative strategy is to keep the total size fixed but continue to allocate recruits in the same proportions. In this case

$$\mathbf{f} = \left(\sum_{i=1}^{k} W_i \right) \mathbf{r},$$

where $\mathbf{r} = \mathbf{n}(\mathbf{I} - \mathbf{P})/\mathbf{nw}'$. It may be shown that this is also an unbiased strategy in the sense that $\bar{\mathbf{n}}(T) = \mathbf{n}$ for all T and also that

$$\mathbf{V}(T+1) = \mathbf{Q}'\mathbf{V}(T)\mathbf{Q} + [\bar{\mathbf{n}}(T)\mathbf{Q}]_\mathrm{d} - \mathbf{Q}'[\bar{\mathbf{n}}(T)]_\mathrm{d}\mathbf{Q} - \bar{\mathbf{n}}(T)\mathbf{w}'\{[\mathbf{r}]_\mathrm{d} - \mathbf{r}'\mathbf{r}\} \tag{6.55}$$

(see Bartholomew, 1975). As one might anticipate, this strategy exercises a tighter control over the structure.

One obvious shortcoming of the 'fixed' strategies is that they take no account of the current structure. Thus there will be occasions when the goal could be maintained precisely but when the fixed strategy fails to do so. This leads naturally to the search for adaptive strategies along the lines of those used for attaining a structure in a deterministic environment. Our second approach proceeds along these lines. At each step we aim to get as near as possible to the goal. Let $N_j = \sum_{i=1}^{k} n_{ij}$ be the total in grade j after wastage and promotion have taken place. We have already seen that the structure will be maintained exactly if $n_j \geq N_j$, for all j. When at least one of these inequalities is violated some modification is required. If we recruit f_j people to grade j the total number there will be $N_j + f_j$. We might then reasonably choose the f_j's to minimize

$$\phi = \sum_{j=1}^{k} (n_j - N_j - f_j)^2 \tag{6.56}$$

subject to $\mathbf{f} \geq \mathbf{0}$. This problem is obviously solved by taking

$$f_j = \max (0, n_j - N_j).$$

In words, this requires each grade which is below target to be made up to that level; those which are already too large are left alone. This strategy has one obvious and serious drawback. Its effect will be that the total size is usually in excess of what is required and never less. It is thus biased upwards. We can ensure that the total size remains fixed by adding to (6.56) the restriction

$$\sum_{j=1}^{k} f_j = \sum_{j=1}^{k} (n_j - N_j).$$

The minimization problem is now essentially the same as that involving the distance function D_2 in Section 6.3, as can be seen by converting all the quantities to proportions.

It remains to consider how the fixed and adaptive strategies compare among themselves. The processes defined by the two adaptive strategies are Markov chains on the set of states consisting of all possible structures. In principle, therefore, such things as means and variances can be obtained. In practice the number of states for systems of a realistic size is prohibitively large. Our comparisons are therefore based on simulation. Table 6.3 gives the average stock vectors obtained from 10,000 trials using the same transition matrix as for

Table 6.3 *Average structures maintained with the adaptive strategies in a sequence of 10,000 trials with transition matrix*

$$P = \begin{pmatrix} 0.7 & 0.2 & 0 \\ 0 & 0.8 & 0.1 \\ 0 & 0 & 0.9 \end{pmatrix}$$

	Maintained by recruitment at the bottom			Maintained by equal recruitment at each level		
Goal	8	8	8	2	5	11
Size free	8.00	9.66	10.67	2.00	5.28	11.45
Size fixed	6.16	8.21	9.63	1.75	4.95	11.30
Goal	36	36	36	12	30	66
Size free	36.00	39.25	41.52	12.00	30.09	66.18
Size fixed	32.16	36.54	39.30	11.88	30.00	66.12

Table 6.4 *Mean square errors of grade sizes with the fixed and adaptive strategies in a sequence of 10,000 trials with transition matrix as in Table 6.3*

		Maintained by recruitment at the bottom			Maintained by equal recruitment at each level		
Goal		8	8	8	2	5	11
Fixed	Size free	3.83	6.36	7.62	1.32	3.70	8.19
	Size fixed	5.30	5.07	5.10	1.25	2.97	3.68
Adaptive	Size free	0.00	6.27	12.26	0.00	0.43	0.90
	Size fixed	5.42	2.38	5.97	0.31	0.59	0.71
Goal		36	36	36	12	30	66
Fixed	Size free	15.61	27.05	33.18	5.57	18.32	47.98
	Size fixed	24.29	24.37	24.45	6.37	16.74	19.10
Adaptive	Size free	0.00	23.18	50.92	0.00	0.19	0.50
	Size fixed	23.66	10.81	25.06	0.17	0.30	0.46

Table 6.2. In the case of those structures near the middle of $\mathcal{M}$, which are maintained by equal recruitment at all levels, the bias in both cases is slight. It is considerably greater for those structures which are maintained by recruitment into the bottom grade only. However, this disadvantage is more than offset when we compare their closeness to the goal as compared with the fixed strategies. This comparison is made in Table 6.4 which gives the mean square errors ($ = \sum_{T=1}^{10,000} (n_i(T) - n_i)^2/10,000$ for each i). Although no single strategy is uniformly best in the cases considered, it is clear that the adaptive strategies are usually much superior especially when the goal is near the centre of $\mathcal{M}$. Even at the extreme point they are quite good, especially as the figures conceal the fact that the variances are quite small, meaning that the variation in structure is small; most of the mean square error is accounted for by the bias.

Strategies for maintaining a structure by promotion control

A similar line of development can be followed for promotion control. Again we find an ambiguity in translating the deterministic fixed strategy into stochastic terms. We showed in (6.20) that the promotion probabilities required to maintain **n*** satisfied

$$n_i^* p_{i,\,i+1} = \sum_{j=1}^{i} r_j \sum_{h=1}^{k} n_h^* w_h - \sum_{j=1}^{i} n_j^* w_j \qquad (i = 1, 2, \ldots, k-1). \quad (6.57)$$

The right-hand side gives the number to be promoted from grade i. The first version (F_1) of the fixed strategy is that which promotes these numbers at each time point. That is we choose

$$n_{i,\,i+1} = n_i p_{i,\,i+1} \qquad (i = 1, 2, \ldots k-1). \quad (6.58)$$

Equation (6.57) can also be regarded as giving the *proportion* of those in grade i to be promoted. As long as the structure is being maintained exactly it is immaterial whether we work with the proportions or numbers. However, as soon as the actual structure departs from **n*** a difference appears. According to the second interpretation the number to be promoted from i would be

$$n_{i,\,i+1} = n_i p_{i,\,i+1} \qquad (i = 1, 2, \ldots k-1) \quad (6.59)$$

where **n** is the current structure. We call this strategy F_2. (Since $n_{i,\,i+1}$ has to be an integer a rounding procedure must be used.) At this stage another complication arises. If the uncontrolled flows occur before the promotions are made there is the possibility that there will not be sufficient promotees available. This will happen whenever $n_{i,\,i+1}$, as given by (6.58) and (6.59), exceeds $n_i - W_i$. The natural modification is to promote all those that are available. This will, of course, introduce a bias into the strategy and for this reason we have given the average stock numbers in the calculations which follow. It would be possible to take the alternative view that promotion takes priority, in which case the problem does not arise, though we should then have to decide whether a person chosen for promotion could also leave.

An adaptive strategy for promotion control can be devised by choosing the promotion numbers to minimize the squared distance between the actual and desired structures. Thus if the current structure is **n** and the goal is **n*** the number in grade i after all flows have taken place is

$$n_i - W_i + f_i - n_{i, i+1} + n_{i-1, i}$$

where n_i, W_i, and f_i are given numbers and $n_{i, i+1}$ and $n_{i-1, i}$ are non-negative integers to be selected. We propose to do this by minimizing

$$\phi = \sum_{i=1}^{k} (n_i{}^* - n_i + W_i - f_i + n_{i, i+1} - n_{i-1, i})^2$$

subject to

$$n_{0, 1} = n_{k, k+1} = 0, \qquad 0 \le n_{i, i+1} \le n_i - W_i \qquad (i = 1, 2, \ldots k-1).$$

This is a straightforward quadratic programming problem which can easily be solved numerically.

Table 6.5 Performance of strategies in 20,000 consecutive trials when $w = (0.1, 0.1, 0.1)$ *and* $r = (1,0,0)$

	Strategy	\(\mathbf{n^*}\)								
		(9,	9,	72)	(9,	81,	0)	(36,	32,	22)
Mean	F_1	12.6	8.8	68.6	12.8	77.2	0	36.1	32.0	21.9
grade	F_2	9.4	9.1	71.6	9.4	80.6	0	36.0	32.0	22.0
size	Adaptive	10.3	9.1	70.5	10.1	79.9	0	36.0	32.0	22.0
Mean	F_1	31.8	2.9	34.6	33.6	33.6	0	27.6	18.0	11.3
square	F_2	7.9	7.2	13.0	7.9	7.9	0	11.5	7.2	9.6
errors	Adaptive	4.8	0.9	6.2	4.4	4.4	0	0	0	0

In order to illustrate the performance of these strategies the results of a simulation exercise are presented in Table 6.5. The three cases relate to a system of size 90 with $w = (0.1, 0.1, 0.1)$ and $r = (1, 0, 0)$. The first two structures are as near to vertices of the maintainable region as the restriction to integers allows; the third is near the centre of the region. On this showing the adaptive strategy is clearly the best and F_2 generally performs better than F_1. These conclusions are supported by other calculations with different values of w and r.

6.5   ATTAINABILITY IN A STOCHASTIC ENVIRONMENT

We have implicitly dealt with attainability in the previous section and little more needs to be added. The various strategies proposed for maintaining a structure can be applied whatever the starting point from which they will converge to a neighbourhood of the goal. It should be remembered that the adaptive strategies were originally arrived at as special cases of T-step strategies in the fixed time treatment of control in a deterministic environment. Although it might be possible to formulate stochastic versions of the fixed time and free time versions of the general control problem this hardly seems necessary in view of the fact that the one-step strategies are both simple and effective.

We can no longer compare the performance of the various strategies by means of averages and mean square errors. Now we are interested in the speed and directness of the approach to the goal. An illustration of how this may be done is provided in Figures 6.4 and 6.5. They relate to the case of control by recruitment

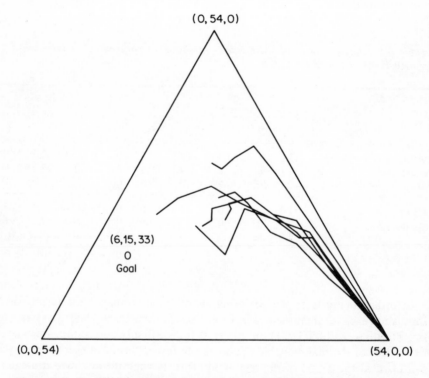

Figure 6.4   Six simulated attempts to reach the goal (6, 15, 33) from (54, 0, 0) using the fixed strategy (first 8 steps).

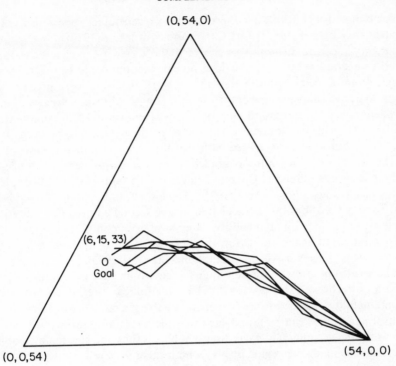

Figure 6.5   Six simulated attempts to reach the goal (6, 15, 33) from (54, 0, 0) using the adaptive strategy (first 8 steps).

for a three-grade system with

$$\mathbf{P} = \begin{pmatrix} 0.7 & 0.2 & 0.0 \\ 0 & 0.8 & 0.1 \\ 0 & 0 & 0.9 \end{pmatrix}.$$

The aim is to reach the goal structure (6, 15, 33) from (54, 0, 0). Figure 6.4 shows six repetitions of the 'sized fixed' fixed strategy and Figure 6.5 has similar results for the adaptive strategy which minimizes the squared distance function. As one might have anticipated, the adaptive strategy takes a more direct path and the various simulations lie closer together. In an obvious sense, therefore, the adaptive strategy exerts a tighter control.

6.6   COMPLEMENTS

Control theory is a large subject, but much of the existing work has been done in an engineering context where the relationships between the variables are expressed by differential equations. There appears to be very little of the theory which is directly applicable to our problems. Two general references on stochastic control are Kushner (1967, 1971). Canon and co-workers (1970) have expounded the theory of control in terms very similar to those used in this chapter, though at a more general level. In fact, the deterministic sections could be regarded as an application of their approach though the work described here predates their book. We have adapted some of their terminology. In the social sciences, Tintner and Sengupta (1972) discussed some problems of control and programming in the context of stochastic economics. The use of mathematical programming in the control of Markov chains is treated in Kushner and Kleinman (1971).

The treatment of the control of graded systems developed here had its origin in a short section at the end of Chapter 3 in the first edition. This was subsequently developed by the author, Forbes (1971c), Davies (1973), Vajda (1975), and others and became the subject of a full chapter in the second edition. For this edition most of the material on attainability, which is of less importance in practice, has been omitted to make room for the new work on control in a stochastic environment. This is based on Bartholomew (1977b, 1979).

The work on control in a deterministic environment has been extended in several directions. Davies (1975) has continued the study of n-step maintainability. A structure is n-step maintainable if a sequence of values for the control parameters can be found such that the structure returns to the goal at every nth step. Maintainability as discussed here is thus one-step maintainability. The n-step maintainable region ($n > 1$) contains that for $n = 1$ so a wider range of structures can be maintained in this wider sense. Grinold and Marshall (1977) also discussed maintainability and attainability. Vajda (1975 and 1978) has given a detailed discussion of extensions of the control problem described in this chapter. He also introduced weaker forms of maintainability in which, for example, it is sufficient to maintain the total size of some subset of the grades. Bartholomew and Forbes (1979) provided an elementary, and largely arithmetical, account of some of the theory of this chapter in the context of practical manpower planning.

Our brief treatment of maintainability in continuous time is based on Hassani (1980). Not all problems in continuous time can be handled by a simple limiting operation on the discrete model. For example, in continuous time, control may be exercised either discretely or continuously. There is clearly scope for further research here, especially in the stochastic environment.

The calculation of the normal approximation to P_M requires a program for the computation of volumes under the normal surface. We have found the method of Milton (1972) satisfactory though, occasionally, it can be very lengthy.

Our treatment of control in a stochastic environment is brief and the calculations given are purely illustrative. Further details are in Bartholomew (1977b, 1979) where the numerical examples cover a wider range of policies. In particular, control strategies which do not require knowledge of the uncontrolled flows are discussed. Additional calculations are presented graphically in Bartholomew (1976c). An exact expression for the variance–covariance matrix for a fixed size system is given in Bartholomew (1975) along with some further examples.

There has been an extensive use of mathematical programming methods to find optimum strategies in education and manpower planning. If the flow numbers are assumed to be proportional to the stocks from which they come the problem is essentially the same as the deterministic formulation given here. Greater prominence is usually given in such treatments to the cost structure and the computational aspects than in our development. Rowe and co-workers (1970) formulated the control problem in terms which require the minimization of a discounted cost function over a finite planning horizon. Charnes, Cooper, Niehaus, and others have developed an extensive theory based on goal programming techniques with an embedded Markov chain. Their approach is some what similar to our fixed time control problem though their models involve various budgetary constraints. All of this work related to a deterministic environment. An early paper which prepared the ground is reproduced in Bartholomew (1976b) and a collection of their earlier papers is in Charnes and co-workers (1970). Morgan (1970) also formulated a control problem as a linear programme in which the variables were recruitment and promotion numbers and costs were attached to recruitment, redundancy, and overmanning. In a subsequent paper (Morgan 1971) the author considers the control of an age structure in which age groups take the place of grades and he proposed a strategy which is similar to those discussed in Section 6.3.

Mehlmann (1980) provides a method of solving the deterministic control problem using dynamic programming. He allows both the recruitment vector and the transition matrix to be control variables but, in addition, he requires that they shall be as near as possible to some preferred values.

The statistical prerequistes for this chapter are minimal. On the mathematical side, the theory of mathematical programming together with the associated ideas of convex sets covers most of what is required. These topics are covered in many texts including the *Handbook of Applicable Mathematics*, Volumes I and III.

CHAPTER 7

Models for Duration and Size

7.1 BACKGROUND

In many social contexts the duration of some activity or the size of some quantity has substantive interest. Thus, for example, the length of human life, the age at marriage, or the spacing between successive births in a family are all of fundamental interest to the demographer. The length of time that a patient survives after treatment for illness, such as cancer, is often a measure of the efficacy of the treatment. The duration of a strike, the length of time a firm remains in business, and the time that an individual spends in his job, at his address, in absence from work, or in unemployment are all further examples of social or economic interest. Likewise, the size of wealth holdings or of incomes are key variables in the study of inequality and the size of firms and towns are features of industrial and social organization which interest economists and geographers. All of these diverse quantities have two things in common which make them appropriate subjects for a stochastic modelling exercise. First, they are all subject to a high degree of variability which means that they must be treated probabilistically. Secondly, empirical work has shown that the frequency distributions of many such variables can be graduated by a few simple curves which appear to be remarkably stable over time. Such regularities are rather uncommon in the social sciences and when we find them it is highly desirable to try to discover how they come about. Not only do we hope thereby to gain some insight into the underlying springs of human behaviour but also to provide a firm basis for techniques of statistical analysis.

The widespread occurrence of a particular form of frequency distribution may come about in several ways. For example, normal distributions are supposed to occur frequently because the values of many variables are the result of a large number of independent small influences. The central limit theorem then ensures their approximate normality. Extreme value distributions likewise have forms which are largely independent of the distributions of the constituent random variables. A closely related fact, which may account for the widespread occurrence of certain distributions, is that certain families are closed under operations such as convolution or mixing. This means, for example, that if samples from two members of the family are mixed the resulting distribution will also belong to that family. A third explanation for the persistence of a particular distributional form arises out of the treatment of social processes as stochastic processes. We have frequently noted in previous chapters that processes

186

approach steady states in which, for example, a class structure reaches a stable, dynamic equilibrium. If the classes of such a process were to represent size categories (e.g. income groups) the steady-state distribution could be interpreted as a frequency distribution. The persistence of a certain distributional form might then be explained by supposing that the underlying process had reached its steady state.

We shall meet models generated in each of these ways. We begin in Section 7.2 with models that have been developed for duration distributions in a variety of applications. For the most part these involve stochastic processes which have already occurred in other contexts. Section 7.3 is a transitional stage in which we consider the lognormal distribution which has served to describe distributions of both duration and size. This leads on to Section 7.4 where we consider models for size distributions. A duration can, of course, be thought of as the size of a time interval and, as Section 7.3 demonstrates, there is no clear dividing line between models for size and duration. All of our models can be criticized as being oversimplified and various extensions in the direction of greater realism are mentioned in the text and Complements section. Nevertheless we would emphasize again that oversimplification is not necessarily a vice. By focusing first on the broad qualitative features of the process we gain considerable insight and provide a basic model to which further refinements can be added later.

7.2 MODELS FOR DURATION

Durations have been studied in other areas outside the social sciences. The best known and most developed is in reliability theory and industrial life-testing. There is a close parallel between much of the work that has been done in this field and the subject-matter of this chapter, but there are important differences between the social and industrial applications. In life-testing the exponential distribution has a central role because the failure risk of many types of equipment is approximately independent of age. Where the exponential distribution fails the Weibull or the gamma distributions are often used. None of these distributions plays a significant part in the applications we shall consider, although the exponential distribution is our point of departure in the next section. In other words, the kinds of model which have been found satisfactorily to describe the life of industrial components are rarely suitable for the durations of social processes. Our exposition will largely be based on modelling the length of time a person spends in a job (their CLS), but we shall indicate other applications as we go. One model is particularly concerned with the duration of a strike.

We shall require the following terminology and notation. Let T denote the duration of the variable of interest; no distinction in notation will be made between the random variable and the values which it takes; T will be continuous. The probability distribution of T is the main object of interest, and this can be expressed in three equivalent ways as follows:

(a) **The survivor function.** This is the probability that an individual survives for length of time T, and is denoted by $G(T)$. The survivor function is the complement of the distribution function, $F(T)$, but $G(T)$ is of more direct relevance in practice.

(b) **The completed length of service (CLS) density function.** This is the density function associated with $F(T)$, related to the survivor function by

$$f(T) = -\frac{dG(T)}{dT}.$$

(c) **The loss intensity (or rate).** This is denoted by $\lambda(T)$ and is defined as follows:

$$Pr\{\text{loss in } (T, T + \delta T)|\text{survival to } T\} = \lambda(T)\delta T.$$

This function is known by many different names appropriate to different applications. In reliability theory and life-testing it is known as the hazard function, age specific failure rate, or, more briefly, as the failure rate. Actuaries call it the force of mortality and in manpower applications it is commonly referred to as the force of separation or the propensity to leave. It appeared in Chapter 4 and 5 as the loss intensity and we shall emphasize the continuity by preserving the same terminology here. The relationship between $\lambda(T)$, $G(T)$, and $f(T)$ is easily deduced as follows:

$$f(T)\delta T = Pr\{\text{loss in } (T, T + \delta T)\}$$
$$= Pr\{\text{survival to } T\} Pr\{\text{loss in } (T, T + \delta T)|\text{survival to } T\}$$
$$= G(T)\lambda(T)\delta T.$$

Hence

$$\lambda(T) = f(T)/G(T) = -\frac{d \ln G(T)}{dT}. \tag{7.1}$$

Conversely,

$$G(T) = \exp\left[-\int_0^T \lambda(x)\,dx\right]. \tag{7.2}$$

In applications to labour wastage the function $\lambda(T)$ provides a concise description of the leaving process, enabling one to identify times in an individual's service when the risk of leaving is particularly high or low.

In practice it is sometimes easier to estimate one of these functions rather than another, but we can easily pass from one to the other using the relations set out above. Most of the subsequent model-building will focus on $f(T)$ or $G(T)$ but the interpretation of the model is often made most easily in terms of $\lambda(T)$. The data from which empirical estimates of these functions have to be made are rarely in the form of a simple random sample. More often they consist of data from a cohort censored at its upper end or of census data relating to the stocks and flows over a short interval of time.

Exponential and mixed exponential distributions

Rice and co-workers (1950) published several empirical CLS distributions obtained in their studies at the Glacier Metal Company. They observed that the distributions could be graduated by smooth J-shaped curves of hyperbolic form, but they did not put forward any theoretical model to account for this phenomenon. Silcock (1954) reviewed the literature on turnover up to that date and proposed two models to account for observed CLS distributions. His first model was based on a remark of Rice and co-workers (1950) to the effect that they had found a regularity in the turnover pattern which appeared to be characteristic of the firm and independent of economic and social forces operating outside the firm. Silcock interpreted this as implying a constant loss intensity and observed that this, in turn, would yield an exponential CLS distribution. It is doubtful whether Silcock's interpretation is justified since the loss intensity could certainly depend on factors internal to the firm, including the individual's length of service. Nevertheless, the exponential does provide a testable assumption about the leaving process and it paves the way for more adequate models. The exponential distribution was fitted by Silcock to several CLS distributions and, in every case, the fit was very poor. Two examples are given in Table 7.1, from which it can be seen that the observed distribution is always more highly skew than the fitted exponential. The censoring at 21 months makes it impossible to investigate the form of the upper tail, but it is clear from the data given that the exponential hypothesis is not tenable. Silcock's experience has been borne out in almost all subsequent investigations.

The second model proposed by Silcock (1954) is a generalization of the first. It retains the simple assumption of constant loss rate for individuals, but this rate is now supposed to vary in the population from which employees are drawn. This is a very plausible hypothesis. Individuals differ in almost all other aspects of their behaviour and it would be surprising if propensity to leave their jobs was an exception. Denoting the constant loss intensity by λ any individual will have the CLS density

$$f(T) = \lambda \, e^{-\lambda T} \qquad (\lambda > 0; T \geq 0).$$

Suppose now that λ is a random variable with distribution function $H(\lambda)$. The CLS distribution for a sample of employees will then have the density function

$$f(T) = \int_0^\infty \lambda e^{-\lambda T} \, dH(\lambda) \qquad (T \geq 0). \tag{7.3}$$

This distribution is always more skew than the exponential distribution with the same mean and it therefore has the main characteristic required by the data. A

Table 7.1 Observed and fitted CLS distributions for two firms

Length of completed service	Glacier Metal Co. (1944–47)				J. Bibby & Sons Ltd. (Males, 1950)			
	Actual number of leavers	Exponential fit	Type XI fit	Mixed exponential fit	Actual number of leavers	Exponential fit	Type XI fit	Mixed exponential fit
Under 3 months	242	160.2	242.0	242.0[a]	182	103.9	195.4	182.0[a]
3 months	152	138.9	150.3	152.0[a]	103	86.8	87.5	103.0[a]
6 months	104	120.4	103.8	101.4	60	72.4	51.8	60.7
9 months	73	104.5	76.5	72.7	29	60.5	35.0	38.0
12 months	52	90.6	59.2	55.8	31	50.5	25.6	25.5
15 months	47	78.5	47.4	45.7	23	42.1	19.7	18.6
18 months	49	68.1	38.8	39.2	10	35.2	15.8	14.7
21 months and over	487	444.8	488.0	497.2	191	177.6	198.2	186.5
Total	1206	1206.0	1206.0	1206.0	629	629.0	629.0	629.0

[a] These figures agree exactly with those observed because the distribution was fitted by equating percentage points. For further details see Bartholomew (1959).

partial justification of this statement can be seen by considering the ratio

$$r = f(T)/\mu^{-1}\,e^{-T/\mu}, \tag{7.4}$$

where

$$\mu = \int_0^\infty Tf(T)\,dT = \int_0^\infty \frac{1}{\lambda}\,dH(\lambda) = E\!\left(\frac{1}{\lambda}\right).$$

For small T,

$$r \to E(\lambda)\,E\!\left(\frac{1}{\lambda}\right) \ge 1,$$

since the arithmetic mean is always at least as great as the harmonic mean. Equality occurs only in the degenerate case when $H(\lambda)$ places all the probability at one point. For large T, write

$$r = \mu \int_0^\infty \lambda e^{-(\lambda - 1/\mu)T}\,dH(\lambda).$$

For $\lambda > \mu^{-1}$ the integrand tends to zero; for $\lambda < \mu^{-1}$ it tends to infinity. Therefore, provided that $H(\lambda)$ assigns some probability to values of λ in excess of $\mu^{-1}, r \to \infty$. This is bound to be the case except when $H(\lambda)$ degenerates to a single point.

In order to fit (7.3) to data we must specify $H(\lambda)$. Silcock (1954) chose a gamma distribution with density function

$$\frac{dH(\lambda)}{d\lambda} = \frac{c^v}{\Gamma(v)}\lambda^{v-1}e^{-c\lambda} \qquad (v > 0; c > 0; \lambda \ge 0)$$

and showed that

$$f(T) = \frac{v}{c}\left(1 + \frac{T}{c}\right)^{-(v+1)} \qquad (T \ge 0). \tag{7.5}$$

The corresponding expressions for the survivor function and the loss intensity are

$$G(T) = (1 + T/c)^{-v} \quad \text{and} \quad \lambda(T) = \frac{v}{c}(1 + T/c)^{-1}. \tag{7.6}$$

This is a J-shaped distribution of the Pearson family, in which it is classified as type XI. The loss intensity has a particularly simple form, indicating a monotonic decline in the propensity to leave with increasing length of service. It may, at first sight, appear paradoxical to have arrived at a decreasing $\lambda(T)$ having started with the assumption that the loss rate was constant for any individual. However, the group of individuals we are observing do not all have the same constant loss rate. The longer members of the group survive the more likely it is that they have low λ's and so we get the apparent decline in λ with time. When Silcock (1954) fitted this distribution to his data the agreement was much improved. Some of Silcock's calculations are given in Table 7.1 under the heading 'Type XI fit'.

The gamma distribution for λ used above is quite flexible, depending as it does on two parameters. Its main advantage in the present context is its mathematical tractability. However, the particular functional form which we adopt does not appear to be critical. An explicit expression for $f(T)$ can be obtained in some cases using an inverse gamma distribution, and an interesting curiosity is provided by the case

$$\frac{dH(\lambda)}{d\lambda} = \frac{c^\rho}{\Gamma(\rho)\Gamma(1-\rho)} \lambda^{-1}(\lambda-c)^{-\rho} \qquad (\lambda \ge c; 0 < \rho < 1), \qquad (7.7)$$

when $f(T)$ itself turns out to be a J-shaped gamma distribution. For some purposes it is simpler to suppose that λ can take only two values, say λ_1 and λ_2. If the associated probabilities are p and $1-p$ respectively, the CLS distribution will be

$$f(T) = p\lambda_1 e^{-\lambda_1 T} + (1-p)\lambda_2 e^{-\lambda_2 T} \qquad (0 < p < 1, \lambda_1, \lambda_2 > 0; T \ge 0). \qquad (7.8)$$

This is the two-term mixed exponential distribution used in Chapter 4. It was fitted by Bartholomew (1959) to the data used by Silcock and some of the results are given in Table 7.1. There is very little to choose between the mixed exponential distribution and the type XI in these examples, in spite of the radical difference in the form of $H(\lambda)$ in the two cases.

It cannot be concluded from the foregoing analysis that the 'mixture' models proposed by Silcock (1954) provide a true explanation of the leaving process. There are other models which give equally good agreement with the data and at least one of them gives the same functional form for $f(T)$. If the model does contain any degree of truth then there are obvious implications for the personnel manager. Since, according to the model, there are innate differences between individuals, the way to influence the leaving rate is by screening the intake. If we could identify people with small or big λ's wastage could be decreased or increased by appropriate selection. The model therefore suggests that we ought to seek ways of discriminating between entrants according to their propensity to leave. This is a problem for the psychologist, but the success of his methods can certainly be put to a statistical test.

The two-term mixed exponential can clearly be extended to three and more terms, but the number of parameters increases rapidly and the effort is hardly worth while unless large amounts of data are available. In the next subsection and again in Chapter 8 we shall meet CLS distributions having the form

$$f(T) = \sum_{i=1}^{k} p_i \lambda_i e^{-\lambda_i T} \qquad \left(T \ge 0; \sum_{i=1}^{k} p_i = 1 \right). \qquad (7.9)$$

If the p_i's are non-negative this belongs to the family (7.3), but this need not necessarily be the case. When these models do lead to a CLS distribution with positive p_i's there is no statistical means of discriminating between mixture and other models using data on length of service alone.

Network models

An interesting class of models for the leaving process can be developed using a Markov process to represent movements between internal states (of mind) through which an individual passes before leaving. Herbst (1963) pioneered this approach, using what he called a decision process model. Similar models, though with important differences of interpretation, have since been discussed by Hoem (1971) in continuous time and by Young (1971) and Feichtinger (1971) in discrete time. A very similar model was discussed in relation to survival after cancer in Chapter 5, but there the interest was not directly in the survival time. The distinctive thing about Herbst's model is that the intermediate states are latent in the sense that they cannot be directly observed. Herbst gives reasons for believing that the states in his model do correspond to what happens in practice, but their existence can at best only be inferred indirectly from the shape of the observed CLS distribution.

The model is most easily explained by reference to Figure 7.1. Each individual is supposed to enter the system in the undecided state and then to pursue a path through the network until he arrives at one or other of the terminal states S_4 and

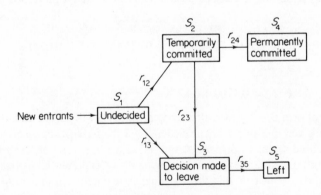

Figure 7.1   The states of the Markov process model for the
leaving process.

S_5. All that we can observe in practice is the time taken to reach S_5; all else about the system must be inferred from the form of this CLS distribution. Movements through the network are governed by a set of transition rates denoted by r's in the diagram. The system is thus being modelled by a continuous time Markov process of the kind discussed in Chapter 4 with transition matrix

$$\mathbf{R} = \begin{pmatrix} r_{11} & r_{12} & r_{13} & 0 & 0 \\ 0 & r_{22} & r_{23} & r_{24} & 0 \\ 0 & 0 & r_{33} & 0 & r_{35} \\ 0 & 0 & 0 & 0 & 0 \\ 0 & 0 & 0 & 0 & 0 \end{pmatrix} \tag{7.10}$$

where $r_{11} = -r_{12} - r_{13}$, $r_{22} = -r_{23} - r_{24}$, $r_{33} = -r_{35}$. Because the matrix is triangular its eigenvalues are equal to its diagonal elements. The $\bar{n}_i(T)$'s of (4.16) may be interpreted as probabilities of being in i at time T if there is one person in 1 to start with. The survivor function is then given by

$$1 - \bar{n}_5(T) = G(T) = d_{51} + c_{53}e^{r_{11}T} + c_{54}e^{r_{22}T} + c_{55}e^{r_{33}T}. \tag{7.11}$$

Herbst was interested in discovering whether the model provides an adequate description of organizational commitment. He therefore compared (7.11) with length of service distributions from two firms collected by Hedberg (1961). Firm A had a high loss of entrants and firm B a low loss. There are seven parameters to be estimated and this would be overambitious unless the data were very extensive. Fortunately, Herbst had very large samples and the parameters were estimated by equating percentage points. The following estimates were obtained:

Firm A

$$G(T) = 0.1493 + 0.4544\,e^{-0.4270T} + 0.4194\,e^{-0.09667} - 0.0231\,e^{-10.15T} \tag{7.12}$$

Firm B

$$G(T) = 0.4700 + 0.1274\,e^{-0.3680T} + 0.4025\,e^{-0.1142T} \tag{7.13}$$

In the case of firm B, $-r_{33}$ was effectively infinite. The observed and fitted values of $G(T)$ are given in Table 7.2. The closeness of the fit with such large samples is remarkable and seems to provide strong evidence in favour of Herbst's theory. The distributions could have arisen as mixtures of exponential distributions, as described above. (It is true that there is a negative coefficient in firm A's survivor function but its contribution is negligible.) Notice that the first term is a constant; this can be thought of as proportional to the survivor function of a degenerate exponential distribution with infinite mean. It relates to the people who eventually reach the permanently committed category.

The statistical problem of estimating the parameters efficiently clearly requires further investigation. Sverdrup (1965) and Hoem (1971) have considered the problem when data are available on individual transitions. In the case of the present example the fit is so good that more efficient procedures would be superfluous.

Table 7.2   Observed and estimated values of $100\{1 - \bar{n}_5(T)\}$
for Hedberg's data

Month	Firm A 7628 entrants		Firm B 968 entrants	
	Actual	Theoretical	Actual	Theoretical
0	100.00	100.00	100.0	100.0
1	82.66	82.66	91.7	91.7
2	68.85	68.85	85.1	85.1
3	58.95	58.94	79.8	79.8
4	52.10	51.66	75.6	75.4
5	46.61	46.17	71.7	71.8
6	42.32	41.92	67.7	68.7
7	38.73	38.53	64.9	66.1
8	35.74	35.77	63.0	63.8
9	33.52	33.46	61.3	61.9
10	31.78	31.56	59.5	60.2
11	30.19	29.81	58.5	58.7
12	28.46	28.33	57.2	57.4
13	27.20	27.03	55.9	56.2
14	25.85	25.87	55.4	55.2
15	24.79	24.83	54.8	54.3
18	22.25	22.30	53.1	52.2
21	20.36	20.43	51.4	50.7
24	19.06	19.04	49.7	49.6
27	18.01	18.00	48.8	48.9
30	17.31	17.23	48.2	48.3
36	16.45	16.22	47.7	47.7
42	15.77	15.65	47.4	47.3
48	15.35	15.33	47.2	47.2
54	15.18	15.16	47.1	47.1
60	15.05	15.06	—	—
66	15.00	15.00	—	—

The method used above to derive the CLS distribution is perfectly general and works with an **R** matrix of any size or structure.

So far we have concentrated on models in which there is only one exit from the system. In many applications there are several, as, for example, when leavers are classified by reason for leaving. In actuarial terminology this would be described as a multiple decrement problem. If we are interested only in the total time in the system, regardless of the reason for leaving, then all that is required is to give all the terminal states the same label. On the other hand, if we can identify leavers by their reason for leaving we may be interested in the length of service distributions conditional on each reason.

To illustrate the method, suppose that the state 4 in Herbst's model had been 'declared redundant' instead of 'permanently committed'. Then we might have been interested in the two conditional length of service distributions for redundant and voluntary leavers. For the voluntary leavers a simple conditional probability argument gives the survivor function

$$G_L(T) = \left\{G(T) - \frac{r_{12}r_{24}}{r_{11}r_{22}}\right\} \bigg/ \left(1 - \frac{r_{12}r_{24}}{r_{11}r_{22}}\right) \qquad (7.14)$$

where $(1 - r_{12}r_{24}/r_{11}r_{22})$ is the probability of ultimate loss in state 5.

Random walk models

Another type of model arises by considering the time taken by a random walk to reach an absorbing barrier. The idea was first put forward by Lancaster (1972) as the basis of a model for the duration of a strike. We shall base our exposition on this application, referring to other uses of the model as we proceed.

Strikes arise as an action of last resort when the two parties to a dispute cannot reach agreement. Let us visualize them at the outset of a strike as separated by a 'distance' d. It is natural to think of this distance as a sum of money if the dispute is over pay, but the issues will usually be more complicated. As the strike progresses the gap between the parties changes as new factors enter the situation and as the hardship and inconvenience caused by the strike makes itself felt. Lancaster's model is a stochastic description of the way in which the gap is closed.

The progress of the negotiations can be represented graphically as shown in Figure 7.2. Points on the graph represent intermediate stages in the negotiations. When the path reaches the horizontal line at d the gap is closed and the duration

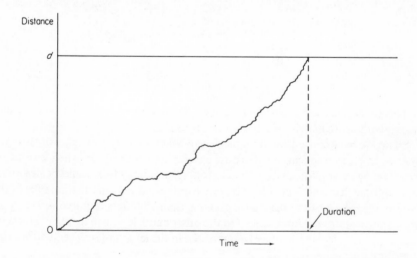

Figure 7.2   Illustration of how the duration of a random walk is related to absorption at a boundary.

of the strike is then the horizontal distance from the origin to the point of absorption. Lancaster supposed that the path of the negotiations was a Wiener process. That is, in moving from T to $T + \delta T$ the path moves up or down by an amount which is normally distributed with mean $\mu \delta T$ and variance $\sigma^2 \delta T$ independently of previous increments. The properties of this process are well known; see, for example, Cox and Miller (1965). Absorption is only certain if $\mu \geq 0$, that is, if the process has a non-negative 'drift'. The density function of the time to absorption then has the density function

$$f(T) = \frac{d}{\sqrt{2\pi}\sigma T^{\frac{3}{2}}} \exp\left[-\frac{1}{2}\left(\frac{d - T\mu}{\sigma \sqrt{T}} \right)^2 \right] \qquad (d, \sigma, \mu > 0). \qquad (7.15)$$

This is known as the inverse Gaussian distribution; its properties are given in Chhikara and Folks (1978). If $\mu < 0$ absorption is not certain and there is thus a

Table 7.3   The fit of the inverse Gaussian distribution to the duration of strikes beginning in 1965 in two sectors of British industry (Lancaster, 1972)

Duration (days)	All stoppages in vehicles and cycles		All stoppages in metal manufacturing	
	Observed	Fitted	Observed	Fitted
2	34	34.0	43	47.1
3	19	18.2	37	30.8
4	10	11.3	21	21.0
5	8	7.7	19	15.3
6	6	5.6	11	11.6
7	5	4.2	8	9.2
8	2	3.3	8	7.4
9	3	2.6	9	6.1
10	2	2.1	3	5.1
11–15	6	6.5	16	16.7
16–20	4	3.1	4	9.1
21–25	4[a]	4.3[a]	4	5.6
26–30	—	—	3	3.7
31–40	—	—	3	4.3
41–50	—	—	5	2.3
> 50	—	—	5	3.7
Total	103	102.9	199	199.0

[a]These frequencies relate to all strikes lasting longer than 20 days. The parameters were estimated by maximum likelihood. The distribution was truncated by the omission of strikes of one day's duration (which are not recorded); it is the truncated distribution which is fitted here (see Lancaster, 1972, for details).

non-zero probability equal to $1 - \exp(2d/\sigma^2)$ that T is infinite. The distribution is then described as defective. Although three parameters appear in (7.15) the inverse Gaussian is essentially a two-parameter distribution because its density can be expressed in terms of d/σ and μ/σ. Lancaster (1972) fitted the distribution to British strike data with considerable success as the illustration in Table 7.3 shows.

Eaton and Whitmore (1977) proposed the same model for the stay of schizophrenic patients in a mental hospital though van Korff (1979) has since shown that the two-term exponential distribution gives a somewhat better fit to their data. In this application d represents the amount of 'healing' that the patient requires before being fit for discharge. According to the model, progress towards that end will fluctuate in the random manner of a Wiener process. A negative drift might be expected for diseases which do not always respond to treatment and end in death or permanent hospitalization. The model can also be given an interpretation in the application to labour wastage. Here one can imagine the accumulation of dislikes which eventually reach a critical level at which the person leaves. Whitmore (1979) spoke of the erosion of 'attachment' to the organization —which comes to the same thing. He fitted the distribution to the data given by Silcock (1954), including the Bibby distribution given in Table 7.1. The fit was excellent, being rather better than the various mixed exponential models.

The Wiener process is only one of many random walks that could be used in continuous or discrete time. As Whitmore (1979) has shown, mixture models can be generated by treating the parameters as random variables.

7.3 THE LOGNORMAL MODEL FOR SIZE AND DURATION

If the logarithm of a positive random variable, X, has a normal distribution, then X is said to have a lognormal distribution. The distribution is positively skewed to a degree that depends on the standard deviation of $\log X$. The lognormal distribution has been successfully fitted to many kinds of duration and size distribution. Lognormality offers substantial statistical advantages since, by working with the logarithm of the variable, the whole range of 'normal theory' methods becomes available. It is, therefore, of great practical interest to discover what kinds of process lead to the lognormal distribution. A full account of the distribution including both its properties and some applications is given in Aitchison and Brown (1957).

Lane and Andrew (1955) appear to have been the first to discover that CLS distributions can be successfully graduated by lognormal distributions. Since then their findings have been confirmed many times at all levels of skill and in many countries. The distribution has the useful feature of a non-zero mode, which, in many applications, gives it an advantage over mixed exponential distributions. Although when the data are coarsely grouped the mode may not be evident, finer groupings often reveal one and it is more satisfactory to have a model which shares this feature. Size distributions have also been found to be close to the lognormal in form. Aitchison and Brown (1957) discuss the use of the distribution to graduate income distributions and they give a number of examples. Steindhl (1965) gives examples from economics among which are the distribution of manufacturing plants according to number of employees, and firms by turnover. In the physical sciences the lognormal is well established as a breakage distribution giving the size distribution of particles after repeated breakage. Although this is of no direct social interest, the mechanism involved provides a useful analogy for certain kinds of social size distribution.

As already noted the ubiquity of the normal distribution derives from the operation of the central limit theorem on sums of random quantities. If a variable is lognormal, one might, therefore, ask whether it could be regarded as a product of random variables since then its logarithm would be a sum. This leads us to consider whether such an interpretation is possible in the various fields of application which we have outlined above. Since all the models we shall mention have the same basic structure it is convenient to begin by setting out the essentials in abstract terms.

Consider a sequence of random variables $T_1, T_2, T_3, \ldots$ generated as follows. T_1 is given and subsequent members of the series are formed by taking

$$T_{j+1} = T_j u_{j+1}, \qquad j = 1, 2, 3, \ldots \tag{7.16}$$

where $u_2, u_3, \ldots$ are random variables with a known joint distribution which does not depend on any of the T's. Another way of expressing (7.16) is to say that the proportionate change in T between j and $j+1$, $(T_{j+1} - T_j)/T_j$, has a distribution which does not depend on the level of T at j. It follows from (7.16) that

$$T_{j+1} = T_1 \prod_{i=2}^{j+1} u_i, \qquad j \geq 1$$

or

$$\log T_{j+1} = \log T_1 + \sum_{i=2}^{j+1} \log u_i. \tag{7.17}$$

Provided that the joint distribution of the u's is such that the central limit theorem applies to their sum then, as j increases, $\log T_{j+1}$ will tend to normality and hence T_{j+1} to lognormality. This is certainly true if the u's are independently and identically distributed, but a modest degree of dependence and inequality of variance can be admitted without destroying the asymptotic normality.

We must now consider how this general model can be interpreted in the various applications. We shall not claim that the model is fully realistic in any case, but rather that it indicates what features of a social process tend to induce lognormality. The first model is suggested by the analogy with repeated breaking of materials. Suppose that a large landowner divides his estate among his heirs. In due course they divide their inheritance among their heirs—and so on through succeeding generations. What will be the distribution of the size of the holdings after many generations? The average holdings will, of course, get smaller, but it is with the shape of the distribution that we are concerned. If T_j is interpreted as the random variable of size of holding in the jth generation then u_{j+1} will be the fraction of his father's estate which descends to a randomly chosen heir of the next generation. If the proportions into which each estate is divided are, to some extent, random and if there is the requisite degree of independence between generations then the reasoning given above would lead us to expect approximate lognormality after many generations. Similar arguments would apply to the repeated subdivision of any other quantity. For example, a society subject to repeated secession might ultimately result in a collection of small communities whose size distribution was roughly lognormal. As pointed out by Aitchison and Brown (1957), the ideas carry over into classification theory where categories are formed by repeated subdivision. For example, workers might first be divided by skill level, then sex, whether manual or white collar, and so on. The distribution of category size which results is likely to be approximately lognormal.

Another way of interpreting the model which is sometimes used to explain wealth distributions is as follows. A person's initial wealth T_1, is subject to change over intervals of time bounded by the points $j = 1, 2, 3, \ldots$. During any interval wealth will change by varying amounts. It is likely that the bigger the starting value the bigger the change. If we suppose that the proportionate change has the same distribution, whatever the size of an individual's wealth and that individual changes are independent, the general model will apply with the u's being independently and identically distributed. In this case the model has one property which is often held to be a disadvantage. It follows from (7.17) that the variance of $\log T_{j+1}$ is proportional to $(j-1)$ which means that wealth distributions should show increasing variability through time. Aitchison and Brown (1957) refer to some empirical evidence for this but, in the main, this does not seem to be the case. In practice other factors will operate, especially at the extremes of the range to counteract the increasing dispersion. This may justify Steinhdl's (1965, p. 11) observation that the fit of the lognormal is often good in the middle of the range but poor at the extremes. Similar arguments would apply to city sizes to the extent that growth, caused by excess of births over deaths (or vice versa) would produce a change in size proportional to present size.

A third interpretation can be given to the basic model by supposing that the changes do not occur in sequence but more or less simultaneously. According to this view a person's income, for example, is determined by a large number of independent factors which act multiplicatively. Thus a price rise in a commodity would raise the income of all producers in the same proportion; a tax levied at a fixed rate would reduce them all in the same proportion. If a large number of factors act in this way the conditions for lognormality might be met.

The interpretation of the lognormal model for the case of duration of employment can be made in a similar fashion though more speculatively. Consider a sequence of jobs held by an individual and let T_j denote time spent in the jth job. The linear equation (7.16) would then be a simple description of how current experience determines future behaviour. It implies that successive lengths of stay are positively correlated so that staying a long time in one job predisposes one to stay a long time in the next and vice versa. This would explain lognormality after many jobs, but it offers no explanation for the observed lognormality of the stay of people in their first job. To deal with this case we must follow the third interpretation for size distributions given above. We now suppose that T_1 is the time that a person expects to stay when he joins. During the early stages of the job he will be subject to the influence of many factors. Some of these will be favourable and incline him to stay longer than he originally expected. Others will be unfavourable and will tend to precipitate leaving. In this way the employee's original expectation will be modified by random perturbations. If we assume that the effect of each factor is to multiply the current expectation of service by a random factor, u, then T_j can be interpreted as the actual length of service for a person subject to j factors. The same stochastic assumption about the u's then ensures the approximate lognormality of the actual length of service.

According to this version of the model, individual differences will still be reflected in T_1, but this is now something which is no longer open to modification in the job since the individual brings it with him. Influence may be brought to bear by trying to change the mean values of the u's in such a way as to increase (or decrease, if necessary) the actual length of service.

In Section 7.2 we gave reasons for introducing into the model variability between individuals. The same argument might be made for many size distributions. To do this here suppose that $\ln T \sim N(\omega, \sigma^2)$. In the wastage application, at least, there are empirical grounds for believing that heterogeneity is likely to produce variation in ω. If we let this have distribution function $H(\omega)$ the mixed lognormal density becomes

$$f(T) = \int_{-\infty}^{+\infty} \frac{1}{\sqrt{2\pi}\sigma T} \exp\left[-\frac{1}{2}\left(\frac{\ln T - \omega}{\sigma} \right)^2 \right] dH(\omega). \qquad (7.18)$$

Note the range of ω over the whole real line. If we suppose that $\omega \sim N(\mu, \tau^2)$ then it is easy to show that

$$\ln T \sim N(\mu, \sigma^2 + \tau^2). \qquad (7.19)$$

In other words, the distribution has the same form but will be more skew than the original because of the larger variance of $\ln T$. Thus, provided that the mixing distribution is reasonably normal (and many human variables do vary in this way), the lognormal form will survive heterogeneity in the scale parameter.

Taken together, the arguments and models set out in this section provide reasons for expecting the lognormal distribution to arise and persist. They do not provide a fully satisfying explanation. For example, even if the first version of the model were correct it offers no explanation of why the relation between T_{j+1} and T_j should have the simple linear form postulated by (7.16).

7.4 MODELS FOR SIZE

Pareto is usually credited with the discovery of the fact that the logarithm of the proportion of people with incomes in excess of x, say, is often approximately proportional to $-\log x$ when x is large. Distributions with this property are often called Pareto distributions and the constant of proportionality is known as the Pareto parameter. From an empirical point of view this makes it very easy to compare income distributions by graphical means. Zipf (1949) found that the same feature occurred with other size distributions including that of city size in the United States and most other Western countries. In this context it has become known as Zipf's law. Zipf, himself, attempted to show that the law was a consequence of what he called the principle of least effort which he supposed lay at the root of all human behaviour. Others, starting with Simon (1955), have tried to find stochastic models of growth which lead to steady-state distributions having the Pareto property. It turns out that a large class of processes lead to distributions having the Pareto form, and in this section we shall describe some of the simplest. Such models describe, in probabilistic terms, how size changes over time. Time and size may both be treated as continuous or discrete. Throughout the section we shall regard size as a discrete variable taking the values 1, 2, 3, ... Time will be treated as continuous in the first two models and as discrete in the third. Simon (1955) stated the law in slightly more general terms than given above to the effect that for many size distributions the upper tail of the probability function has the form

$$p_i \propto i^{-\alpha}\beta^i, \tag{7.20}$$

where β is so close to unity that β^i only has a significant effect on p_i for very large values of i and $\alpha > 0$.

Shorrocks' model

This model is based on a continuous time Markov process and is equivalent to the birth and death process with immigration discussed in most texts on stochastic processes. Since there is no upper limit to size (in the theory, at least) the process has a countable infinity of states unlike the finite processes we have considered hitherto. This poses the problem of whether a steady state exists since the location and spread of the distribution could increase indefinitely. We shall treat these matters heuristically, but for full justification of the steps involved reference must be made to the basic texts mentioned in the Complements to Chapter 5. When a Markov process is used as a model for growth $r_{ij}\delta T$ is interpreted as the probability that an individual moves from size category i to size category j in $(T, T + \delta T)$. The state probabilities $p_i(T)$ then give the size distribution at time T. Our interest will be in the limiting form of this distribution.

Shorrocks (1975) proposed such a model for the distribution of wealth with

$$\left.\begin{array}{ll} r_{i,i+1} = (v + \lambda i), & i = 1, 2, \ldots \\ r_{i,i-1} = \mu i & i = 2, 3, \ldots \\ \quad r_{ij} = 0 & \text{otherwise when } i \neq j \end{array}\right\}. \qquad (7.21)$$

As defined, this model relates to a closed population, but open populations can be accommodated by supposing that anyone who leaves is replaced by a new person who takes over the wealth of the leaver. The state probabilities for such a process satisfy (see (4.4))

$$\frac{\mathrm{d}p_i(T)}{\mathrm{d}T} = -\{v + (\lambda + \mu)i\}p_i(T) + \{v + \lambda(i-1)\}p_{i-1}(T) + \mu(i+1)p_{i+1}(T) \quad i > 1$$

$$\frac{\mathrm{d}p_i(T)}{\mathrm{d}T} = -vp_1(T) + \mu p_2(T). \qquad (7.22)$$

If a steady-state distribution exists it can be found from (7.22) by setting the derivative equal to zero and suppressing the T. Solving recursively and using the fact that $\sum_{i=1}^{\infty} p_i = 1$ we find that if $\lambda < \mu$,

$$p_i = \frac{\Gamma(i + \rho - 1)}{\Gamma(\rho)\Gamma(i)}(1 - \beta)^\rho \beta^{i-1}, \qquad i = 1, 2, \ldots, \qquad (7.23)$$

where $\rho = v/\lambda$ and $\beta = \lambda/\mu$. The condition for the steady state to exist is $\lambda < \mu$ which implies that the 'birth-rate', λ, must be less than the death-rate. Equation (7.23) defines a negative binomial distribution whose behaviour when i is large can be deduced using the fact that

$$\Gamma(x)/\Gamma(x + a) \sim x^{-a} \quad \text{as } x \to \infty.$$

We thus find

$$p_i \sim \text{constant} \times i^{\rho - 1}\beta^i$$

which is of the form (7.20) with $\alpha = 1 - \rho$. For the distribution to agree with (7.20) the birth and death-rates must be nearly equal (giving $\beta = 1$) and $0 < \rho < 1$.

We must now consider the rationale for choosing the rates as in (7.21). The restriction of movement to adjacent classes implies that growth is continuous, which is a reasonable assumption in many applications. If we put $v = 0$, the form of the transition rates is consistent with the law of 'proportionate effect' since the larger the present wealth, say, the more likely is an increase. This, in turn, implies that the mean change for those in class i in $(T, T + \delta T)$ is proportional to i. This version of the model seems especially appropriate for city sizes because, in the absence of migration one would expect growth (positive or negative) to be proportional to size. With birth and death rates very nearly equal, Zipf's law with $\alpha = 1$ then emerges. If migration from outside the system is allowed, $v > 0$ and the Pareto parameter is then less than one. In the wealth application a positive v

implies that wealth accrues from some sources regardless of present wealth. The model also allows us to have $\lambda = 0$, meaning that growth is unrelated to present size. In that case the solution of (7.22) yields a Poisson distribution with

$$p_i = \frac{\rho^{i-1}}{(i-1)!} e^{-\rho} \qquad i = 1, 2, \ldots. \tag{7.24}$$

The foregoing analysis involves the homogeneity assumption that all members of the population are subject to the same law of growth. Previous experience suggests that we ought to allow for individual differences. This might be done by treating one or more of the parameters as random variables. However, there is a lack of identifiability already built in to the model as we now show. Let us suppose that the law of change embodied in (7.21) applies to any given individual but with $\lambda = 0$ and that it varies continuously with a gamma distribution with density

$$f(\rho) = c^\gamma \rho^{\gamma-1} e^{-c\rho}/\Gamma(\gamma), \qquad \rho \geq 0$$

then

$$\begin{aligned} p_i &= c^\gamma \int_0^\infty \rho^{i+\gamma-2} e^{-\rho(1+c)} \, d\rho/\Gamma(\gamma)(i-1)! \\ &= \frac{\Gamma(\gamma+i-1)}{\Gamma(\gamma)\Gamma(i)} \left(\frac{c}{1+c}\right)^\rho \left(\frac{1}{1+c}\right)^{i-1} \qquad i = 1, 2, \ldots \end{aligned} \tag{7.25}$$

which has the same form as (7.23). The occurrence of a negative binomial distribution could therefore result either from a homogeneous system with $\lambda > 0$ or from a mixture of systems with $\lambda = 0$.

Although the negative binomial does belong to the family of distributions satisfying (7.20) the value of its Pareto parameter cannot exceed one. In practice distributions occur with larger values, commonly in the range 1 to 2. We therefore go on to consider a second model which is not subject to this limitation.

The Yule–Simon model

This is also a Markov model but with rates

$$r_{i,i+1} = (v + \lambda i) \quad (i = 1, 2, \ldots), \qquad r_{i,1} = \mu, \quad (i = 2, 3, \ldots) \tag{7.26}$$

$$r_{ij} = 0 \quad \text{otherwise when } i \neq j.$$

In its growth aspect this model is identical with that of (7.21). However, it is designed to describe an open system of fixed size in which losses are replaced by new entrants at the lowest level. The choice of a constant loss rate, μ, implies the same risk of loss at all levels. It is clear that λ must be non-negative and so we have a system in which members grow at a rate proportional to their size and are subject to a constant risk of removal. Simon (1955) proposed a discrete time version of the model and pointed out that Yule had also derived it to describe the distribution of the number of genera of plants having i species.

It is not immediately clear how this model could account for city sizes since although growth might well be approximated by a linear growth rate, cities do not vanish. Simon (1955) did, however, give an alternative interpretation of the model which does have more plausibility in this context. He showed that the same distribution would arise as a quasi-steady state in a growing population of cities fed by a flow of new cities reaching some minimum size.

The steady-state equations for the model of (7.26) are easily shown to be

$$- \{v + i\lambda + \mu\} p_i + \{v + (i-1)\lambda\} p_{i-1} = 0, \qquad i = 2, 3, \ldots$$
$$p_1 = \mu/(v + \lambda + \mu).$$

Thus

$$p_i = \frac{\prod_{j=1}^{i-1} (\rho + j)}{\prod_{j=1}^{i} (\rho + \eta + j)} \eta, \qquad i = 2, 3, \ldots$$

$$p_1 = \eta/(1 + \rho + \eta), \qquad (7.27)$$

where $\eta = \mu/\lambda$. Simon (1955) considered only the special case $v = \rho = 0$ for which

$$p_i = \eta B(i, \eta + 1), \qquad i = 1, 2, \ldots \qquad (7.28)$$

For large i we may express (7.27) in terms of gamma functions and use the same approximation as before. By this means we find

$$p_i \sim \text{constant} \times i^{-(1+\eta)} \qquad (7.29)$$

so that $1 + \eta$ is the Pareto parameter which can now take any value greater than one.

In addition to the applications already mentioned Simon (1955) fitted the distribution to word frequencies and to the distribution of the number of contributions made by authors to various learned journals over a period of years. The result for papers and abstracts in *Econometrica* over a 20-year period are given in Table 7.4. The model is clearly too simplistic to describe the situation fully, but it is remarkable how much has been captured by it. The estimated value of η is 0.69 which implies that an author is 69 per cent as likely to cease activity as to produce a new paper.

As with Shorrock's model, the distribution in (7.27) can also be derived as a mixture. As before, we first set $\lambda = 0$ so that growth no longer depends on size. In that case (7.27) reduces to the geometric distribution

$$p_i = \left(\frac{v}{v+\mu}\right)^{i-1} \left(1 - \frac{v}{v+\mu}\right) \qquad i = 1, 2, \ldots \qquad (7.30)$$

Now let $x = v/(v+\mu)$ have a beta distribution with density function

$$f(x) \propto x^{\rho-1}(1-x)^{\eta-1}$$

Table 7.4   Numbers of authors contributing i papers to Econometrica over a 20-year period[a]

i	1	2	3	4	5	6	7	8	9	10	11 or more	Total
Actual frequency	436	107	61	40	14	23	6	11	1	0	22	721
Fitted frequency using (7.28)	453	119	51	27	16	11	7	5	4	3	25	721

[a] From Simon (1955).

then

$$p_i = \frac{B(i + \rho - 1, \eta + 1)}{B(\rho, \eta)} \qquad i = 1, 2, \ldots \tag{7.31}$$

which is another way of writing (7.27). This leaves us in the same dilemma as before. The good fit of Table 7.4 could mean that the more papers an author has contributed the more likely he is to contribute another; or it could mean that authors have different propensities to contribute without past record having any influence on present performance.

The Markov process obviously provides a framework within which numerous other models of this kind can be generated. We shall not pursue this course (though the reader will benefit by doing so). Instead we shall do essentially the same thing but in discrete time.

Discrete time models for size

In principle the only change to be made in passing to discrete time is to replace the transition rates by transition probabilities. However, there are complications which need consideration. If we wish to retain the law of proportionate effect we can no longer limit transitions to adjacent classes. If we did, the only changes possible would represent different proportions of the current size and so the same proportionate change could not be given the same probability. In any case we might wish to allow for several changes if the discrete time interval is of reasonable length. Even allowing for transitions between non-adjacent categories, the construction of a transition matrix which gives the same proportionate change of size the same probability is impossible with a small number of parameters. However, these difficulties can be circumvented by regarding size as a continuous variable with the states of the chain formed by grouping. If we form the groups so that the boundaries (and, hence, the mid-points) form a geometric progression then a move from group i to $i + h$ involves the same proportionate change of size (on average) for all i. Thus the whole transition matrix can be specified in terms of the elements of a single row. Modifications have to be made at the lower end of the range because, for example, no move down is possible from the lowest group. In practice, the number of groups will be finite with all individuals above a certain level in the highest group. However, for the theory, we shall suppose there to be an unlimited number of groups. In spite of 'end effects' it will be possible to see by inspection of an empirical transition matrix whether it is plausible to assume the law of proportionate effect. The following estimated matrix for an income distribution, quoted by Shorrocks (1976), does conform broadly to the pattern required.

$$\begin{pmatrix} 0.64 & 0.29 & 0.04 & 0.03 & 0.00 \\ 0.14 & 0.56 & 0.26 & 0.03 & 0.01 \\ 0.02 & 0.22 & 0.54 & 0.21 & 0.01 \\ 0.01 & 0.04 & 0.27 & 0.54 & 0.14 \\ 0.00 & 0.01 & 0.05 & 0.27 & 0.67 \end{pmatrix}$$

Champernowne (1953 and 1973) proposed a number of models for incomes embodying the law of proportionate effect within the discrete time Markov framework. According to the simplest of these an individual can move up one class, stay where he is, or move down up to r classes. If such a downward move would take him beyond the lowest class he is assumed to move into the lowest class. Since the transition probabilities do not depend on the current state they may be written as

$$\left.\begin{aligned} i > r \quad & p_{ij} = p_{-(i-j)}, \quad j = i-r, i-r+1, \ldots, i+1 \\ & = 0 \text{ otherwise} \\ i \le r \quad & p_{ij} = p_{-(i-j)}, \quad j = 2, 3, \ldots, i+1 \\ & p_{i1} = \sum_{h=i-1}^{r} p_{-h}, \quad p_{ij} = 0 \text{ otherwise} \end{aligned}\right\} . \qquad (7.32)$$

If a steady-state structure $\mathbf{q}$, say, exists it will satisfy $\mathbf{q} = \mathbf{qP}$ where the elements of $\mathbf{P}$ are given by (7.32). Written out in full, these equations are

$$q_i = \sum_{j=i-1}^{i+r} q_j p_{i-j} \qquad i = 2, 3, \ldots \qquad (7.33)$$

The first equation, with $i = 1$, can be ignored for the moment. Let us consider a possible solution of the form $q_i = Ax^i$. If such a solution exists substitution in (7.33) shows that it must satisfy

$$x = p_1 + xp_0 + x^2 p_{-1} + \ldots + x^{r+1} p_{-r}$$
$$= \phi(x) \text{ say.} \qquad (7.34)$$

For $\{q_i\}$ to be a probability distribution x must satisfy $0 < x < 1$. Whether or not there are any eligible roots can be investigated graphically by considering the intersections of $y = x$ and $y = \phi(x)$. Here, $\phi(x)$ increases monotonically (since its coefficients are all non-negative) from $y = p_1$ at $x = 0$. It intersects with $y = x$ at $x = 1$ *because* $\sum_{h=-r}^{1} p_h = 1$. For there to be a root in $(0, 1)$ the slope of $\phi(x)$ at $x = 1$ must clearly be strictly greater than 1. That is,

$$p_0 + 2p_{-1} + \ldots + (r+1)p_{-r} = 1 + p_{-1} + 2p_{-2} + \ldots + rp_{-r} = 1 + \mu > 1$$

where μ will be recognized as the mean number of steps taken to the left at any stage. The condition for the process to approach an equilibrium is thus that $\mu \geq 0$. If $\mu < 0$ the equilibrium distribution will be geometric since there is at most one root in $(0, 1)$, that is,

$$q_i = (1-x)x^{i-1} \qquad i = 1, 2, \ldots, \tag{7.35}$$

where x is the root of $\phi(x) = 0$ in $(0, 1)$. We must remember that i is measured on a logarithmic scale. The lower boundary of the ith group will be of the form a^{i-1} on an arithmetic scale and so if S denotes size the upper tail probability has the form

$$Pr\{S \geq S_0\} \propto S_0^{-\alpha}; \tag{7.36}$$

the distribution thus has the exact Pareto form for all S and not merely for large S. As Champernowne (1973) points out, the assumptions of the model are grossly oversimplified when applied to incomes, but he is able to show that the Pareto form of the upper tail is preserved under various generalizations incorporating the law of proportionate effect.

The same geometric form arises from a discrete version of the Yule–Simon model. For this we have

$$\begin{aligned} p_{11} &= 1 - p_1, & p_{ii} &= p_0, & i &> 1, \\ p_{i,i+1} &= p_1, & p_{i1} &= 1 - p_0 - p_1, & \text{all } i. \end{aligned} \tag{7.37}$$

In this case the equilibrium distribution always exists and is

$$q_i = \frac{(1-p_0-p_1)}{(1-p_1)}\left(\frac{p_0}{1-p_1}\right)^{i-1}, \qquad i = 1, 2, \ldots \tag{7.38}$$

One common modification which can be made to both of the above models is to introduce a lower limit on size—a minimum income or threshold size which has to be exceeded before the individual enters the population. This does not affect the behaviour of the distribution in the region of the upper tail.

It is worth asking why the operation of a law of proportionate effect does not produce a lognormal distribution for size with these models as the argument of Section 7.3 might suggest that it should. In the Yule–Simon model the distribution of proportionate income change clearly depends on i and the conditions for the central limit theorem to apply are not met, likewise with Champernowne's model.

Although the number and range of models for size that we have discussed here is limited we have taken the discussion far enough to show that the widespread occurrence of positively skew distributions of lognormal or Pareto form is in no way surprising and that it is intimately bound up with the operation of the law of proportionate effect.

7.5 COMPLEMENTS

The theory on which the models of this chapter depend is largely that of Markov processes to which references have been given in the Complements sections of Chapters 3 and 5. The extensions used here are to countably infinite state spaces and, in the case of the Wiener process, to continuous state spaces; both are covered in the standard treatments. Conditions for linear functions of non-identically distributed random variables to be asymptotically normal will be found in advanced treatments of probability theory, for example Feller (1966, Theorem 3, p. 256). The standard treatment of the lognormal distribution is Aitchison and Brown (1957). The lognormal is sometimes known as the Gibrat distribution (or law) after the economist who proposed it for graduating distributions of income and wealth.

We have reviewed a wide range of models for the leaving process without exhausting all the possibilities, even of the models we have formulated. In spite of this, some of the commonest distributions used in reliability theory and life-testing have found no place in our discussion. Among these are the gamma distribution and the various extreme value distributions—in particular the Weibull distribution. These distributions are capable of producing the long upper tail required only for very small values of their shape parameters. We have already noted that a J-shaped gamma distribution can arise as a mixture of exponentials, but the unimodal gamma distribution has a monotonic *increasing* failure rate.

Weibull distributions have been successfully fitted to data on the distribution of wars and strikes in American industry by Horvath (1968). He proposes an extreme value model analogous to Weibull's original model for the strength of materials as follows. During the course of a conflict, whether a war or a strike, there are many obstacles to a settlement. The time taken to reach agreement on any one of these issues may be thought of as a random variable and the conflict is supposed to be ended as soon as the first obstacle is overcome. This is thus a 'weakest point' theory according to which one probes all the potential points of breakthrough until success is achieved at one of them. If there are many such points and if the distribution of the time to breakthrough on each is the same then the extreme value theory shows that the appropriate limiting form of the distribution of duration is the Weibull. The attraction of this model lies in its generality since it is independent of the actual form of the component distribution. However, its plausibility on other grounds may be questioned. In many bargaining situations it is not reasonable to suppose that the conflict will end when one of the obstacles to agreement has been overcome. It may be necessary to reach a settlement on several or, indeed, all of the points at issue. Even if the model is accepted as realistic up to this point it is doubtful whether the time to reach agreement on each issue would have the same distribution or even whether there would be enough issues to make the limit a good approximation. Nevertheless, the model does fit the data and it should clearly be entertained as a possible explanation.

A different kind of model for the duration of a war was proposed by Weiss (1963). His first model relates the chance of ending the conflict at a given time to the number of deaths at that time. In a second model he allows the chance of termination to be a function of both time and the number of deaths. These models are more general than those discussed in this chapter in that they make the loss intensity a function of something other than time (though, in this case, the number of deaths is a monotonic function of time). Such a generalization is clearly a step in the direction of realism, but the added complexity which it brings is a serious obstacle. It is difficult enough to distinguish among the simpler models of this chapter and it therefore seems unlikely that further elaboration will be useful until much more detailed data are available.

The two-term mixed exponential distribution has been fitted to distributions of the duration of unemployment where the two populations have been tentatively identified with the 'hard core' and 'transient' unemployed (see *IMS Monitor*, Vol. 2 (1973) pp. 29–32, unsigned). Continuous mixtures in the same field appear in Lancaster and Nickell (1980). We have already noted the use of a mixed exponential for the duration of stay in hospital by Van Korff (1979). In later chapters we shall meet further examples of distributions of duration. Chapter 8 deals with multistage renewal processes and we shall find there the distribution of stay in such a system and find that this, too, has the mixed exponential form.

The inverse Gaussian distribution has considerable statistical advantages and Whitmore (1979) notes that it is similar in shape to the lognormal if $d|\mu|/\sigma^2$ is 10 or more and if $\mu > 0$. The properties of the distribution and a full bibliography are set out in Chhikara and Folks (1978). Whitmore (1976) reviewed a number of applications which, apart from those covered here, include the distribution of the realization time for fixed security price changes. This is based on the fact that security prices in an efficient market follow a random walk (see Cootner, 1964). A brief exploration of discrete time random walk models with arbitrary distributed increments was made in the second edition of this book (Chapter 6, pp. 201–2).

The use of mixed exponential and lognormal curves for graduating CLS distributions is illustrated in Chapter 3 of Bartholomew and Forbes (1979). This includes (p. 75) reference to unpublished work by D. Cronin on the log-logistic distribution. The logistic distribution is very similar in shape to the normal and hence the same is true for the log-logistic and lognormal distributions. The log-logistic distribution can arise as a mixture of Weibull distributions which, in turn, can occur in the theory of extreme values. An interesting feature of the log-logistic distribution is evident from its survivor function which is

$$G(x) = \{1 + Ax^\alpha\}^{-1}$$
$$\sim 1/Ax^\alpha$$

if x is large, so the distribution has a Pareto tail.

We thus have the various mixed exponential models, the lognormal, inverse Gaussian, and log-logistic models which all lead to duration distributions of similar shape. This is an advantage if we are looking for a convenient distributional form for some statistical purpose since we can choose whichever is the most tractable for the object in view. If, on the other hand, we are seeking for an understanding of the social process it is clear that, on purely empirical grounds, it will be almost impossible to distinguish between the various models. There is thus a limit to what can be learnt from duration alone; other aspects of the process must be observed if further progress is to be made.

The basic references on size distributions are Champernowne (1973), Steindhl (1965), Ijiri and Simon (1977), and Zipf (1949) though the latter does not discuss stochastic models. Cliff and co-workers (1975, Chapter 2) consider several models, including Zipf's, for graduating size distributions in a geographical context. Kendall (1961) gave several further examples of the Yule–Simon distribution. Recent work on income and wealth distributions is contained in Lillard and Willis (1978) and Pestieau and Possen (1979).

An alternative approach to the derivation of Zipf's law was used by Hill (1970 and 1974), Hill and Woodroofe (1975), and Chen (1980). They start with the classical occupancy problem involving the allocation of balls to boxes. Wold and Whittle (1957) gave a further model, based on a birth and death process, for the distribution of wealth. Among other things they assume that the heirs to an estate divide the inheritance into equal parts. A useful review of the stochastic theory of industrial size distributions together with several numerical examples and many references was given by Collins (1973). Shorrocks (1975) emphasized the limitations of confining the analysis of the models to the steady state and, in his paper, he also considered the transient behaviour.

One of the recurring problems in stochastic modelling is that two or more models turn out to be empirically indistinguishable. This phenomenon has occurred several times in previous chapters and has been particularly acute with models for duration and size. Lancaster and Nickell (1980) give a lucid account of the difficulty in connection with durations but the problem is not a new one. It has long been recognized in the study of accident proneness where, by looking at the frequency distribution of number of accidents alone, one cannot determine whether the apparently increasing risk with number of accidents already incurred is real or simply due to individual variation in proneness. In a formal sense this is identical with the dilemma we noted in connection with Shorrock's and Simon's models for size distributions. The existence of competing models serves to show what kinds of data would be needed to resolve the ambiguity.

CHAPTER 8

Models for Social Systems with Fixed Class Sizes

8.1 INTRODUCTION

In this chapter we return to the study of the dynamics of social systems whose members move among a set of classes of some kind. Here, however, we assume that the sizes of those classes are given. This represents a complete change of viewpoint from that which we adopted for Markov models. There the flow probabilities were given and the class sizes were random variables. Here the class sizes are given for each time and the flow numbers are the random variables in which we are interested. We have already moved some way in this direction when, in Chapters 3 and 5, we considered Markov models with total size fixed. Now we go further and fix all the class sizes.

There are several fields of application which call for this kind of model. One is in manpower planning where the classes represent grades whose sizes are fixed by the budget or amount of work to be done at each level. Recruitment and promotion can only occur when vacancies arise through leaving or expansion. In studies of residential mobility a fixed stock of housing in different localities imposes constraints on movement of the kind which make a fixed size model more appropriate. White (1970a) has used models of this kind to study the flows of clergy of several large American denominations. Incumbencies were classified into groups on the basis of prestige or status among which individuals move as vacancies arise. White was not interested in the manpower planning aspects but in the sociological implications of different mobility patterns. More recently Stewman (1975a) has applied White's methods to the study of a state police force.

The stochastic element in such processes occurs principally in the loss mechanism; individuals leaving or moving create vacancies which generate sequences of internal moves. There may also be randomness in the method by which vacancies are filled. The branch of stochastic theory on which these models depend is called renewal (or replacement) theory. It appears first to have arisen in connection with the renewal of human populations through death and birth. More recently the main applications have been in the context of industrial replacement and reliability theory. There are obvious similarities between the renewal of components in such things as computers and aircraft and the replacement of employees or the occupation of houses. The key random variable in all cases is the length of time an entity remains active. In Chapter 7 we prepared the ground for this by discussing models for duration. In this chapter we shall

212

introduce some of these models into the more general framework of fixed size systems.

One might expect to develop the treatment of such models along lines parallel to those which we followed for Markov models. There we made use of the closed/open and discrete time/continuous time dichotomies. However, it is clear that a fixed class system is necessarily open otherwise there could be no vacancies. It is possible to develop the models in discrete or continuous time. For numerical work in manpower planning the former has many advantages and it was the course followed in Bartholomew and Forbes (1979). The continuous time treatment has the advantage here of arising naturally out of the analysis of Chapter 7 and of relating more readily to the standard treatments of renewal theory.

8.2 SYSTEMS CONSISTING OF A SINGLE CLASS

We begin with the simplest type of renewal system in which there is no internal differentiation into classes. Our aim will be to see how the inflow, outflow, and age distribution depend on the life distribution, which we assume to be given. In order to convey the basic ideas let us consider a highly simplified and rather unrealistic situation. Consider a new house whose first occupant remains there for length of time t_1. He is followed, immediately, by someone whose length of stay is t_2, and so on. The times at which changes occur are thus t_1, $t_1 + t_2$, $t_1 + t_2 + t_3$ . . . . If the t's are assumed to be independent and identically distributed then we have a renewal process. The theory of such processes enables us to find, for example, the distribution of the number of renewals in any time interval. In practice we often observe such systems at a single point in time and then we might be interested in how long the present occupant had been in residence and how long it would be before he moves. The former will be called the 'age' of the occupant and we shall show how to find its distribution. Only rarely shall we be interested in a single house or job, but rather in the population of a region or the employees of a firm. The aggregate behaviour of such systems can be found by 'adding up' the constituent processes.

A key function, on which almost everything else depends, is called the renewal function. It is denoted by $H(T)$ and is the expected number of renewals in $(0, T)$. Its derivative, $h(T)$ is known as the renewal density; $h(T)\delta T$ is the expected number of renewals in $(T, T + \delta T)$, when δT is sufficiently small. Under the same condition the chance of two or more renewals in this interval may be neglected and so $h(T)\delta T$ may also be interpreted as the probability of a renewal in $(T, T + \delta T)$. We shall work throughout with continuous life distributions for which $h(T)$ exists and it will be more convenient to use $h(T)$ rather than $H(T)$.

The renewal equation

We consider the event that a replacement is required in $(T, T + \delta T)$. This event has probability $h(T)\delta T$. It can also be regarded as the union of the following mutually exclusive and exhaustive events, the sum of whose probabilities must therefore also be equal to $h(T)\delta T$.

	Event	Probability
(a)	First renewal in $(T, T + \delta T)$	$f(T)\delta T$
(b)	First renewal in $(x, x + \delta x)$ and another in $(T, T + \delta T)$	$f(x)\delta x h(T - x)\delta T$

$f(t)$ denotes the probability density of the life. In arriving at the second probability we have made use of the fact that at the time of the first renewal the process can be thought of as beginning again with a new incumbent. The probability of a renewal at time $T - x$ later is thus $h(T - x)\delta T$. Under (b) we have imagined the interval $(0, T)$ divided into lengths of δx so there is one such probability for each interval. The total probability obtained by integration over x is then

$$f(T)\delta T + \delta T \int_0^T f(x)h(T - x)\mathrm{d}x.$$

Equating to $h(T)\,\delta T$ and cancelling δT we have the renewal equation

$$h(T) = f(T) + \int_0^T f(x)h(T - x)\mathrm{d}x. \tag{8.1a}$$

An obvious change of variable gives the equivalent equation

$$h(T) = f(T) + \int_0^T h(x)f(T - x)\mathrm{d}x. \tag{8.1b}$$

This equation can sometimes be solved by taking the Laplace transform of each side because the integral on the right-hand side transforms into the product of the transforms, thus

$$h^*(s) = f^*(s) + f^*(s)h^*(s),$$

whence

$$h^*(s) = f^*(s)/\{1 - f^*(s)\}. \tag{8.2}$$

For this method to work it must be possible to find $f^*(s)$ and to invert the right-hand side of (8.2).

If the life distribution is exponential with parameter λ the solution is simple. In this case $f^*(s) = \lambda/(\lambda + s)$ which gives $h^*(s) = \lambda/s$ implying that $h(T) = \lambda = \mu^{-1}$, where μ is the mean life. This is intuitively reasonable; if people stay for four years, on average, we would expect 25 per cent to leave per year. However, this result is true, in general, only for the exponential distribution and we have already seen that many life distributions occurring in the social sciences depart considerably from this form. In such cases intuition will not be a reliable guide and further analysis is necessary. The qualification 'in general' is important because it is true, under very general conditions, that

$$\lim_{T \to \infty} h(T) = \mu^{-1}. \tag{8.3}$$

Thus the replacement rate will settle down to a fixed level after a sufficiently long time. It remains to investigate how large T must be for (8.3) to be an adequate approximation and what the behaviour of the system is in the short term. We shall do this below in the context of manpower planning where life becomes completed length of service (CLS).

Solution of the renewal equation for the mixed exponential CLS distribution

When the CLS distribution has the form

$$f(T) = p\lambda_1 e^{-\lambda_1 T} + (1-p)\lambda_2 e^{-\lambda_2 T}$$

its Laplace transform is

$$f^*(s) = p\left(\frac{\lambda_1}{\lambda_1 + s}\right) + (1-p)\left(\frac{\lambda_2}{\lambda_2 + s}\right). \tag{8.4}$$

When this expression is substituted in (8.2), $h^*(s)$ may be inverted by standard methods, as in Bartholomew (1959), to give

$$h(T) = \mu^{-1} + \{p\lambda_1 + (1-p)\lambda_2 - \mu^{-1}\} \exp\{-(p\lambda_2 + (1-p)\lambda_1 T)\}, \tag{8.5}$$

where $\mu = p/\lambda_1 + (1-p)/\lambda_2$ is the mean of $f(T)$. This formula shows that $h(T)$ approaches its limit in an exponential curve. Further, since $p\lambda_1 + (1-p)\lambda_2 \geq \mu^{-1}$, it follows that the number of recruits required will always be in excess of the number predicted by equilibrium theory. The expected number of recruits needed in any time interval can be found by integrating $h(T)$. Some illustrative calculations are given in Table 8.1, using two of the mixed exponential distributions fitted in Table 7.1.

We have not carried the calculations beyond the end of the second year because it is doubtful whether the fit of the mixed exponential distribution is adequate in the upper tail.

The figures in Table 8.1 show, in a striking fashion, the rapid decline in recruitment which can occur in a new organization. In the examples we have chosen, the figure is roughly half in the second year what it was in the first. This conclusion has obvious implications both for recruitment planning and the interpretation of wastage figures. It suggests that the high initial wastage might be reduced by careful selection of employees with a view to rejecting those with short service prospects. It must be emphasized that we have assumed that the CLS distribution does not change with time. This would be a questionable assumption under the conditions likely to exist when a new organization is established.

Table 8.1 *Percentage recruitment (wastage) figures in successive quarters for a new firm with mixed exponential CLS distribution*

| Quarter | Recruitment using the CLS distributions fitted to data from: | |
	Glacier Metal Co.	J. Bibby & Sons (males)
1	22.6 ⎫	34.6 ⎫
2	18.2 ⎬ 69.5	28.6 ⎬ 108.3
3	15.3 ⎪	24.2 ⎪
4	13.4 ⎭	20.9 ⎭
5	12.1 ⎫	18.5 ⎫
6	11.3 ⎬ 44.6	16.8 ⎬ 65.4
7	10.8 ⎪	15.5 ⎪
8	10.4 ⎭	14.6 ⎭

For the purpose of interpreting crude wastage figures[†] the meaning of Table 8.1 is clear. A change in the wastage rate does not necessarily indicate a change in those factors which precipitate leaving. It may simply reflect a change in the length of service structure. This fact makes it quite meaningless, for example, to compare the crude wastage rate of a new firm with that of an old one.

An approximate solution of the renewal equation

We have seen that some empirical CLS distributions can be satisfactorily graduated by a mixed exponential curve—at least in the interval 0 to 21 months. However, we have also pointed out that, over longer periods, the lognormal is more satisfactory because it has a longer upper tail. The same is true of the type XI distribution obtained from Silcock's model. Unfortunately it is not possible to obtain a simple explicit expression for $h(T)$ for either of these distributions. In order to investigate the form of $h(T)$ over longer periods we therefore require an approximate method for obtaining a solution to the renewal equation.

The approximation which we will use was derived by Bartholomew (1963b). It was intended for use when the CLS distribution is extremely skew and has the following form:

$$h^0(T) = f(T) + F^2(T) \left/ \int_0^T G(x)\mathrm{d}x \right. \tag{8.6}$$

---

† Crude wastage is defined as the number of leavers in the period divided by the average number of employees during the same period (see Section 8.5).

where $G(T) = 1 - F(T)$. The approximation has the following properties in common with the exact solution of the renewal equation.

(a) If $f(T) = \lambda e^{-\lambda T}$ then $h^0(T) = h(T) = \lambda$ for all T.

(b) $\lim_{T \to \infty} h^0 T = \lim_{T \to \infty} h(T) = \mu^{-1}$.

(c) $h^0(0) = h(0) = f(0)$.

(d) $\left. \dfrac{d^i h^0(T)}{dT^i} \right|_{T=0} = \left. \dfrac{d^i h(T)}{dT^i} \right|_{T=0}$ $(i = 1, 2)$.

These properties suggest that the approximation is likely to be good everywhere if $f(T)$ is close to the exponential and will always be good near $T = 0$ and when T is large. The method of derivation used in Bartholomew (1963b), also suggests that the approximation will be good if $f(T)$ has a long upper tail. Calculations by Butler (1970) confirmed this for the case of the lognormal distribution. A further useful property of the approximation is that $h^0(T)$ is an upper bound for $h(T)$ if the loss intensity associated with $f(T)$ is non-increasing. Such distributions have been studied in reliability theory where they are known as DFR (decreasing failure rate) distributions (see, for example, Barlow and co-workers, 1972, Chapter 5).

The simplicity of the approximation is apparent when we consider its form for Silcock's model. In this case

$$f(T) = \frac{v}{c}\left(1 + \frac{T}{c}\right)^{-(v+1)}, \qquad G(T) = \left(1 + \frac{T}{c}\right)^{-v}$$

and

$$\int_0^T G(x)\,dx = \mu\left\{1 - \left(1 + \frac{T}{c}\right)^{-v+1}\right\}.$$

Hence, from (8.5),

$$\left. \begin{aligned} h^0(T) &= \frac{v}{c}\left(1 + \frac{T}{c}\right)^{-(v+1)} + \frac{\left\{1 - \left(1 + \frac{T}{c}\right)^{-v}\right\}^2}{\mu\left\{1 - \left(1 + \frac{T}{c}\right)^{-v+1}\right\}} \\[2mm] &\sim \frac{1}{\mu}\left\{1 - \left(\frac{c}{T}\right)^{v-1}\right\} \quad \text{if } v > 1 \end{aligned} \right\} \qquad (8.7)$$

When v is a little greater than one, the approach to equilibrium is very slow. If $0 < v < 1$ the mean is infinite and $h(T)$ approaches zero like T^{v-1}. In the case $v = 1$ the zero limit is approached like $(\log T)^{-1}$. The calculations made by Silcock (1954) for eight distributions gave six out of eight values of v between 0.5 and 1.

We would therefore expect the wastage rate to go on declining slowly over all periods likely to be of practical interest.

The approximation also takes a simple form for the lognormal CLS distribution. Thus we have

$$
\left.
\begin{aligned}
f(T) &= \frac{1}{\sqrt{2\pi}\,\sigma T} \exp\left\{ -\frac{1}{2}\left(\frac{\ln T - \omega}{\sigma}\right)^2 \right\}, \\[2mm]
F(T) &= \Phi\left(\frac{\ln T - \omega}{\sigma}\right)
\end{aligned}
\right\}
\tag{8.8}
$$

and

$$
\int_0^T G(x)\,\mathrm{d}x = T\left\{ 1 - \Phi\left(\frac{\ln T - \omega}{\sigma}\right) \right\} + \mu\Phi\left(\frac{\ln T - \omega}{\sigma} - \sigma\right)
$$

where $\mu = e^{\omega + \frac{1}{2}\sigma^2}$ and $\Phi(.)$ is the standard normal distribution function. The approximation can therefore be calculated using tables of the normal probability integral. It is clear from (8.8) that equilibrium will not be reached until $X = (\ln T - \omega)/\sigma$ is large enough for $\Phi(X)$ to be near one. To investigate this point in more detail let us examine the form of $h^0(T)$ for large T. If X is large we may write:

$$
\Phi(X) \sim 1 - \frac{1}{\sqrt{2\pi}\,X} e^{-\frac{1}{2}X^2}.
$$

Straightforward manipulation then gives

$$
h^0(T) \sim \mu^{-1}\left\{ 1 - \frac{1}{\sqrt{2\pi}} \frac{\sigma}{X(X-\sigma)} e^{-\frac{1}{2}(X-\sigma)^2} \right\}^{-1}.
\tag{8.9}
$$

If $\sigma = 2$ and $X = 4$, say, then

$$
h^0(T)|_{X=4} = \frac{1.02}{\mu}.
$$

When $X = 4, T = \mu e^\sigma = 400\,\mu$. This means that it would take about 400 times the average CLS to get within 2 per cent of the equilibrium value. Bearing in mind that the lognormal distribution cannot represent the true state of affairs beyond about 40 or 45 years, it is clear that equilibrium behaviour will be of limited practical interest.

The transient behaviour of the renewal model is illustrated in Table 8.2 assuming a lognormal CLS distribution. Calculations have been made for a typical case by taking $\omega = 0$ and $\sigma = 2$. If the unit of time is one year then $\omega = 0$ corresponds to a 'half-life' of one year. For this distribution $\mu = e^2 = 7.389$ so that $h(\infty) = 0.1353$.

The figures in this table show the extreme slowness with which $h^0(T)$ approaches its limit. Even after 20 years the renewal density is still roughly twice its equilibrium value. The fact that $h(T)$ for a comparable mixed exponential

Table 8.2   Approximation to the renewal density for a lognormal CLS distribution with parameters $\omega = 0$ and $\sigma = 2$

$\ln T$	0	1	2	3	4	5	10	∞
T	1	2.71	7.39	20.1	54.6	148.4	22.026	∞
$h^0(T)$	0.574	0.424	0.318	0.244	0.194	0.164	0.125	0.135

distribution reaches equilibrium in a matter of a few years serves to emphasize the need for accurate graduation of the tail of the distribution if long-term predictions are required.

Distribution of the number of recruits

Standard renewal theory provides methods for finding the distribution of the number of recruits as well as its expectation. Let us begin by considering an organization with $N = 1$. The expected number of recruits required in the interval $(0, T)$ is then

$$\bar{n}(T) = H(T) = \int_0^T h(x)\,dx.$$

For large T it is known (see Cox, 1962, p. 40) that $n(T)$ is approximately normally distributed with mean $T\mu^{-1}$ and variance $V^2 T\mu^{-1}$, where V^2 is the square of the coefficient of variation of the CLS distribution. For the lognormal distribution $V^2 = e^{\sigma^2} - 1$ which, for $\sigma = 2$, is equal to 53.6. Long-term predictions of recruiting needs are thus subject to a high degree of uncertainty. However, we have already seen in the case of the mean that the limit is approached so slowly that asymptotic results are practically useless. The same is true of the variance, so we must use exact results when dealing with this kind of CLS distribution.

The exact value of the variance of $n(T)$ can be found by using the following equation:

$$E\{n^2(T)\} = \bar{n}(T) + 2\int_0^T \bar{n}(T - x)h(x)\,dx \tag{8.10}$$

(see Parzen, 1962, p. 179). Since we have the means of finding $\bar{n}(T)$ and $h(T)$ either exactly or approximately for any CLS distribution we can compute $E\{n^2(T)\}$ and hence the variance of $n(T)$. We illustrate the calculations using the mixed exponential distribution for which $h(T)$ was found earlier. If we let

$$a = p\lambda_1 + (1 - p)\lambda_2 - \mu^{-1}$$
$$b = p\lambda_2 + (1 - p)\lambda_1$$

then

$$h(T) = \mu^{-1} + a e^{-bT}$$

and

$$\bar{n}(T) = T\mu^{-1} + \frac{a}{b}(1 - e^{-bT})$$

(8.11)

Making the necessary substitutions in (8.10) and subtracting $\bar{n}^2(T)$ we find the following expression for the variance:

$$\text{var}\{n(T)\} = \frac{T}{\mu}\left\{1 + \frac{2a}{b}\right\} + \frac{a}{b}\left\{1 - \frac{4}{\mu b} + \frac{a}{b}\right\} + \frac{2a}{b}\left\{\frac{1}{\mu} - \frac{a}{b}\right\}T e^{-bT}$$

$$- \frac{a}{b}\left\{1 - \frac{4}{\mu b}\right\} e^{-bT} - \frac{a^2}{b^2} e^{-2bT}.$$

(8.12)

The corresponding results for an organization of size N are simply obtained by multiplying the mean and variance given above by N. Since $n(T)$ for a large organization can be regarded as the sum of the numbers for single-member systems, the central limit theorem ensures its approximate normality. Some numerical values of expectations and standard errors are given in Table 8.3. For the purposes of this calculation we have considered two hypothetical firms with 1000 employees. We assume the CLS distribution to be mixed exponential in both cases. In the first case we have used the parameter values obtained by fitting the curve to the Glacier Metal Co. data; in the second we have used a more skew member of the mixed exponential family suggested by data relating to the United Steel Cos. given in Silcock (1954).

Table 8.3 *Expectations and standard deviations of numbers of recruits required in various time intervals for two mixed exponential CLS distributions when* $N = 100$

		Time interval			
		0–3 months	0–6 months	0–12 months	0–24 months
$p = 0.6513$	Mean	226	408	695	1142
$\lambda_1 = 0.2684$	S.E.	16.2	21.8	29.0	39.2
$\lambda_2 = 2.4228$	Approx. S.E.	18.4	22.6	29.2	39.2
$p = 0.5377$	Mean	458	738	1072	1508
$\lambda_1 = 0.2187$	S.E.	27.0	34.3	41.8	52.5
$\lambda_2 = 4.8940$	Approx. S.E.	32.9	36.3	42.4	52.6

The rows labelled 'Approx, S.E.' were obtained using only the first two terms of the variance as given by (8.12). The approximation overestimates the true value but the difference is negligible for all but the shortest time periods. It is clear from the table that predictions for even short periods are subject to considerable uncertainty.

The foregoing results relate to the total number of replacements in $(0, T)$. More often we require to know about the number of replacements in some fairly short interval (T_1, T_2) at a distance from the origin. If this distance is such that the system has reached equilibrium then it is well known that the numbers of renewals in non-overlapping intervals are independent Poisson variables. This result is due to Khintchine (1960, Chapter 5). In view of the slowness which manpower replacement systems approach their equilibrium state it is highly desirable to have some results on the distribution of the number of renewals in the transient state. A limiting result relevant to this requirement was given by Grigelionis (1964). It was rederived by Butler (1970) and its adequacy as an approximation was investigated by Bartholomew and Butler (1971).

The result concerns the number of replacements required in a system of size N in an interval of time (T_1, T_2) as $T_2 \to T_1$ and $N \to \infty$, with the expected number of leavers held constant. Let $T_2 - T_1 = \delta T$ be sufficiently small for the chance of two or more replacements in the interval to be negligible. Then

$$\left. \begin{array}{l} Pr\{\text{one replacement in the } i\text{th job}\} = h(T_1)\delta T + O(\delta T^2) \\ Pr\{\text{no replacement in the } i\text{th job}\} = 1 - h(T_1)\delta T + O(\delta T^2) \end{array} \right\}. \quad (8.13)$$

The expected number of replacements in $(T_1, T_1 + \delta T)$ for the whole system is thus $Nh(T_1)\delta T$. For this to remain fixed as $\delta T \to 0$ and $N \to \infty$ we must have $\delta T = K/N$ for some fixed K. The total number of replacements is thus a sum of Bernoulli variables whose probabilities of being non-zero are tending to zero as their number tends to infinity. Under the conditions specified the limiting distribution is Poisson in form with mean $h(T)K$ (see Feller, 1968, p. 282). We note in passing that the result remains true if the probability of replacement associated with the ith job depends on i (Feller's result is stated for this more general case). This means that the different jobs could have different CLS distributions or that the time origins for the job streams could be different.

The practical value of limit theorems lies in their usefulness as approximations before the limit is reached. Bartholomew and Butler (1971) investigated the adequacy of the Poisson approximation both theoretically and by simulation, using a mixed exponential CLS distribution. They made exact calculations of the probabilities in (8.13) and simulated a small system with $N = 33$. Their conclusions can be summarized by saying that the approximation in only likely to be good if $(T_2 - T_1)$ is less than about one-tenth of the average CLS. In practice, average CLS's are of the order of a few years so the approximation should be reasonable if the prediction interval is no longer than a few months.

Butler (1971) also tested the adequacy of the theory using data collected for employees in a steelworks. He found that the flow numbers in small homogeneous groups over fairly short time periods were approximately distributed in the Poisson form. However, there were a few large flows which could not be accounted for by the Poisson hypothesis and it seemed likely that these arose due to the transfer of groups of individuals between departments. Overall, the agreement with the theory was encouraging.

The derivation of the Poisson distribution given above does not imply that the numbers leaving in non-overlapping intervals will be independent. In general they will not be and Bartholomew and Butler (1971) investigated this point also. Their calculations suggest that the correlation between the numbers of renewals in contiguous intervals is negligible if the intervals are short enough for the Poisson approximation itself to be adequate.

In spite of the qualifications with which the Poisson approximation must be surrounded it provides a very useful rule-of-thumb in practice. It means that approximate standard errors can be calculated by simply taking the square root of the expected values. The latter are needed in any case, so the extra labour of finding measures of error is minimal.

The age distribution

A useful way of learning about the past of an organization and identifying possible problems in the future is to look at the age distribution. As already noted, age in this context refers to age in the system and we use the term here instead of the more natural 'length of service' to avoid confusion with CLS. The identification of abnormal characteristics in an age distribution presupposes some knowledge of what is normal. Renewal theory enables us to determine what form the age distribution will take under various assumptions about past history. To illustrate the approach we take the case of an organization of fixed size set up with a full complement at time zero and operating according to the assumptions of the renewal model.

Some founder members may still be in post at time T and any such must obviously have age T. The probability that a randomly selected person has age T is thus the proportion who will be expected to survive to T. That is,

$$Pr\{x = T\} = G(T), \tag{8.14}$$

where x denotes age. For values of $x < T$ we have

$Pr\{$individual at T has age in $(x, x + \delta x)\} = a(x|T)\delta x$, say,

$= Pr\{$he joined in $(T - x, T - x + \delta x)$ and survived for $x\}$

$= Pr\{$joined in $(T - x, T - x + \delta x)\}$ $Pr\{$survived for $x\}$

since the survival time is independent of the time of joining. Thus

$$a(x|T)\delta x = h(T - x)\delta x G(x) \qquad (0 \le x < T)$$

or

$$a(x|T) = h(T - x)G(x). \tag{8.15}$$

The age distribution thus depends on the CLS distribution, directly through $G(x)$ and indirectly through the renewal density.

For typical CLS distributions we have seen that $h(T)$ is a decreasing function over most, if not all, its range. Hence $h(T - x)$ is an increasing function of x for fixed T. The survivor function $G(x)$ is always non-decreasing in x. The product of the two functions may therefore increase or decrease, but if T is large and x is small $h(T - x)$ will change only slowly so that $a(x|T)$ will be like $G(x)$ near the origin. In the limit as $T \to \infty$ this correspondence occurs for all x. Under these conditions $G(T) \to 0$ so that the discrete lump of probability at $x = T$ vanishes. Also, for fixed x,

$$\lim_{T \to \infty} h(T - x) = \mu^{-1}.$$

Hence

$$a(x|\infty) = \mu^{-1}G(x) \qquad (0 \le x < \infty). \tag{8.16}$$

The steady-state age distribution is thus proportional to the survivor function and so is a non-decreasing function of age.

This result provides a means of estimating a survivor function in the absence of any information about leavers. It requires, of course, the steady-state assumption which is unlikely to be exactly satisfied. Nevertheless, there are circumstances where flow data are not available and an estimate of the survivor function, however rough, is well worth having. If, on the other hand, $G(x)$ can be estimated from flow data the estimate can be compared with the age distribution to see whether the system is near its steady state.

Age distributions can be calculated under more complicated assumptions about past history and in this way the shape of the age distribution can be related to this history. We have seen, for example, how a system maintained at a constant size will eventually have an age distribution whose density decreases with increasing age. A 'hump' in one age group of an observed distribution would thus provide clear evidence of a happening in the past giving rise to an excess number in that group.

Expanding systems

The foregoing analysis can easily be extended to cope with an expanding system. All that is necessary is to aggregate the individual processes using as origin for each the time at which they joined. Thus suppose that the initial size of the system is $N(0)$ and that it increases by an amount $M(T_i)$ at time $T_i (i = 1, 2, \ldots)$. Then the expected number of replacements in $(T, T + \delta T)$ is given by

$$N(0)h(T)\delta T + \sum M(T_i)h(T - T_i), \tag{8.17}$$

where the summation extends over all i for which $T_i < T$. If we wish to treat expansion as a continuous process with $M(T)\delta T$ denoting the increase in $(T, T + \delta T)$ the corresponding expression for the replacement rate would be

$$R(T) = N(0)h(T) + \int_0^T M(x)h(t-x)\,dx \qquad (8.18)$$

where we have used $R(T)\delta T$ to mean the expected number of replacements in $(T, T + \delta T)$. This is a natural extension of the usage of earlier chapters. Again we see that the result depends on the renewal density.

We shall illustrate the use of (8.18) by demonstrating how crude wastage rates depend on the age of the system. This fact renders them unsuitable as indices of wastage as we shall argue in Section 8.3. The crude wastage rate for the interval is defined as the expected number of losses divided by the average size; that is,

$$w(T_1, T_2) = \int_{T_1}^{T_2} R(T)\,dT \Big/ \left\{ \frac{1}{(T_2 - T_1)} \int_{T_1}^{T_2} N(T)\,dT \right\}, \qquad ((8.19)$$

where $N(T) = \int_0^T M(x)\,dx + N(0)$, the total size at T.

In order to illustrate the effect of expansion on crude wastage we shall give two examples. Let us assume that the CLS distribution is mixed exponential; then we have shown that

$$h(T) = \mu^{-1} + a\,e^{-bT}$$

where

$$a = p\lambda_1 + (1-p)\lambda_2 - \mu^{-1}, \qquad b = p\lambda_2 + (1-p)\lambda_1$$

and

$$\mu = p/\lambda_1 + (1-p)/\lambda_2.$$

In our first example we assume that the system is subject to a linear growth law of the form

$$M(T) = M$$

M being the rate of growth. The function $R(T)$ can be obtained from (8.18) by straightforward integration. On substituting the result in (8.19) we find that

$$w(T_1, T_2) = \frac{T_2 - T_1}{\mu} + \frac{2a}{b(T_1 + T_2)} \left\{ T_2 - T_1 - \left(\frac{e^{-bT_1} - e^{-bT_2}}{b} \right) \right\}. \qquad (8.20)$$

Some calculations based on this formula are given in Table 8.4 for three CLS distributions of the mixed exponential family. The parameter values selected were those obtained by fitting the distribution to the data relating to the Glacier Metal Co., J. Bibby & Sons, and the United Steel Cos. which have already been used in this chapter.

Table 8.4 *Percentage wastage in successive quarters for a group expanding at a constant rate*

Quarter	Glacier Metal Co.	J. Bibby & sons	United Steel Cos.
1	23.5	35.8	49.6
2	21.3	32.9	40.5
3	19.5	30.2	33.5
4	18.0	28.0	28.6
5	16.8	26.2	25.2
6	15.9	24.6	22.7
7	15.1	23.3	20.9
8	14.5	22.2	19.4

A comparison of these figures with those in Table 8.1 shows that although the wastage rate still decreases it does not do so as rapidly as before. It can be seen from (8.20) that $w(T_1, T_2)$ does not depend on the value of M and hence the wastage approaches the same equilibrium value regardless of the rate of expansion.

The previous example is based on a continuous growth function. In our second example we shall suppose that $N(T)$ jumps from an initial value of $N(0)$ to a new value of $N(0) + N$ at $T = T_J$. The expected number of losses in this case is clearly

$$\bar{n}(T) = N(0) \int_0^T h(x)\,dx \qquad (T \le T_J)$$

$$= N(0) \int_0^T h(x)\,dx + N \int_0^{T-T_J} h(x)\,dx \qquad (T > T_J).$$

Some quarterly wastage rates for this kind of growth are given in Table 8.5.

Table 8.5 *Percentage wastage in successive quarters for a firm expanded from $N(0)$ to $N(0) + N$ at $T = T_J$ assuming the United Steel Cos. CLS distribution*

	$T_J = \frac{1}{2}$ year		$T_J = 1$ year	
Quarter	$N = \frac{1}{2}N(0)$	$N = N(0)$	$N = \frac{1}{2}N(0)$	$N = N(0)$
1	45.8	45.8	45.8	45.8
2	28.0	28.0	28.0	28.0
3	27.9	32.4	19.0	19.0
4	18.9	21.2	14.4	14.4
5	14.4	15.6	23.3	29.0
6	12.1	12.7	16.7	19.5
7	11.0	11.3	13.3	14.7
8	10.4	10.6	11.5	12.3

Calculations are given for two values each of T_J and N, assuming a CLS distribution like the one for the United Steel Cos.

The effect of an abrupt increase in size is to arrest the decreasing wastage rate. With a large increase in size there is a temporary increase in wastage, but the effect is short-lived and after a year its influence has almost vanished.

Contracting systems

We dealt with expanding systems by aggregating individual processes, but this cannot be done in reverse because the future behaviour would then depend on which processes were deleted. We shall not develop the theory of contracting processes, but instead consider a special but important case from first principles. This concerns the maximum rate of contraction which can be achieved. It is obvious that this occurs when all recruitment ceases. The problem is to determine the expected size of the system as a function of time since recruitment stopped. Let us measure time from this point at which the size of the system is assumed to be $N(0)$. After a time T the expected size will be

$$N(0)\{1 - D(T)\},$$

where $D(T)$ is the distribution function of the time before a member, chosen randomly at $T = 0$, leaves. This is called the residual length of service distribution. For an individual aged $\cdot x$ at $T = 0$ this will have distribution function

$$D(T|x) = \{F(T+x) - F(x)\}/\{1 - F(x)\}, \tag{8.21}$$

where $F(\cdot)$ is the distribution function of CLS. If the initial density function of age is $a(x)$ then

$$D(T) = \int_0^\infty a(x) D(T|x)\mathrm{d}x. \tag{8.22}$$

The age distribution could be deduced as described earlier or estimated empirically. For our immediate purpose we can leave it as an arbitrary function.

It is common to specify a desired rate of contraction in terms of a percentage rate of run-down. In continuous time this amounts to requiring that

$$N(T) = N(0)\mathrm{e}^{-\alpha T} \qquad (0 < \alpha < \infty). \tag{8.23}$$

It is an important practical question to ask how large α can be without the need for redundancies. In other words we require the largest α for which

$$N(0)\mathrm{e}^{-\alpha T} > N(0)\{1 - (D(T)\} \tag{8.24}$$

for all T.

In the case when the CLS distribution is exponential $D(T) = \mathrm{e}^{-\lambda T}$ and the inequality is clearly satisfied if $\alpha \leq \lambda$. Redundancy is thus avoided provided that the rate of contraction is less than the wastage rate. This result might seem obvious at first sight and might be supposed to hold whatever the CLS distribution. This is a common misapprehension as we shall now demonstrate by supposing that the CLS distribution is mixed exponential. Substituting the density into (8.21) and then into (8.22) we find

$$1 - D(T) = e^{-\lambda_1 T} \int_0^\infty a(x) \frac{p e^{-\lambda_1 x}}{p e^{-\lambda_1 x} + (1-p) e^{-\lambda_2 x}} dx + e^{-\lambda_2 T}$$
$$\times \int_0^\infty \frac{a(x)(1-p) e^{-\lambda_2 x}}{p e^{-\lambda_1 x} + (1-p) e^{-\lambda_2 x}} dx$$
$$= A e^{-\lambda_1 T} + B e^{-\lambda_2 T}, \text{ say, where } A + B = 1, A, B \geq 0.$$

The inequality (8.24) will now only be satisfied for all T if $\alpha \leq \min(\lambda_1, \lambda_2)$. We have already seen that the equilibrium wastage rate in this case is $\{p/\lambda_1 + (1-p)/\lambda_2\}^{-1}$ which is always larger than $\min(\lambda_1, \lambda_2)$. Hence it will not be possible to contract the system at a rate as great as might be inferred from the wastage rate. For some other CLS distributions, such as the Silcock type XI, $\{1 - D(T)\}$ decreases more slowly than any exponential and so a fixed rate of contraction would certainly lead to redundancies sooner or later.

8.3 THE MEASUREMENT OF LABOUR TURNOVER

We have pointed out in other contexts the role of stochastic models in constructing meaningful measures of social phenomena. One of the simplest examples of their use in this connection is provided by the problem of measuring labour turnover. Labour turnover refers to the flow of people into and out of an organization. In the context of the renewal models of the present chapter, for which input is simply related to output, turnover is equivalent to wastage. We shall use the two terms synonymously, and practically all measures which have been proposed are based on the wastage flow alone. Some writers speak of measuring stability, which is essentially the converse of turnover, but the two concepts are equivalent—a high stability implies low turnover and vice versa.

The measurement of labour turnover serves two purposes. One is as an important ingredient of manpower models as, for example, when earlier in this chapter we related the crude wastage rate to the age structure of an organization. The second use is as an index of morale or efficiency. It is with this use that we shall be concerned in this section. A high turnover of employees is usually regarded as a sign of poor morale leading to inefficient operation, and there is an extensive literature devoted to the question of how to measure and interpret it. A bewildering variety of measures have been proposed and a good deal of confusion still surrounds the whole question. Some of the literature was listed in Chapter 7 in the course of the discussion of models for the leaving process. To those given there may be added the papers by Bowey (1969), Bibby (1970), Hyman (1970), Forbes (1971b), Van der Merwe and Miller (1971) and Clowes (1972). We shall here aim to clarify the issues by bringing the results of the earlier stochastic analysis to bear. A measure will not, of course, automatically be a good one just because it is based on a stochastic model. An inappropriate model will lead to misleading conclusions, as Stoikov's (1971) model and the subsequent discussion by Bartholomew (1971) shows.

Possibly the earliest, and certainly one of the most widely used, measures of turnover is the crude rate defined by

$$I = \frac{\text{number who leave in a given interval}}{\text{average number employed during the same interval}} \times 100 \text{ per cent.}$$

(8.25)

The inadequacy of such an index is readily apparent from the analysis of Section 8.2 where we used the index to show the influence on loss of the age of a system. A measure which depends so strongly on age structure cannot adequately reflect other attributes—unless the system has reached its steady state. Thus a difference in the crude rate between two firms may simply be a reflection of the fact that one has a much younger age structure. Lane and Andrew (1955) provide a striking illustration of this sort of situation by giving data on two firms of which the one with the higher wastage also had the longer expected length of service.

In demographic work an analogous situation arises in the measurement of mortality, where the pitfalls of using crude death rates are well understood. Salubrious seaside resorts have high crude death rates, not because they are unhealthy but because they attract as residents a disproportionate number of elderly people. Demographers and actuaries deal with this problem by calculating standardized mortality rates which are estimates of what the crude rate would be if the population had a chosen standard age distribution. Exactly the same course can be followed when measuring turnover. Suppose we adopt a standard age distribution with density function $s(x)$. (We return below to the question of how $s(x)$ might be chosen.) The expected number of leavers in $(T, T + \delta T)$ from those with age in $(x, x + \delta x)$ will then be

$$N(T)s(x)\delta x \lambda(x)\delta T,$$

where $N(T)$ is the size of the system at time T and $\lambda(x)$ is the force of separation common to all members of the system. The total expected number of leavers in an interval (T_1, T_2) will thus be

$$\int_{T_1}^{T_2} N(T)\,\mathrm{d}T \int_0^\infty s(x)\lambda(x)\,\mathrm{d}x$$

(8.26)

where, again, we treat $N(T)$ as a continuous variable. The average number present in (T_1, T_2) is

$$\int_{T_1}^{T_2} N(T)\,\mathrm{d}T/(T_2 - T_1),$$

so the crude turnover index for this $s(x)$ is estimated to be

$$I = (T_2 - T_1) \int_0^\infty s(x)\lambda(x)\,dx. \qquad (8.27)$$

This expression demonstrates the way in which the crude index depends on both the age structure and the force of separation. By adopting a standard form for the age distribution we eliminate variation from this source and have a measure depending solely on the force of separation.

A natural choice for $s(x)$ is the steady-state age distribution which would be attained in a constant size system as given by (8.16). This will give an index showing what the crude rate would be in equilibrium. Making the substitution

$$I = (T_2 - T_1) \int_0^\infty \frac{1}{\mu} G(x)\lambda(x)\,dx = \frac{T_2 - T_1}{\mu}$$

$$\times \int_0^\infty f(x)\,dx = \frac{T_2 - T_1}{\mu}. \qquad (8.28)$$

The standardized index thus turns out to be equivalent to the average CLS—a measure proposed by Lane and Andrew (1955).

The foregoing discussion serves to underline what is otherwise obvious, namely that a measure of wastage should depend only on the propensity to leave. This is expressed by $\lambda(x)$ or either of the equivalent functions $f(x)$ and $G(x)$.

The 'best' way of describing the leaving process is thus by giving an estimate of one or other of these functions. Now that graphical output is commonly available on computers, graphs of these functions can easily be displayed and reproduced. A diagram is easy to assimilate and less open to misinterpretation than any index. However, there may continue to be circumstances in which the information conveyed by $\lambda(x)$ has to be summarized into a single number. The mean of the CLS distribution is an obvious possibility which has already arisen in connection with the standardized index. The median, or half-life, has similar appeal with the added advantage of being both easy to interpret and easy to calculate. An equally simple measure is obtained by quoting the proportion who would be expected to survive for a specified period.

In Chapter 7 we failed to find any one-parameter distribution which successfully fitted CLS data. Any attempt to convey all the relevant information contained in $\lambda(x)$ by a single number is therefore bound to be inadequate. For example, two distributions can have the same proportions surviving to one year, say, but be very different at other points. There is less risk of concealing important differences if two indices are used, each highlighting different features of the distribution. The two parameters ω and σ of a fitted lognormal distribution contain most of the information about a CLS distribution, though not in an easily interpreted form. Perhaps the simplest pair of measures is $G(x_1)$ and $G(x_2)$ for two suitably chosen lengths of service x_1 and x_2 (x_1 might be the end of the training period and x_2 the time an average person is ready for promotion).

The treatment here and in Chapter 7 has dealt with the dependence of leaving on the length of service to the exclusion of all else. In reality $\lambda(x)$ is a function not only of the length of service but of many other things such as sex, place of residence, skill level, etc. An apparent difference between two firms as judged by any of our measures may therefore be the result of different sex ratios or some other such factors. This serves to emphasize that the length of service-based indices can only be meaningfully used if the groups involved are homogeneous with respect to other factors which influence leaving. If it is desired to compare heterogeneous groups the indices must be standardized. This can be done in the same way as described for age.

8.4   MODELS FOR SYSTEMS WITH SEVERAL GRADES

In a system with several grades, vacancies need not be filled by direct recruitment. Instead, a member of some other grade may be transferred, in which case another vacancy will be created and so on. Throughout this section we shall consider only simple hierarchies of grades such that a vacancy is filled either by direct recruitment or by promotion from the grade immediately below. The sequence of replacements stops when the vacancy reaches the lowest grade where it must be filled by recruitment. As far as possible the notation and terminology will be consistent with that of previous chapters. As the grade sizes are assumed to be constant we shall drop the argument and write them as $n_1, n_2, \ldots n_k$ with $\Sigma_{i=1}^{k} n_i = N$. The constancy of the grades imposes some restrictions on the flows which apply whatever the model. Thus consider any time interval of unit length and define random variables denoting the flows as follows.

R_i = the number recruited to grade i $\left.\begin{array}{l}\\\\\end{array}\right\}$ $(i = 1, 2, \ldots, k)$
W_i = the number who leave from grade i

P_i = the number who are promoted from grade i $(i = 1, 2, \ldots, k-1)$.

Since the size of each grade is fixed it follows that the total flow in must balance the total flow out. That is

$$P_{i-1} + R_i = P_i + W_i \qquad (i = 1, 2, \ldots, k) \qquad (8.29)$$

with $P_0 = P_k = 0$. It follows also that the expectations of these random variables will satisfy the same equations. In general these expectations will depend on the location and length of the interval chosen. We shall therefore analyse the process by considering an interval of infinitesimal length. The following functions are defined for this purpose.

$n_i w_i(T)\delta T$ = expected number of losses from i in $(T, T + \delta T)$
$n_i h_i(T)\delta T$ = expected number of promotions from i in $(T, T + \delta T)$
$R_i(T)\delta T$ = expected number of recruits into i in $(T, T + \delta T)$.

Defined in this way we may think of $w_i(T)$ and $h_i(T)$ as rates which give an individual's propensity to leave or be promoted at T. As a direct consequence of (8.29) we have

$$n_{i-1} h_{i-1}(T) + R_i(T) = n_i h_i(T) + n_i w_i(T) \qquad (i = 1, 2, \ldots, k) \qquad (8.30)$$

with $h_0(T) = h_k(T) = 0$.

A model with fixed loss rates

Any model has to specify two things: the loss mechanism and the way in which vacancies are filled. In our first model we suppose that each individual has a constant propensity to leave which depends only on the grade. That is, we let

$$w_i(T) = \lambda_i \qquad (i = 1, 2, \ldots, k) \qquad (8.31)$$

and further assume that individuals behave independently. These assumptions are analogous to those we made in Chapter 5 for the continuous time Markov model. When a loss occurs in grade $i(i > 1)$ we shall assume that the vacancy is filled by promotion from the grade below with probability s_i; otherwise it is filled by recruitment. This assumption has some empirical support. All vacancies in grade 1 (the lowest) are filled by recruitment. The wastage assumption together with (8.30) gives

$$R_i(T) = n_i h_i(T) - n_{i-1} h_{i-1} + n_i \lambda_i. \qquad (8.32)$$

According to the second assumption,

$$R_i(T) = (1 - s_i) \{ n_i h_i(T) + n_i \lambda_i \}. \qquad (8.33)$$

Eliminating $R_i(T)$ between (8.32) and (8.33) we find the following recursion formula for the promotion rates:

$$n_{i-1} h_{i-1}(T) = s_i n_i h_i(T) + s_i n_i \lambda_i$$

which gives

$$n_i h_i(T) = \sum_{j=i+1}^{k} n_j \lambda_j \prod_{r=i+1}^{j} s_r. \qquad (8.34)$$

Neither $h_i(T)$ nor $R_i(T)$ depend on T in this model. We illustrate the use of this formula by considering two special cases.

Case I. $\lambda_i = \lambda (i = 1, 2, \ldots, k)$

This case is equivalent to the model for a single grade with exponential life distribution as far as total inflow and outflow are concerned. For the promotion rates we have

$$h_i = \frac{\lambda \sum\limits_{j=i+1}^{k} n_j}{n_i} \prod\limits_{r=i+1}^{j} s_r \qquad (i = 1, 2, \dots, k-1). \qquad (8.35)$$

The effect of the grade sizes on each promotion rate is clear from this expression. They vary directly as the total size of the higher grades and inversely as the size of the current grade. Also, the rate is proportional to each of the probabilities s_r associated with the higher grades. If recruitment is allowed only at the lowest level, then $s_1 = 0$, $s_i = 1$ $(i > 1)$ then

$$h_i = \lambda(N - N_i)/n_i \qquad (i = 1, 2, \dots, k-1) \quad \text{where } N_i = \sum\limits_{j=1}^{i} n_j. \qquad (8.36)$$

Suppose we ask, in this case, what grade structure would be necessary to make the promotion rates equal. Equation (8.36) shows that this requires that

$$n_i \propto (N - N_i) \qquad (i = 1, 2, \dots k-1)$$

implying that

$$n_{k-1} = cn_k, \qquad n_i = n_{i+1}(1 + c), \qquad (i = 1, 2, \dots k-1). \qquad (8.37)$$

Apart from the highest grade, the grades form a geometric series. A somewhat similar result was obtained in Chapter 5 for a system with fixed input and random grade sizes. There it was shown that if the promotion rates were equal the *expected* grade sizes would form a geometric series. This result illustrates the point that Markov and renewal models are sometimes equivalent when interpreted deterministically, that is, when random variables are replaced by their expectations.

Case II. $s_i = s (i = 2, 3, \dots k)$

Here we have

$$n_i h_i = \sum\limits_{j=i+1}^{k} n_j \lambda_j s^{j-i}. \qquad (8.38)$$

Suppose next that $\lambda_j = \lambda s^{-j}$ so that

$$h_i = \lambda(N - N_i)/n_i s^i.$$

If we now had $n_i \propto (N - N_i)$ the promotion rates would increase geometrically as we moved up the hierarchy. This is a direct consequence of the larger wastage rates at the higher levels. If, on the other hand $\lambda_j = \lambda$ then

$$h_i = \lambda \sum\limits_{j=i+1}^{k} n_j s^j / n_i s^i.$$

For such a system the structure needed to make the promotion rates equal would have to satisfy

$$n_i s^i \propto \sum_{j=i+1}^{k} n_j s^j.$$

This again implies a geometric structure though with common ratio $s(1+c)$ instead of $(1+c)$. In other words the structure has to taper off more rapidly than when no one is recruited at higher levels.

The distribution of the flows

It is a straightforward matter to show that all the flows have Poisson distributions when the loss rates are constant. If each of the n_i individuals in grade i is subject to a constant loss rate λ_i then the number of losses from that grade will be Poisson with mean $n_i \lambda_i$. In order to deduce the distribution of the remaining flows we must make use of our assumption about how vacancies are filled. According to this assumption the conditional distribution of P_i, given $P_i + W_i$, is binomial with parameters $P_i + W_i$ and s_i. To complete the determination of the distribution we need the following lemma.

Lemma

If $X \sim Poisson\ (\mu)$ and. $Y|X \sim Binomial\ (X, p)$ then $Y \sim Poisson\ (p\mu)$.

We start with grade k for which $P_{k-1} + R_k = W_k$. Applying the lemma with $X = W_k$, $p = s_k$ we deduce that $P_{k-1} \sim$ Poisson $(n_k \lambda_k s_k)$ and similarly, $R_k \sim$ Poisson $\{\lambda_k (1 - s_k)\}$. Turning next to grade $k-1$, $P_{k-2} + R_{k-1} = W_{k-1} + P_{k-1}$. Putting $X = W_{k-1} + P_{k-1}$, $Y = P_{k-2}$ and noting that $W_{k-1} + P_{k-1}$ will be Poisson because it is the sum of two independent Poisson variables, the lemma yields

$$P_{k-2} \sim \text{Poisson } \{s_{k-1}(n_k \lambda_k s_k + n_{k-1} \lambda_{k-1})\}.$$

Proceeding down the hierarchy in this fashion we eventually find that all of the recruitment and promotion flows have Poisson distributions. Because of the constraints of (8.29), it is not true that they are independent.

The validity of our reasoning does not depend on the length of the time interval over which the flows take place or on the age of the system. This is a direct consequence of the assumption of constant loss rates. In general, if the loss rates were to depend on the age of the individual or seniority in the grades both of these features would be relevant and the argument much more difficult.

Lengths of stay

Another important aspect of renewal systems concerns lengths of stay both in individual grades and in the system as a whole. As already noted, the latter may be compared with other distributions of durations resulting from the models of Chapter 7.

An individual in grade i is subject to two forces of decrement. One is the constant loss rate λ_i and the other is due to promotion. We shall first consider how to find the survivor function for a person entering grade i if he were subject to only one of these forces. Since they operate independently the survivor function for someone subject to both forces is obtained as the product of the two survivor functions thus found. In the absence of promotion, it follows from the assumption of a constant loss rate that the survivor function would be $e^{-\lambda_i \tau}$, where τ denotes length of stay in the grade.

The waiting time for promotion in the absence of loss will clearly depend upon how promotees are selected. We shall consider two promotion rules. The first requires selection at random among those eligible; the second supposes that the most senior of those in the grade below is promoted. In the case of random promotion we first find the survivor function conditional on there being P_i promotions in $(0, \tau)$. On any one occasion when a promotion is made the probability that any particular individual is chosen is $1/n_i$. The probability that the individual is *not* promoted on each of the P_i occasions when promotions are made is therefore

$$\left(1 - \frac{1}{n_i}\right)^{P_i}.$$

The unconditional probability is obtained using the fact that P_i has a Poisson distribution with mean $n_i h_i \tau$, thus

$$Pr\{\text{survive to } \tau \text{ in } i | \text{no loss}\} = e^{-n_i h_i \tau} \sum_{P_i=0}^{\infty} \left(1 - \frac{1}{n_i}\right)^{P_i} \frac{(n_i h_i \tau)^{P_i}}{P_i!} = e^{-h_i \tau}. \quad (8.39)$$

Multiplying the two survivor functions we have

$$Pr\{\text{survive to } \tau \text{ in } i\} = e^{-(h_i + \lambda_i)\tau}. \quad (8.40)$$

That is, the length of stay in grade i is exponential with mean

$$(h_i + \lambda_i)^{-1}.$$

Next suppose that the most senior member of each grade is promoted. On entry to grade i (by whatever route) an individual will have $n_i - 1$ people ahead of him. To be promoted he must first wait until all these have left the grade and, after that, he must wait for the next vacancy to occur. Vacancies occur in grade $i+1$ in a Poisson process with rate parameter $n_i h_i$ and each individual in i is also subject to a loss rate λ_i. The process by which the $n_i - 1$ people leave grade i is thus a pure death process with death rate, when the number remaining is m, given by

$$\mu_0 = 0$$
$$\mu_m = n_i h_i + \lambda_i m \qquad (m = 1, 2, \ldots n_i - 1). \quad (8.41)$$

The theory of the death process (see, for example, Bailey, 1964, or Feller, 1968) enables us to find the probability that there will be m people remaining at time τ. We denote this $P_m(\tau)$. The probability $P_0(\tau)$ is the same as the complement of the survivor function since if no one remains at τ the time for them all to leave must be less than τ. In our case the differential–difference equations for the probability $P_m(\tau)$ are

$$P'_m(\tau) = -(m\lambda_i + n_i h_i)P_m(\tau) + \overline{(m+1\,\lambda_i} + n_i h_i)\,P_{m+1}(\tau), \quad (m = 1, 2, \ldots n_i - 2)$$

$$P'_0(\tau) = (\lambda_i + n_i h_i)P_1(\tau) \tag{8.42}$$

$$P'_{n_i-1}(\tau) = -\{(n_i - 1)\lambda_i + n_i h_i\}P_{n_i-1}(\tau)$$

with initial conditions $P_{n_i-1}(0) = 1$, $P_m(0) = 0$ $(m < n_i - 1)$. We shall work with the Laplace transforms of these probabilities so we first convert the equations by multiplying both sides by e^{-st} and integrating from 0 to infinity. We need the result that the transform of $P'_m(\tau)$ is $sP_m^*(s) - P_m(0)$. It then follows that

$$\{s + m\lambda_i + n_i h_i\}P_m^*(s) = \{\overline{m+1}\,\lambda_i + n_i h_i\}P_{m+1}^*(s) \quad (m = 1, 2, \ldots n_i - 2) \tag{8.43}$$

$$sP_0^*(s) = (\lambda_i + n_i h_i)P_1^*(s), \quad sP_{n_i-1}^*(s) - \{(n_i - 1)\lambda_i + n_i h_i\}P_{n_i-1}^*(s).$$

Starting with $P_{n_i-1}^*(s)$ these equations yield

$$P_0^*(s) = \frac{1}{s}\prod_{j=1}^{n_i-1}\left[\frac{j\lambda_i + n_i h_i}{j\lambda_i + n_i h_i + s}\right]. \tag{8.44}$$

This is the transform of the distribution function of waiting time. The transform of the density function is $sP_0^*(s)$. After the $n_i - 1$ individuals have left we have to wait for the next vacancy. This waiting time has an exponential distribution with parameter $n_i h_i$. The sum of these two waiting times thus has the transform

$$sP_0^*(s)\left(\frac{n_i h_i}{n_i h_i + s}\right).$$

Combining this with (8.44) the survival time, in the absence of loss, has Laplace transform

$$\prod_{j=0}^{n_i-1}\frac{j\lambda_i + n_i h_i}{j\lambda_i + n_i h_i + s}. \tag{8.45}$$

This is the transform of an Erlang distribution and it could be inverted by resolving into partial fractions and inverting term by term. However, if we suppose that the grade sizes are large the density degenerates, showing that the waiting time is approximately constant. Taking the natural logarithm of (8.45) and assuming n_i to be large,

$$-\sum_{j=0}^{n_i-1}\ln\left[1 + \frac{s}{j\lambda_i + n_i h_i}\right] \sim -\frac{s}{\lambda_i}\sum_{j=0}^{n_i-1}\frac{1}{j + n_i h_i/\lambda_i}. \tag{8.46}$$

The last sum is simply related to the digamma function

$$\phi(x) = \sum_{i=1}^{\infty} \frac{x}{i(i+x)}$$

and (8.46) may thus be written as

$$-\frac{s}{\lambda_i}\{\phi(n_i - 1 + n_i h_i/\lambda_i) - \phi(-1 + n_i h_i/\lambda_i)\}.$$

Now for large x, $\phi(x) \sim \ln x + 0.5772\ldots$ and hence (8.46) is approximately

$$-\frac{s}{\lambda_i}\ln\left[\frac{h_i + \lambda_i}{h_i}\right].$$

The waiting time distribution thus has a Laplace transform which is approximately

$$\exp -\frac{s}{\lambda_i}\ln\left[\frac{h_i + \lambda_i}{h_i}\right]. \tag{8.47}$$

This is obviously the transform of a distribution where all the probability is concentrated at the point

$$\tau_i = \frac{1}{\lambda_i}\ln\left[\frac{h_i + \lambda_i}{h_i}\right] \tag{8.48}$$

where h_i is given by (8.34). Under the seniority rule the length of stay, for someone who does not leave, will therefore be approximately constant. The set of values $\{\tau_i\}$ will thus indicate the typical career prospectus of someone who stays with the firm.

The survivor function in grade i for someone subject to both forces of decrement is thus $e^{-\lambda_i \tau}$ for $0 \le \tau < \tau_i$, and 0 for $\tau \ge \tau_i$.

The foregoing results can also be used to find the completed length of service distribution for a person entering the system at the lowest level. (Those who enter higher up are in effect entering an essentially similar system with fewer grades.) Let $G(T)$ denote the survivor function of an individual entering at the lowest level, then we may write:

$$G(T) = \sum_{i=1}^{k} Pr\,\{\text{survives to } T \text{ and is in grade } i \text{ at } T\}. \tag{8.49}$$

To be in grade i at T an individual must have been promoted $i-1$ times and not have left. Suppose that these promotions occurred at times $T_1, T_2, \ldots T_{i-1}$ and let us first find the probability of not leaving conditional on these times. Between T_j and T_{j+1} the individual is subject to the loss rate λ_j. Hence

$$Pr\,\{\text{survives to } T \,|\, \text{no loss}\} = \exp -\sum_{j=0}^{i-1} \lambda_j(T_{j+1} - T_j) \tag{8.50}$$

with $T_0 = 0$, $T_i = T$. To find the unconditional probability we need the joint density function of $T_1, T_2 \ldots T_{i-1}$ and the probability that no further promotion takes place in $T - T_{i-1}$. This is

$$b_1(T_1)b_2(T_2 - T_1) \ldots b_{i-1}(T_{i-1} - T_{i-2})\{1 - B_i(T - T_i)\}, \qquad (8.51)$$

where $b_i(\cdot)$ is the density of length of stay in grade i (already found) and $B_i(\cdot)$ is its distribution function. The required probability is obtained by multiplying (8.50) and (8.51) and then integrating over the region $0 \leq T_1 \leq T_2 \leq \ldots \leq T_i \leq T$. This apparently complicated expression can easily be handled by taking its Laplace transform because it has the form of a convolution integral. If we have an integral of the form

$$\int\int \ldots \int_{0 \leq x_1 \leq x_2 \ldots x_k \leq X} a_1(x_1)a_2(x_2 - x_1) \ldots a_{k-1}(X - x_k)dx_1 \ldots dx_k$$

then its Laplace transform is

$$a_1^*(s)a_2^*(s) \ldots a_i^*(s).$$

In our case,

$$a_j(u) = b_j(u)e^{-\lambda_j u} \qquad (j = 1, 2, \ldots, i-1)$$
$$a_i(u) = \{1 - B_i(u)\}e^{-\lambda_i u}$$

so the required transforms are

$$a_j^*(s) = b_j^*(s + \lambda_j), \qquad (j = 1, 2, \ldots, i-1)$$
$$a_i^*(s) = \{1 - B_i^*(s + \lambda_i)\}/(s + \lambda_i).$$

Finally therefore, the transform of $G(T)$ is

$$G^*(s) = \sum_{i=1}^{k} \left\{\frac{1 - b_i^*(s + \lambda_i)}{s + \lambda_i}\right\} \prod_{j=1}^{i-1} b_j^*(s + \lambda_j). \qquad (8.52)$$

In the case of promotion at random we have from (8.39) that

$$b_j^*(s + \lambda_j) = h_j/(\lambda_j + h_j + s) \quad \text{with } h_k = 0.$$

Substituting in (8.52) and using the fact that

$$f^*(s) = 1 - sG^*(s)$$

we obtain

$$f^*(s) = \frac{\lambda_1}{\lambda_1 + h_1 + s} + \sum_{i=2}^{k}\left(\frac{\lambda_i}{\lambda_i + h_i + s}\right)\prod_{j=1}^{i-1}\left(\frac{h_j}{\lambda_j + h_j + s}\right) \qquad (h_k = 0). \quad (8.53)$$

As a check it may be verified that this transform reduces to $\lambda/(\lambda + s)$ if $\lambda_j = \lambda$ for all

j and to 1 if $s = 0$. The expression on the right-hand side of (8.53) can be resolved into partial fractions and inverted term by term to give a density function of the form

$$f(T) = \sum_{i=1}^{k} p_i(\lambda_i + h_i)e^{-(\lambda_i + h_i)T}, \qquad (8.54)$$

where $h_k = 0$ and

$$\sum_{i=1}^{k} p_i = 1.$$

This is the familiar mixed exponential distribution, although the p_i's need not all be positive as we shall see below.

In the simplest case when $k = 2$,

$$p_1 = (\lambda_1 - \lambda_2)/(\lambda_1 - \lambda_2 + h_1)$$
$$= n_1(\lambda_1 - \lambda_2)/\{n_1(\lambda_1 - \lambda_2) + n_2\lambda_2\},$$

which satisfies $0 \le p_1 \le 1$ if $\lambda_1 \ge \lambda_2$. Thus, whenever the loss intensity is higher in the first grade than in the second, we shall have a mixed exponential distribution of the kind we have met before. When fitting this distribution we found that fairly typical parameter values were $p_1 = \frac{1}{2}$ and $(\lambda_1 + h_1)/\lambda_2 = 10$. To achieve these values with the present model we should need $\lambda_1/\lambda_2 = 5.5$ and $n_2/n_1 = 4.5$. If this model represented the true state of affairs we should not be able to distinguish it from the others if we only observed the total length of service. To do so it would be necessary to identify the two grades, test whether their loss intensities were constant, and, if so, to see whether or not their values agreed with those predicted by the fitted distribution.

For $k > 2$ the CLS distribution can take on a variety of shapes. In order to illustrate some of the possibilities we shall list four examples below for the case $k = 3$.

(a) $n_1 = n_2 = n_3,\quad \lambda_j = \lambda/j \qquad (j = 1, 2, 3)$

$$f(T) = \frac{\lambda}{324}(209\,e^{-11\lambda T/6} + 75\,e^{-5\lambda T/6} + 40\,e^{-\lambda T/3}).$$

(b) $n_1 = n_2 = n_3,\quad \lambda_j = \lambda/(4-j) \qquad (j = 1, 2, 3)$

$$f(T) = \frac{\lambda}{60}(209\,e^{-11\lambda T/6} - 405\,e^{-3\lambda T/2} + 216\,e^{-\lambda T}).$$

(c) $n_1 = 2n_2 = 3n_3,\quad \lambda_j = \lambda/j \qquad (j = 1, 2, 3)$

$$f(T) = \frac{\lambda}{18}(14\,e^{-4\lambda T/3} + 3\,e^{-2\lambda T/3} + e^{-\lambda T/3}).$$

(d) $n_1 = 2n_2 = 4n_3,\quad \lambda_j = \lambda/(4-j) \qquad (j = 1, 2, 3)$

$$f(T) = \frac{\lambda}{6}(20\,e^{-4\lambda T/3} - (18 - 9T)\,e^{-\lambda T}).$$

The term involving the factor T in case (d) arises because, in that example, $\lambda_2 + h_2 = \lambda_3$ so that the inversion leading to (8.54) breaks down. The difficulty can be overcome by finding the limit of $f(T)$ as given in that equation when $\lambda_3 \to \lambda_2 + h_2$.

In those cases where the loss rates decrease as we move up the hierarchy we get a mixed exponential distribution with positive weights. The foregoing can therefore be regarded as a further model capable of explaining observed CLS distributions. When the loss rates increase as we go up the hierarchy the CLS distribution is unimodal and has less skewness.

A similar method can be used when promotion is by seniority. For large grade sizes the approximation represented by (8.48) yields a result immediately without recourse to the Laplace transform. It follows from the fact that the waiting time for promotion is constant in the limit that the hazard function of the CLS distribution is given by

$$\lambda(T) = \lambda_1, \qquad 0 \le T < \tau_1$$

$$\lambda(T) = \lambda_i \quad \text{for} \sum_{j=1}^{i-1} \tau_j \le T < \sum_{j=1}^{i} \tau_j, \qquad (i = 2, 3, \ldots, k).$$

The density function is then obtained by substituting in

$$f(T) = \lambda(T) \exp - \int_0^T \lambda(x)\,dx, \qquad T \ge 0.$$

The curve consists of a series of exponential segments. If the λ's form a decreasing sequence the density will be more skew than the exponential with an excess of frequency over the latter at either end of the range. In this respect it is similar in shape to the mixed exponential which resulted in these circumstances in the case of random promotion. In general terms, therefore, the promotion rule does not have a major influence on the broad shape of the distribution. It does, of course, strongly affect the career expectations of the individual.

Age distributions

As with the single grade system we can also determine the age distribution of the members of the system at any time. It will be time dependent, but will approach an equilibrium when it can easily be expressed in terms of the CLS distribution already found. We shall carry out the calculation for both promotion rules in order to compare their effects on the age structure.

Let $a_i(t)$ denote the equilibrium probability density function of age, measured from entry to the system, for members of grade i; $a(t)$ denotes the corresponding density for the whole system. Clearly

$$Na(t) = \sum_{i=1}^{k} n_i a_i(t). \qquad (8.55)$$

When promotion is by seniority the solution of our problem is almost immediate.

Under the assumption of large grade sizes promotion takes place from grades i to $i+1$ at time

$$t_i = \sum_{j=1}^{i} \tau_j, \qquad (i = 1, 2, \ldots, k-1).$$

All members of grade i will therefore have age between t_i and t_{i+1} and hence

$$a_i(t) = a(t) \Big/ \int_{t_i}^{t_{i+1}} a(x)\,dx, \qquad t_i \le t < t_{i+1}$$

$$= Na(t)/n_i, \qquad (i = 1, 2, \ldots, k), \tag{8.56}$$

where $a(t) = \mu^{-1}G(t)$, $t_0 = 0$ and $t_k = \infty$.

When promotion is at random the derivation starts with the observation that

$$a_i(t) \propto a(t)Pr\{\text{individual of age } t \text{ is in } i\}. \tag{8.57}$$

A person will be in grade i at time t if, and only if, he has been promoted $i-1$ times in $(0, t)$. The number of promotions in $(0, t)$ is a pure birth process with birth rate h_j when j promotions have occurred. The probability that $i-1$ promotions take place in $(0, t)$ can be written down from general theory (see also Chapter 9, equation (9.6) or Bartlett, 1955, Section 3.2). In the present notation this probability is

$$P(i-1|t) = \prod_{j=1}^{i-1} h_j \sum_{j=1}^{i} e^{-h_j t} \prod_{\substack{r=1 \\ r \ne j}}^{i} \frac{1}{(h_r - h_i)} \qquad (i = 1, 2, \ldots, k-1). \tag{8.58}$$

In our example below we shall take the case of equal h's, in which case the number of promotions has distribution

$$P(i-1|t) = \frac{(ht)^{i-1}}{(i-1)!} e^{-ht} \qquad (i = 1, 2, \ldots, k-2)$$

$$P(k-1|t) = 1 - \sum_{j=0}^{k-2} P(i|t). \tag{8.59}$$

Returning to (8.57) we thus have

$$a_i(t) = a(t)P(i-1|t) \Big/ \int_0^{\infty} a(x)P(i-1|x)\,dx \qquad (i = 1, 2, \ldots, k-1)$$

and

$$a_k(t) = \frac{1}{n_k}\left\{ Na(t) - \sum_{j=1}^{k-1} n_j a_j(t) \right\}. \tag{8.60}$$

As an example, consider a system with $\lambda_i = \lambda$ for all i and

$$n_{k-1} = n_k, \qquad n_i = n_{k-1} 2^{k-i-1} \qquad (i = 1, 2, \ldots, k-2) \tag{8.61}$$

with all n's large. This system has the property that the wastage and promotion densities are all equal to λ. The time of promotion out of grade i is from (8.48)

$$t_i = \sum_{j=1}^{i} \tau_j = \frac{i}{\lambda} \ln 2 = \frac{0.6932i}{\lambda}.$$

The times at which successive promotions take place thus increase in arithmetic progression.

The average ages for the members of each grade can easily be calculated. Thus we find

$$\left.\begin{aligned}
\xi_i &= \frac{\lambda N}{n_i} \int_{((i-1)\ln 2)/\lambda}^{(i \ln 2)/\lambda} t e^{-\lambda t} \, dt = \frac{1}{\lambda}\{1 + (i-2)\ln 2\} \\
&\qquad\qquad (i = 1, 2, \ldots, k-1) \\
\xi_k &= \frac{1}{\lambda}\{1 + (k-1)\ln 2\}
\end{aligned}\right\} \qquad (8.62)$$

The average length of time which a member has spent in a given grade is

$$\xi_i - t_{i-1} = \frac{1 - \ln 2}{\lambda} = \frac{0.3068}{\lambda} \qquad (i = 1, 2, \ldots, k-1; t_0 = 0)$$

$$\xi_k - t_{k-1} = \frac{1}{\lambda}.$$

This system has the rather interesting property that the average experience within a particular grade is the same for all grades except the highest.

When promotion is at random we substitute (8.59) into (8.60) with $a(t) = \lambda e^{-\lambda t}$ and find

$$a_i(t) = \frac{2\lambda}{(i-1)!}(2\lambda t)^{i-1} e^{-2\lambda t} \qquad (i = 1, 2, \ldots, k-1)$$

and hence

$$\left.\begin{aligned}
\xi_i &= i/2\lambda \qquad (i = 1, 2, \ldots, k-1) \\
\xi_k &= (k+1)/2\lambda
\end{aligned}\right\} \qquad (8.63)$$

On comparing the results given in this equation with those of (8.62) we see that ξ_2 is the same in both cases. For $i > 2$ the average age is greater when promotion is by seniority but when $i = 1$ is smaller. In the random case the average age in each grade is given by

$$\left.\begin{aligned}
\xi_i - \xi_{i-1} &= 1/2\lambda \qquad (i = 1, 2, \ldots, k-1; \xi_0 = 0) \\
\xi_k - \xi_{k-1} &= 1/\lambda
\end{aligned}\right\} \qquad (8.64)$$

As in the previous example we find that random promotion leads to greater average experience within each grade.

This example shows that there may well be practical advantages in adopting promotion policies which are not too rigidly tied to seniority.

8.5   GRADED SYSTEMS WITH MORE GENERAL LOSS MECHANISMS

The analysis of the last section allowed the loss rates to depend on grade but not on age or seniority within the grade. Both kinds of dependence can be incorporated into the models, but we increasingly have to rely on equilibrium or 'large grade size' approximations to get explicit results. We shall not pursue this line of development here though some further details are given in the Complements section. Instead we shall illustrate, by means of a single example, how the results already obtained are likely to be affected by departures from the assumption of constant propensity to leave.

We consider the case of a simple hierarchy with recruitment into the lowest grade only. Propensity to leave will be made a function of total length of service but not of grade or seniority in grade. It can thus be described in terms of its CLS distribution or hazard function. Discussion is limited to the case where vacancies are filled by promoting the most senior in the grade below. This rule has a simple consequence which makes it easy to find an approximation to the expected flow rates. It is this. *Under the operation of this rule every member of grade $i + 1$ must have served at least as long as any member of grade i $(i = 1, 2, \ldots, k - 1)$.* This implies that at any time T there will be a critical age, $t_i(T)$ say, at which people are being promoted from i to $i + 1$. This, of course, will fluctuate, but with large grade sizes we should expect its variation to be small. (The results of the last section demonstrated this feature for the model considered there.) It then follows that the proportion in grade i and below will be the same as the proportion of the population with age less than $t_i(T)$. Equating these two proportions we have

$$\int_0^{t_i(T)} a(x|T)\,dx = N_i/N, \qquad (i = 1, 2, \ldots, k - 1), \qquad (8.65)$$

where $a(x|T)$ is the density function of the age distribution. If $a(x|T)$ has a probability mass at $x = T$, as in (8.15), (8.65) requires slight modification. The reader may check that when $a(x|T) = e^{-\lambda x}$, then (8.65) is consistent with (8.48).

For a system in equilibrium $a(x|T) = \mu^{-1} G(x)$ and then (8.65) becomes

$$\mu^{-1} \int_0^{t_i} G(x)\,dx = N_i/N \qquad (i = 1, 2, \ldots, k - 1). \qquad (8.66)$$

This equation enables us to find the ages at which promotions take place. To find the equilibrium wastage rates we note that

$$\sum_{j=1}^{i} n_j w_j = \frac{N}{\mu} \int_0^{t_i} f(x)\,dx = \frac{N}{\mu} F(t_i) \qquad (i = 1, 2, \ldots, k-1). \qquad (8.67)$$

These equations determine the wastage rates; the remaining rates follow from (8.30) with $R_i(T) = 0$ for $i > 1$.

In order to investigate the effect of making this generalization let us take

$$f(T) = \frac{v}{c}(1 + T/c)^{-(v+1)}$$

for which

$$\mu = c/(v-1), \qquad G(T) = (1 + T/c)^{-v}$$

and

$$\int_T^{\infty} G(x)\,dx = \mu(1 + T/c)^{-v+1}.$$

On substituting these expressions into (8.67) and using (8.66) to eliminate t_i we find that

$$\sum_{j=1}^{i} n_j w_j \sim \frac{N}{\mu}\left\{ 1 - \left(\frac{N-N_i}{N}\right)^{v/(v-1)} \right\} \qquad (i = 1, 2, \ldots, k-1). \qquad (8.68)$$

By allowing v to tend to infinity in equation (8.68) we recover the solution for the exponential distribution.

To illustrate the theory we have given in Table 8.6 some values of w_1 when $k = 2$ for a type XI CLS distribution with $\mu = 1$. The variance of the distribution is given in the last row of the table; it does not exist for $v \leq 2$. These

Table 8.6 *Values of w_1 when $k = 2$ for a type XI CLS distribution with shape parameter v*

	v				
Structure	Limit as $v \to 1$	2	3	11	∞
$n_1 = 2n_2$	1.50	1.33	1.21	1.05	1.00
$n_1 = n_2$	2.00	1.50	1.29	1.07	1.00
$n_1 = \frac{1}{2}n_2$	3.00	1.67	1.37	1.08	1.00
Variance $\{v/(v-2)\}$	—	—	3.00	1.22	1.00

calculations show that the effect of increasing the dispersion of the CLS distribution is to increase the wastage rate of the lowest grade relative to that of the system as a whole. This implies a decrease in w_2 and hence a reduction in the promotion rate between grades 1 and 2. The amount of this change depends upon the structure of the organization, being greatest when the higher grade is larger than the lower. Put the other way round this conclusion states that the equilibrium promotion chances would be increased if the variability in completed length of service could be decreased. In a sense this is the converse of the conclusion reached in Section 5.4. There we found that a decrease in the variability of CLS would tend to reduce the relative size of the lowest grade.

8.6   VACANCY CHAINS

White (1970a) introduced a novel and fruitful way of studying mobility in systems with fixed grade sizes in which it is the flows of vacancies, rather than people, which are modelled. When a person moves from A to B a vacancy moves from B to A. Results about one kind of move are thus readily converted to the other. The concept of a vacancy has no meaning in a Markov chain model where class sizes are not fixed. It is only when a move creates a vacancy which has to be filled that we can speak of a vacancy arising. A loss from the system introduces a vacancy whose path through the system can be traced until it finally makes its exit at the point where someone is recruited. The length and path of such a sequence of moves can throw considerable light on the functioning of the system. We use the term *vacancy chain* to refer to the set of moves initiated by the loss of a person and ending with the entry of a new one.

In a manpower system, where promotion is by one step at a time and recruitment is limited to a few levels, the length of the chain is the difference between the levels of the grades at which the loss and entry take place. As such it is of rather limited practical interest. In more complex systems such as those relating to the national churches considered by White there is much more to be learnt from the study of vacancy chains.

Any model which seeks to model the flow of vacancies needs two elements. First, a chain must be initiated and this may be done by defining a loss mechanism for people. Secondly, we must specify how a vacancy moves. White (1970a) assumed that vacancies moved independently according to a Markov chain. We have already used a special case of this assumption when, earlier in the chapter, we supposed that a vacancy in a hierarchial system was filled from within with a fixed probability. We now extend this to allow the vacancy to move to any state by introducing transition probabilities $s_{ij}(i, j = 1, 2, \ldots k)$ for movements between i and j. Since a vacancy may also leave the system we introduce $s_{i,k+1}$ to denote the probability of leaving from grade i. The passage of a vacancy through the system can thus be described in terms of an absorbing Markov chain with transition matrix.

S	$\mathbf{s}'_{k+1}$
0	**1**'

where $\mathbf{S} = \{s_{ij},\}\mathbf{s}_{k+1} = \{s_{i,k+1}\}$ and $\mathbf{1} = (1, 1, \ldots, 1)$. In order to study the paths of individual vacancies we do not need to specify the means by which they enter. The theory already given in Chapter 3 for a cohort of individuals can be applied immediately to vacancies. We supplement the results given there by an expression for the distribution of the length of the vacancy chain. Let $s_i(T)$ be the probability that a vacancy which enters class i leaves the system after T steps and let $\mathbf{s}(T)$ be the vector of these probabilities. Then

$$s_i(T) = \sum_{r=1}^{k} Pr \{\text{moves from } i \text{ to } r \text{ in } T-1 \text{ steps}\} Pr\{\text{leaves from } r\}$$

$$= \sum_{r=1}^{k} s_{ih}^{(T-1)} s_{h,k+1}$$

from which

$$\mathbf{s}(T) = \mathbf{S}^{T-1}\mathbf{s}_{k+1} = \mathbf{S}^{T-1}(\mathbf{I} - \mathbf{S})\mathbf{1}'. \tag{8.69}$$

The last step follows from the fact that

$$\sum_{j=1}^{k} s_{ij} + s_{i,k+1} = 1 \qquad \text{for all } i.$$

White compared the distributions of the lengths of observed chains with the theoretical distributions given by (8.69). The vector of mean chain lengths is given by

$$\mu = \sum_{T=1}^{\infty} T\mathbf{s}(T) = (\mathbf{I} - \mathbf{S})^{-1}\mathbf{1}'. \tag{8.70}$$

This is an alternative derivation of the result of (3.30) showing that the means are given by the row sums of the fundamental matrix.

One can apply this analysis equally to discrete and continuous time systems because, as vacancies are filled instantaneously, 'real' time does not enter the analysis. The T which appears above simply counts the number of steps which a vacancy takes. An extension to the case when vacancies are not filled instantaneously has been given for discrete time in Bartholomew and Forbes (1979).

8.7 COMPLEMENTS

We have drawn extensively on renewal theory which is treated in most textbooks on stochastic process, for example, Bartlett (1955), Feller (1968), and Cox and Miller (1965). An early review paper is Smith (1958) and the more recent review of Markov renewal processes by Cinlar (1975), in which renewal processes are a special case, is also relevant. Perhaps the most convenient and accessible treatment with an applied emphasis is the monograph by Cox (1962).

The principal mathematical tool we have used is the Laplace transform. Widder (1946) is a basic reference, but many texts on applied and engineering mathematics deal with the subject and often give tables of transforms. The derivation of the result that $\lim_{T \to \infty} h(T) = \mu^{-1}$ can be made via the Laplace transform using what are known as Tauberian theorems.

Renewal models for social process have been treated in this chapter in a radically different way from the earlier editions. There the models were introduced in a very general form. The rather simpler results given here were then deduced as special cases valid for constant wastage rates, large grade sizes, or as limits when $T \to \infty$. In the present treatment we have aimed to deal with renewal models in a way which is more nearly parallel with that given for Markov models in earlier chapters. Thus, for the most part we have allowed loss rates to be grade specific but not depending on age or seniority. By this means we hope the essential simplicity of many of the results will become more apparent and so encourage their wider use. Readers who require a fuller treatment will need to consult the original papers.

The accuracy of the approximation $h^0(T)$ to the renewal density was investigated by Bartholomew (1963b) in a discrete time version and by Butler (1970) for continuous time. These investigations showed that the approximation is usually very good for life distributions more skew than the exponential. It usually turns out to be an upper bound for $h(T)$ but the conditions for this to be the case are not known. The renewal equation can immediately be generalized to the case where the life distribution is a function of the age of the system. The approximation $h^0(T)$ also generalizes and, again, proves to be satisfactory. This is a particularly useful result since there appear to be no general methods of solving the renewal equation in this case.

The exact treatment for the k-grade system when propensity to leave is a function of age is given in Bartholomew (1963a). It is shown there that the approximate method used in Section 8.5 is accurate enough for practical purposes for quite small systems. It may also be shown that the more general systems approach steady states in which the promotion and recruitment densities attain limiting values not depending on T. In the steady state, therefore, they appear, in an average sense at least, like the basic model of this chapter for which we assumed constant grade-specific loss rates. The arguments we used for deriving age and CLS distributions in Section 8.4 go through provided that the flows are Poisson processes. This is approximately true for large k-grade systems, as may be shown using arguments similar to that used in Section 8.2 for a single grade system.

A parallel treatment to that given here can be developed in discrete time. Minor complications arise concerning the order in which flows take place, and the rule for promoting the most senior has to be modified to allow for there normally being several with the same seniority. For practical work in manpower planning discrete time renewal models have computational advantages. A treatment which utilizes the duality of person and vacancy flows is given in Bartholomew and Forbes (1979). They also give a version in which there may be delays in filling vacancies. This is clearly more reasonable in many applications.

Renewal models have found many applications in manpower planning, but usually in a more general form than those discussed here. They are normally expressed as computer algorithms and analysed numerically. An account of one such model known as the KENT model is given by Hopes in Bartholomew (1976b) and by Wishart in Smith (1976). Applications of the theory of k-grade renewal processes to industrial replacement are discussed in Bartholomew (1963c) and to manpower planning in Robinson (1974). The vacancy chain idea has been applied to housing markets by White (1971).

CHAPTER 9

The Simple Epidemic Model for the Diffusion of News, Rumours and Ideas

9.1 INTRODUCTION

The diffusion of information in a social group is a phenomenon of considerable interest and importance. Ideas, rumours, fashions, news, and innovations are all examples of things which spread from person to person in a population in a more or less haphazard fashion. A large amount of research has been devoted to this subject and some of this has involved the construction of stochastic models. Our object in this and the following chapter is to describe some of these models and to use them to gain insight into the nature of the phenomenon.

The system which we shall study may be described as follows. There is a population of N units which we shall usually describe as people, but which may be groups of people as, for example, families. Information is transmitted to members of the group from a source either at an initial point in time or continuously. For example, the source may be a television commercial, a newspaper, a roadside advertisement hoarding, or a group of people introduced into the population from outside. Persons who receive the information may become 'spreaders' themselves by transmitting the information to others whom they meet. The process of diffusion continues until all have heard the news or until transmission ceases.

In many cases a stochastic model will be required to provide an adequate description of the process. The chance element enters at two points. Whether or not a given person hears the information will depend on (a) his coming into contact with the source or a spreader, and (b) on the information being transmitted when contact is established. In social systems such as the armed forces where there are well-defined channels of communication the chance factor is negligible. In less rigidly organized systems neither (a) nor (b) is a certain event and so the development of the process is unpredictable. Hence it can only be described stochastically.

In spite of the obvious stochastic nature of the process much of the existing theory is deterministic. The reason for this is that the mathematics of the full stochastic version of many of the models is so intractable. We shall often find that we have to fall back on the deterministic version of a model to make progress. However, the deterministic model will be regarded as an approximation and models will always be formulated stochastically in the first instance.

The progress of the spread of information may be described in a variety of ways. Most work, empirical and theoretical, has centred on the growth of the number of people who have received the information at any time—called hearers. There are two random variables relevant to this particular aspect of the process which we shall study. The first is the number of hearers at time T which will be denoted by $n(T)$ or, where there is no risk of confusion, by n. The second way of describing the growth in the number of hearers is by the time taken for n people to hear. This time is denoted by T_n. The two random variables, T_n and $n(T)$, stand in an inverse relation to one another and for many practical purposes either will serve. Our choice between them will be governed chiefly by the mathematical advantages which each offers in a particular circumstance.

When observed values of $n(T)$ are plotted against T we have an empirical growth curve which often turns out to be sigmoid in shape. This indicates an initial accelerating rate of spread followed, towards the end, by a deceleration. One of the main objects of theoretical research has been to construct models to provide a satisfactory explanation for this feature. It will come as no surprise to learn that growth curves with this characteristic can be generated by at least two different models.

Much of the empirical work relates to the diffusion of innovations. Here it is important to distinguish between the time at which a potential innovator hears of the innovation and the time at which he adopts it. In many cases adoption may be almost immediate so that the distinction is of no practical importance. In other cases this is not so. The time to adoption must be treated as the sum of the time to hearing and the subsequent delay before adoption. Ryan and Gross (1943) give data on both aspects of a study of the spread of hybrid corn in two Iowa farming communities. In this and the following chapter we shall concentrate on the hearing aspect of the total process.

A second class of random variables relates to the number of intermediaries between the original source of information and the hearer. Someone who receives the news direct from the source is called a first-generation hearer. Those who first hear from a member of the first generation are second-generation hearers, and so on. The number of hearers in the gth generation at time T will be denoted by $n_g(T)$, where g takes integer values from 1 to N. We shall be mainly interested in the ultimate number of hearers in each generation for which we shall use the notation $n_g(= n_g(\infty))$. These random variables have a particular interest because news which spreads from person to person is liable to distortion. The distribution of the n_g's therefore gives us some idea of the extent to which a message may be distorted in the course of its diffusion through a social group.

In the case of a population consisting of a school or club it may be realistic to think of all the members as located at a single point since diffusion will tend to take place when the members of the population are together. Similarly, if the means of communication is by telephone or radio, location may be irrelevant over small areas at least. In many other situations, however, the spatial spread of the information will be of prime interest. This is well illustrated by Hägerstrand's

(1967) pioneering study of the diffusion of innovations in Swedish communities and with Morrill's (1965) study of the growth of Negro ghettos in U.S. cities. Two aspects of the process are of interest in the study of spatial diffusion. One is the rate at which the news spreads throughout the area. That is, we are still interested in the random variable $n(T)$, but we shall now wish to regard it as a function of location as well as of time. The other aspect is the pattern of the state of knowledge at any given point in the course of the diffusion. Pattern is not an easy concept to quantify in this context, but it refers to regularities in the location of hearers in relation to the original source. For example, a clustering of hearers in a north–south direction or along a main road would constitute a pattern indicative of the manner in which the diffusion had taken place.

There are obvious similarities between epidemics and social diffusion, but there are also important differences. For example, the random variables $\{n_g\}$ have no particular relevance in epidemic theory and it is only recently that they have received any attention at all. The word 'epidemic' provides a convenient description of either process and we shall continue to use it without meaning to imply a medical application. A comprehensive account of epidemic theory up to 1974 is given in Bailey (1975).

The plausibility of the various assumptions that we shall have to make depends partly on the kind of information being transmitted. The term 'information' is being used here in a neutral sense to cover such diverse things as rumours, news, advertising material, and public announcements. Our models should not, therefore, be regarded as of universal application but rather as pointers to what might happen under various specified sets of conditions.

Models for diffusion can be conveniently classified according to whether or not they include any mechanism for stopping the act of spreading the information. If everyone who acquires the information continues to pass it on indefinitely, then it will usually be the case that the whole population will ultimately be informed. In these circumstances the main object of the theory is to study the approach to that limiting state. The remainder of this chapter is devoted to such models which, following a well-established usage in epidemic theory, we call *simple epidemic models*. In practice, however, there are many reasons why spreaders may cease this activity, in which case it is possible for the process to die out before the whole population has been informed. This possibility raises a range of questions about the extent to which any significant diffusion will take place at all. Models incorporating this feature will be called *general epidemic models* and they form the subject of Chapter 10.

9.2 THE BASIC MODEL

Description of the model

Our basic model is a special case of the pure birth process. Let E_s denote the transmission of the information from a source to any given member of the population. This is assumed to be a random event with

$$Pr\{E_s \text{ in } (T, T + \delta T)\} = \alpha \delta T \qquad (\alpha > 0), \tag{9.1}$$

where α is described as the *intensity of transmission of the source*. In this model we treat contact with the source and reception of information from the source as a single event. The above assumption may thus be expressed by saying that all members are equally exposed to the source. It would be a plausible assumption if the source were a television commercial and if the population consisted of regular viewers. It would not be realistic if the population also included people who rarely view the programme. Our model can be generalized to include variable exposure. However, we shall show that, in most circumstances, α plays a minor role in the development of the process; a simple assumption will therefore suffice.

Let us denote the transmission of news between any given pair of individuals by E_I. Our second assumption about the process is that

$$Pr\{E_I \text{in} (T, T + \delta T)\} = \beta \delta T \qquad (\beta > 0), \tag{9.2}$$

where β is the *intensity of transmission between individuals*. We assume that this probability is the same for all pairs of individuals. This, in turn, implies that we have a homogeneously mixing population. In such a population any uninformed member is equally likely to receive the news from any of the n persons who are active spreaders. The assumption of homogeneous mixing seems plausible only in very small groups and experimental evidence supports this view. Nevertheless, there are advantages in studying the simple model first and then introducing greater realism by way of appropriate generalizations. Finally, we assume that all transmissions, whether from the source or between pairs of individuals, are independent of each other.

We are now in a position to relate the process that we have described to the pure birth process. When exactly n people have received the information we shall say that the system is in state n. A stochastic process is a time-homogeneous birth process if the probability of a transition from state n to state $n + 1$ is given by

$$Pr\{n \to n + 1 \text{ in } (T, T + \delta T)\} = \lambda_n \delta T \qquad (\lambda_n \geq 0) \tag{9.3}$$

and if no other types of transition (for example, $n \to n - 1$) are possible. It is obvious that n can only increase and the identification of the two processes will be complete when we have expressed λ_n in terms of the parameters of our model. The number who have heard can be increased in one of two ways. Either the next person to hear will receive the information from the source or from another person. As there are $N - n$ persons who have not heard, the total contribution to

λ_n from the source is $(N - n)\alpha\,\delta T$. The contribution from communication between persons is obtained as follows. Of all the possible pairs which could be formed there are $n(N - n)$ which consist of one 'knower' and one 'ignorant'. These are the only pairs which can give rise to the transition $n \to n + 1$ and the total probability associated with them is $n(N - n)\beta\,\delta T$. Combining these results we have

$$\lambda_n = (N - n)(\alpha + \beta n) \qquad (n = 0, 1, \ldots, N - 1). \tag{9.4}$$

The theory associated with out model can thus be developed from that for the birth process with quadratic birth rate, given by (9.4).

The model which we have described was proposed by Taga and Isii (1959), but it is almost identical with the simple epidemic model discussed, for example, in Bailey (1975). In epidemic theory the source consists of one or more persons who introduce the infection to the group. Thus if one person starts the epidemic we have to put $\alpha = \beta$ when we find

$$\lambda_n = \beta(n + 1)(N - n) \qquad (n = 0, 1, \ldots, N - 1). \tag{9.5}$$

This particular case has received the greatest attention and we shall return to it later. In the application to diffusion of news it is not necessary that $\alpha = \beta$ or that β should be a multiple of α as in epidemic theory. It is also worth drawing attention to the fact that the assumptions of the model seem more reasonable when it is applied to the diffusion of news. For example, the application to the epidemic requires that we ignore the incubation period of the disease and that each infected individual remains infectious until the epidemic is over. Both assumptions are unrealistic for many infectious diseases but are quite reasonable for the diffusion of some kinds of information. They then require that the information is transmitted instantaneously and that it is not forgotten.

Analysis of the model

We shall now use the birth process model to make deductions about the development of the diffusion process in time. In view of the fact that our model is a pure birth process it is natural to begin by studying the distribution of $n(T)$. Historically this was the course followed, and we shall begin by briefly describing some of the results which have been obtained. However, it is now clear that the approach via the random variables $\{T_n\}$ is capable of yielding more information about the process in a much simpler fashion.

The expression for the distribution of $n(T)$ may be found in Bartlett (1955, Section 3.2). It is given by

$$Pr\{n(T) = 0\} = e^{-\lambda_0 T},$$

$$Pr\{n(T) = n\} = \prod_{i=0}^{n-1} \lambda_i \sum_{i=0}^{n} \frac{e^{-\lambda_i T}}{\prod_{\substack{j=0 \\ j \neq i}}^{n} (\lambda_j - \lambda_i)}. \tag{9.6}$$

Since λ_i can be found from (9.5) in terms of α, β, and N the problem is solved in principle. Even for small values of N the computation of the distribution is formidable; it is given by Bailey (1975, Table 5.1) for $N = 10$ and $\alpha = \beta$. For large values of N the task is not practicable. The feature of the distribution which is of greatest interest is the mean, $\bar{n}(T)$. When plotted as a function of T it gives a visual representation of the expected development of the process. If we are primarily interested in the rate at which the news is spreading at T we would wish to plot the derivative of $\bar{n}(T)$. This latter curve is often called the 'epidemic curve' and provides a clearer picture of the growth and subsequent decline of the epidemic. The expressions for $\bar{n}(T)$ and its derivative were obtained by Haskey (1954) and are given by Bailey (1975). The formulae are rather complicated but computations have been carried out by Haskey (1954), Bailey (1975), and Mansfield and Hensley (1960) for $N \leq 40$. Two epidemic curves, plotted from their calculations, are given in Figure 9.1 for the case $\alpha = \beta$. Bailey (1963) gave a complete solution in terms of known functions and later, Bailey (1968b), an approximation, valid in large populations, based on a perturbation of the deterministic theory to be discussed later.

The abscissa on Figure 9.1 is plotted in units of β^{-1}. This is the expected time taken for any given pair of people to meet—a fact which follows directly from

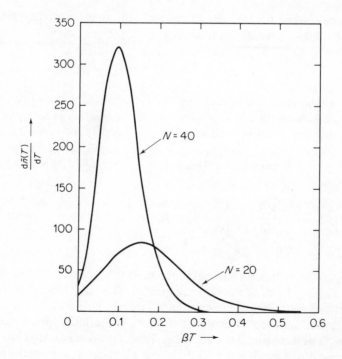

Figure 9.1   The epidemic curve for the simple epidemic when $\alpha = \beta$.

(9.2). Thus, for example, it is clear that the diffusion is completed in the case $N = 40$ in about one-third of the average time taken for two given persons to meet. It will be noted that the spread is more rapid in the larger of the two populations shown in the figure. We shall encounter this phenomenon again below, where the reason for it will be made clear.

The investigation of the form of the epidemic curve for larger N is facilitated by the following observation. Let T' denote the time taken for the news to reach a specified member of the population and let $F(T')$ be its distribution function. Then it is clear that

$$\bar{n}(T) = NF(T) \tag{9.7}$$

and

$$\frac{d\bar{n}(T)}{dT} = Nf(T). \tag{9.8}$$

Equation (9.8) shows that the epidemic curve is proportional to the probability density function of the time taken to inform a given member of the population. Williams (1965) exploited this relationship to deduce the moments and asymptotic form of the epidemic curve. We shall derive similar results by considering the random variables $\{T_n\}$.

The time at which the nth person receives the information can be represented as a sum of random variables as follows:

$$T_n = \sum_{i=1}^{n} \tau_i, \tag{9.9}$$

where τ_i is the time interval between the reception of the information by the $(i-1)$th and ith persons. It now follows from the Markov property of the birth process that the random variables $\{\tau_i\}$ are independently and exponentially distributed with parameters λ_{i-1} $(i = 1, 2, \ldots, N-1)$. Using this fact the exact distribution can, in principle, be found. It was investigated by Kendall (1957) for $n = N$ who found that when $\alpha = \beta$,

$$W = (N+1)T_N - 2 \ln N$$

has the limiting density function

$$f(W) = 2K_0(2e^{-\frac{1}{2}W})e^{-W} \qquad (-\infty < W < \infty) \tag{9.10}$$

where $K_0(\cdot)$ is the modified Bessel function of the second kind and zero order. This is a unimodal density with some positive skewness. Note that the limiting distribution is not normal because the τ_i's do not satisfy the conditions of the Lindberg–Feller central limit theorem for non-identically distributed random variables. (For these conditions see Feller, 1966, Theorem 3, p. 256.) The result has been rediscovered by McNeil (1972). For our purposes we shall find it possible to obtain all the information we require from the cumulants of T_n.

The rth cumulant of τ_i is

$$\kappa_r = (r-1)!/\lambda_{i-1}^r.$$

Since the τ_i's are independent the rth cumulant of their sum is

$$\kappa_r(T_n) = (r-1)! \sum_{i=1}^{n} \lambda_{i-1}^{-r}, \qquad (9.11)$$

where λ_{i-1} is given by equation (9.4). Our main interest is in the expectation of T_n, which is obtained by putting $r = 1$ in (9.11). This gives

$$E(T_n) = \kappa_1(T_n) = \sum_{i=0}^{n-1} \frac{1}{(N-i)(\alpha+\beta i)}$$

$$= \frac{1}{\alpha+\beta N} \sum_{i=0}^{n-1} \frac{1}{N-i} + \frac{\beta}{\alpha+\beta N} \sum_{i=0}^{n-1} \frac{1}{\alpha+\beta i}$$

$$= \frac{1}{\beta(N+\omega)} \{\phi(N) - \phi(N-n) - \phi(\omega-1) + \phi(n+\omega-1)\} \qquad (9.12)$$

where $\omega = \alpha/\beta$ and

$$\phi(x) = \sum_{i=1}^{\infty} \frac{x}{i(i+x)} = \sum_{i=1}^{\infty} \left(\frac{1}{i} - \frac{1}{i+x}\right)$$

is the digamma function. This function is tabulated in the *British Association Mathematical Tables*, Volume I (1951) for $x = 0.0(0.01)1.0$ and $10.0(0.1)60.0$. For large x

$$\phi(x) \sim \ln x + \gamma,$$

where $\gamma = 0.5772 \ldots$ is Euler's constant.

We can now use (9.12) to study the expected development of the diffusion process. Nothing essential will be lost if we suppose that N is large. The limiting behaviour of $E(T_n)$ depends on whether or not n is near to zero or N. Initially let us suppose that n/N is fixed and denoted by p. Then if N is large and $p \neq 0$ or 1

$$E(T_{Np}) \sim \frac{1}{\beta(N+\omega)} \left\{\ln \frac{pN+\omega-1}{1-p} + \gamma - \phi(\omega-1)\right\}. \qquad (9.13)$$

If ω is fixed we have approximately that

$$E(T_{Np}) \sim \frac{\ln N}{N\beta} + \frac{1}{N\beta} \ln \frac{p}{1-p}. \qquad (9.14)$$

Two important conclusions follow from these formulae. First, it is clear that the parameter β is much more important than ω, and hence α, in determining the rate of diffusion. Secondly, a proportion p will be reached faster in a large than in a small population. Both of these phenomena can be explained by reference to the form of λ_n. Except near the start and finish, the coefficient of β in λ_n is an order of magnitude larger than that of α. Also λ_n is an increasing function of N.

The above formulae do not hold if p is zero or one because, in deriving them, we had to assume that N, $N - n$ and n were large. The case $p = 1$, or $n = N$, is of particular interest because T_N is then the *duration* of the epidemic. Proceeding to the limit we find the expression comparable to (9.13) to be

$$E(T_N) \simeq \frac{2 \ln N - \phi(\omega - 1) + \gamma}{\beta(N + \omega)}. \tag{9.15}$$

Thus (9.14) shows that half of the population will have heard in time $(\ln N)/N\beta$ and (9.15) that all will have heard in twice that time.

Equation (9.13) also enables us to find the asymptotic form of the function $\bar{n}(T)$. This follows from the fact that, when n and N are both large with $0 < p < 1$,

$$\frac{n}{N} = p \doteq F(T)|_{T = E(T_n)}. \tag{9.16}$$

Hence the function on the right-hand side of (9.13) is $F^{-1}(p)$. It follows that, for large N and $\bar{n}(T)$ not near 0 or N,

$$\bar{n}(T) \sim N\left(\frac{e^X - \omega + 1}{e^X + N}\right) \tag{9.17}$$

where $X = \{\beta(N + \omega)T - \gamma + \phi(\omega - 1)\}$. For small values of ω/N this is an S-shaped function, similar to the normal ogive. Under the conditions discussed below, $\bar{n}(T)/N$ becomes the same as the probability integral of a logistic distribution. The epidemic curve is easily obtained by differentiating (9.17). It is a unimodal curve with some degree of positive skewness. An illustrative diagram is given in Williams (1965).

We have already found an expression for the median of the epidemic curve. It is a simple matter to find its mean, which is the time that a randomly selected individual can expect to wait before hearing the news. If we write this quantity as $E(T')$ it is obvious that

$$E(T') = \frac{1}{N} \sum_{n=1}^{N} E(T_n).$$

The summation over n on the right-hand side of (9.12) involves the manipulation of double sums but readily yields

$$E(T') = \frac{1}{\beta N}\{\phi(N + \omega - 1) - \phi(\omega - 1)\}. \tag{9.18}$$

This equation is a slight generalization of equation (21) in Williams (1965), whose analysis was restricted to the case when ω is an integer. As we should expect from the result for the median, the mean of the epidemic curve is of order $(\log N)/N$.

The consideration of mean values does not tell us how far an actual realization of the process may depart from our expectation. To investigate this we may consider the distribution of the random variables $\{T_n\}$. We already have a general expression for the rth cumulant of T_n in (9.11). This result will now be used to show that the time taken for n people to hear is subject to considerable variation. Using the identity

$$\frac{1}{x^r y^r} \equiv \frac{1}{(x+y)^{2r}} \sum_{i=1}^{r} \binom{2r-i-1}{r-1}(x+y)^i \left(\frac{1}{x^i}+\frac{1}{y^i}\right)$$

the cumulants given by (9.11) in conjunction with (9.4) may be written

$$\kappa_r(T_n) = \frac{(r-1)!}{\beta^r} \sum_{i=1}^{r} \binom{2r-i-1}{r-1} \frac{1}{(N+\omega)^{2r-i}}$$
$$\left\{\sum_{j=1}^{n} \left(\frac{1}{(N+1-j)^r}+\frac{1}{(j+\omega-1)^r}\right)\right\} \qquad (r>1). \quad (9.19)$$

Introducing the polygamma functions defined by

$$\phi^{(s)}(x) = (-1)^s(s-1)! \sum_{i=1}^{\infty} \frac{1}{(i+x)^s} \qquad (s>1)$$

the cumulants may be written:

$$\kappa_r(T_n) = \frac{(r-1)!}{\beta^r} \sum_{i=1}^{r} \binom{2r-i-1}{r-1} \frac{(-1)^{i-1}}{(i-1)!(N+\omega)^{2r-i}}$$
$$\times \{\phi^{(i)}(N-n) - \phi^{(i)}(N) + \phi^{(i)}(\omega-1) - \phi^{(i)}(\omega+n-1)\}. \quad (9.20)$$

In the limit as $N \to \infty$, with $n/N = p$ held fixed with $0 < p < 1$,

$$\kappa_r(T_n) \sim \frac{(-1)^r}{\beta^r(N+\omega)^r} \phi^{(r)}(\omega-1) \qquad (r>1). \quad (9.21)$$

On the other hand, if $(N-n)$ is fixed so that $p \to 1$ as $N \to \infty$ a different limiting form is obtained. For example, in the case $n = N$,

$$\kappa_r(T_N) \sim \frac{(-1)^r}{\beta^r(N+\omega)^r} \{\phi^{(r)}(\omega-1) + \phi^{(r)}(0)\} \qquad (r>1). \quad (9.22)$$

A notable feature of (9.21) is that it does not depend on n. In particular, the asymptotic variance of T_n is independent of n. The polygamma functions are tabulated in the British Association Tables referred to above so that numerical values for the cumulants are readily obtainable. It follows from (9.21) that the variance of T_n is a decreasing function of ω, which means that its variability goes down as the 'strength' of the source is increased. The extent of this reduction in variance can be gauged from the fact that $\phi^{(2)}(0) = 1.6449$ and $\phi^{(2)}(1) = 0.6449$. Even when n is large the distribution of T_n may have considerable skewness and kurtosis. Some illustrative calculations are given in Table 9.1.

Table 9.1 *Asympotic values of the skewness,* $\sqrt{\beta_1}$, *and the kurtosis,*
β_2, *of* T_n

		ω				
		1	2	5	10	∞
$0 < p < 1$	$\sqrt{\beta_1}$	1.14	0.73	0.47	0.32	0.00
	β_2	5.40	4.19	3.44	3.21	3.00
$p = 1$	$\sqrt{\beta_1}$	0.81	0.81	0.97	1.04	1.14
	β_2	4.20	4.33	4.90	5.12	5.40

The covariance of any pair of T_n's can be found at once from the fact that

$$\text{cov}(T_n, T_{n+j}) = \text{var}(T_n) \qquad (j \geq 0). \qquad (9.23)$$

The deterministic approximation

With the simple birth process model there is no need to have recourse to deterministic methods. All that we required to know about the expected behaviour of the process can be determined from the stochastic model. Nevertheless, it is instructive to consider a deterministic version of the model as a preparation for the analysis of the more intractable models which occur later. We treat $n(T)$ as a continuous function. According to (9.4) the expected amount by which it will increase in $(T, T + \delta T)$ is $(N - n)(\alpha + \beta n)\delta T$. In the deterministic treatment we suppose that it increases by exactly this amount in each small increment of time. This implies that $n(T)$ satisfies the differential equation

$$\frac{dn(T)}{dT} = (N - n(T))(\alpha + \beta n(T)). \qquad (9.24)$$

Solving this with the boundary condition $n(0) = 0$ we have for $T > 0$,

$$n(T) = N \frac{\exp[\beta(N + \omega)T - 1]}{\exp[\beta(N + \omega)T] + N/\omega}. \qquad (9.25)$$

A comparison of this with (9.17) shows that the stochastic solution and the deterministic approximation are not the same. The deterministic curve lags behind the stochastic curve by an amount which varies with both N and ω. The position for small N is illustrated by Bailey (1975, Figures 5.1 and 5.2). For large N the same conclusion follows from a comparison of the means of the two epidemic curves. In the deterministic case,

$$E(T') = \frac{1}{N\beta} \ln\left(\frac{N + \omega}{\omega}\right). \qquad (9.26)$$

Williams (1965) showed that this is always less than the stochastic version given by (9.18). In order for the deterministic solution to be equivalent to the asymptotic stochastic solution it is necessary for ω to be large. It may easily be verified that (9.17) and (9.25) become identical in the limit as $\omega \to \infty$. For fixed ω the difference between the deterministic and stochastic means is of the order of N^{-1}. This may appear to be insignificant until it is recalled that the duration of the whole epidemic is of order $(\log N)/N$. On this time scale a difference of order N^{-1} can be of practical importance for moderate values of N. We must therefore be on the alert for this kind of occurrence when we come to more complex models.

Models with imperfect mixing

The principal assumptions of the birth process model are (a) that all members are equally exposed to the source, and (b) that all pairs of members have equal likelihood of communicating. Assumption (a) is not crucial unless α is large compared to β because the main contribution to λ_n, once the process has started, comes from the term $\beta(N - n)n$. The second assumption is certainly invalid in most human populations. It is therefore necessary to investigate the effect of relaxing (b).

We begin by going to an extreme and suppose that there is no communication at all between members of the population. The diffusion is thus entirely attributable to the source. This case is, in fact, covered by our model and is obtained by setting $\beta = 0$. Expressions for the epidemic curve and duration can be obtained from those already given by letting $\omega \to \infty$ with N fixed.[†]However, they can easily be obtained from first principles along with other results which cannot be found in the general case. When $\beta = 0$ we have what is called the pure death process (see Bailey, 1964, Section 8.5). The exact distribution of $n(T)$ turns out to be binomial with

$$Pr\{n(T) = n\} = \binom{N}{n} e^{-\alpha(N-n)T} (1 - e^{-\alpha T})^n \qquad (n = 0, 1, \ldots, N). \quad (9.27)$$

Hence

$$\bar{n}(T) = N(1 - e^{-\alpha T}) \quad (9.28)$$

and the epidemic curve is

$$\frac{d\bar{n}(T)}{dT} = N\alpha e^{-\alpha T}. \quad (9.29)$$

There is thus a marked qualitative difference between this case and that illustrated on Figure 9.1. The rate of diffusion declines continuously with time instead of

---

† Note that this is not the same set of conditions which led to the equivalence of the deterministic and stochastic epidemic curves in the last section. In that case N was allowed to tend to infinity.

first rising to a maximum. It should therefore be possible, in practice, to form some judgement about the relative importance of interpersonal and source-personal diffusion by an inspection of the empirical epidemic curve. In a study of the diffusion of information about a new drug reported in Coleman (1964b, Figure 17.2) it was found that the growth of knowledge was roughly exponential. This can be interpreted to mean that advertising rather than personal recommendation led to adoption of the new drug.

The time taken for a proportion p of the population to hear the news is also easy to obtain. Since the T_n's are partial sums of the τ's it follows at once that

$$E(T_n) = \frac{1}{\alpha} \sum_{i=1}^{n} \frac{1}{N-i+1}. \tag{9.30}$$

If N is large and $p(=n/N)$ is not near to one

$$E(T_{Np}) \sim -\frac{1}{\alpha}\ln(1-p). \tag{9.31}$$

This result stands in marked contrast to (9.13) and (9.14). In the present model the time taken to reach a given proportion does not depend on N. By combining the results obtained from our two extreme assumptions we may conclude, in general, that the time taken to reach a proportion p cannot be an increasing function of population size. This argument does not cover the limiting case $p = 1$. In this case (9.31) must be replaced by

$$E(T_N) \sim \frac{1}{\alpha}\ln N. \tag{9.32}$$

Thus the total duration does increase with size but only slowly.

In order to chart the territory between the two extreme degrees of mixing we shall consider the case of a stratified population. One such model was discussed by Haskey (1954). He supposed that the population was composed of two strata with different rates of contact between and within groups. A deterministic model with k strata was analysed by Rushton and Mautner (1955); we return to this later. A similar kind of model for a population with $N/3$ strata was solved semi-deterministically by Coleman (1964b, Chapter 17). We shall first consider a simpler model of the same kind.

Suppose that the population is made up of k strata of equal size. The members of all strata are equally exposed to the source and the rate of contact between members of the *same* stratum is β, as before. However, there is no contact at all between the members of different strata. Under these conditions the diffusion in each stratum develops according to the theory of the preceding sections. The diffusion in the system as a whole is then found by pooling the results; no new theory is required.

Coleman (1964b) developed the theory for the case $N = 2k$. In this case each stratum is of size 2 and there is no difficulty in obtaining the exact distribution of

$n(T)$ from (9.6). Let $n_i(T)$ denote the number who have heard at time T in the ith stratum $(i = 1, 2, \ldots, N/2)$; then

$$\lambda_0 = 2\alpha, \qquad \lambda_1 = \alpha + \beta, \qquad \lambda_2 = 0.$$

Hence

$$\left. \begin{aligned} Pr\{n_i(T) = 0\} &= e^{-2\alpha T} \\ Pr\{n_i(T) = 1\} &= \frac{2\alpha(e^{-2\alpha T} - e^{-(\alpha + \beta)T})}{\beta - \alpha} \\ Pr\{n_i(T) = 2\} &= 1 - Pr\{n_i(T) = 0 \text{ or } n_i(T) = 1\} \end{aligned} \right\}. \tag{9.33}$$

The limiting forms appropriate when $\alpha = \beta$ are easily deduced. Since

$$n(T) = \sum_{i=1}^{N/2} n_i(T)$$

it follows that

$$\bar{n}(T) = \frac{N}{2}\bar{n}_i(T) \quad \text{and} \quad \operatorname{var} n(T) = \frac{N}{2}\operatorname{var} n_i(T),$$

and that $n(T)$ is approximately normal. The moments of $n_i(T)$ are readily found from (9.33); in particular,

$$\left. \begin{aligned} \bar{n}_i(T) &= 2 - \frac{2(\omega e^{-(\alpha + \beta)T} - e^{-2\alpha T})}{\omega - 1} & \omega \neq 1 \\ \bar{n}_i(T) &= 2\{1 - (1 + \beta T)e^{-2\beta T}\} & \omega = 1 & \quad (\text{i.e. } \alpha = \beta) \end{aligned} \right\}. \tag{9.34}$$

The epidemic curve in the case $\alpha = \beta$ has the form

$$\frac{1}{N}\frac{d\bar{n}(T)}{dT} = \beta\{1 + 2\beta T\}e^{-2\beta T}. \tag{9.35}$$

The density on the right-hand side of (9.35) is decreasing with mean $3/4\beta$. This may be compared with the approximate value of $(\ln N)/N\beta$ for a homogeneously mixing population. The mean for a population with no mixing at all and transmission intensity $\alpha = \beta$ from the source is $1/\beta$. With the limited degree of communication permitted by our stratified model the expected time to hear is reduced but is still independent of the population size. If $\beta = \infty$ the second member of each stratum automatically receives the news at the same time as the first member. Our model is then equivalent to a freely mixing population made up of $N/2$ pairs.

The foregoing analysis expresses in quantitative form the obvious conclusion that incomplete mixing reduces the rate of diffusion—at least in a population of small non-communicating strata. This conclusion can be strengthened by considering the duration of the diffusion for general k. Let us denote by $T_{(i)}$ the duration for the ith stratum. Then

$$T_N = \max_i T_{(i)}.$$

The probability distribution of T_N is then that of the largest member of a sample of size k from the distribution of the duration in a stratum. It is possible to make progress with the general theory of the distribution of T_N, but an inequality due to Gumbel (1958) provides sufficient information for our purposes. For any random variable x with finite mean μ and variance σ^2 he states that

$$E(x_{\max}) \leq \mu + \sigma \frac{k-1}{\sqrt{2k-1}}$$

where k is the sample size. Applying this result to the case of T_N, μ is the average duration for a stratum and σ^2 is its variance. If N/k is large we may use the asymptotic forms and obtain

$$E(T_N) \leq \frac{2\ln(N/k)}{\beta(N/k+\omega)} + \frac{\sqrt{\phi^{(1)}(\omega-1)}}{\beta(N/k+\omega)} \frac{(k-1)}{\sqrt{2k-1}}$$

$$= k\frac{2\ln N}{N\beta} + O(N^{-1}) \tag{9.36}$$

for fixed ω. This result suggests but does not prove, because of the inequality, that division into k strata multiplies the duration by a factor k.

The question of imperfect mixing can be investigated in somewhat greater generality by reverting to the deterministic approximation and following Rushton and Mautner (1955). Suppose there are k strata; let the size of the ith stratum be N_i and denote by $n_i(T)$ the number of hearers at time T ($i = 1, 2, \ldots, k$). As before, let α be the intensity of transmission of the source; let β_i be the transmission intensity within the ith stratum and let γ_{ij} be the corresponding quantity between members of the ith and jth strata. The set of differential equations corresponding to (9.24) is then

$$\frac{dn_i(T)}{dT} = \{N_i - n_i(T)\}\{\alpha + \beta_i n_i(T)\} + \{N_i - n_i(T)\} \sum_{j\neq i} \gamma_{ij} n_j(T)$$

$$(i = 1, 2, \ldots, k). \tag{9.37}$$

The first term on the right-hand side of (9.37) is the contribution to the rate of increase of $n_i(T)$ from the source and contact within the ith stratum. The second term accounts for the change resulting from between-stratum contact. The initial conditions will be $n_i(0)=0$ for all i. Rushton and Mautner (1955) provided a general method of solving the system (9.37), but considerable insight into the effects of imperfect mixing can be had by considering the following special case. Let $N_i = N/k$, $\beta_i = \beta$ and $\gamma_{ij} = \gamma$ for all i and j; then, because of the deterministic nature of the process, $n_i(T) = n(T)/k$. Under these simplifying assumptions all of the equations in (9.37) have the same form and can therefore be treated as a single equation, since

$$\frac{d}{dT}\left\{\frac{n(T)}{k}\right\} = \left\{\frac{N}{k} - \frac{n(T)}{k}\right\}\left\{\alpha + \frac{\beta}{k}n(T)\right\} + \left\{\frac{N}{k} - \frac{n(T)}{k}\right\}\gamma\left\{n(T) - \frac{n(T)}{k}\right\}$$

or

$$\frac{dn(T)}{dT} = \{N - n(T)\}\left\{\alpha + \left(\frac{\beta + (k-1)\gamma}{k}\right)n(T)\right\}. \tag{9.38}$$

In order to see the effect of imperfect mixing we must compare this equation with (9.24). The difference lies in the coefficient $\{\beta + (k-1)\gamma\}/k$ which replaces the β of the original equation. The epidemic therefore develops at the same rate as in a population with homogeneous mixing and with transmission intensity $\{\beta + (k-1)\gamma\}/k$. This expression is a weighted average of β and γ and since $\gamma < \beta$ in any meaningful application it will be less than β. In the extreme case when $\gamma = \beta$, (9.38) reduces to (9.24). When $\gamma = 0$, meaning that there is no communication at all between strata, the epidemic spreads as in a homogeneously mixing population with transmission intensity β/k. In effect this amounts to a scaling of time by a factor k so that it takes k times as long to attain a given degree of spread in the stratified population as in a homogenously mixing one of the same total size. This conclusion provides an interesting confirmation of the stochastic result reached in (9.36).

The foregoing conclusions emphasize that the assumption of homogenous mixing is crucial and they provide a caution against undue reliance on the model in cases where it is known to be suspect. In spite of this rather severe limitation a careful analysis of the pure birth model yields valuable insight into the mechanism of diffusion and so provides a preparation for the general epidemic models of Chapter 10.

9.3 THE DISTRIBUTION OF HEARERS BY GENERATION

Introduction

As we have already observed, the random variables $\{n_g\}$ have a particular interest in the study of the diffusion of rumours which are subject to distortion as they pass from person to person. In this section we shall investigate the distribution of hearers by generation when diffusion takes place according to the pure birth model. The case $g = 1$ is of particular importance as it relates to those who hear directly from the source and so the distribution of n_1 will be treated in some detail. As before, we shall proceed as far as possible with the stochastic version of the model and then revert to the deterministic approximation.

The joint distribution of $\{n_g\}$

Consider, first, the problem of finding the joint distribution of the variables $\{n_g(T)\}$, giving the generation distribution at time T. Denote the number of ignorants (the term used for those who have not heard) at time T by $m(T)$. Then if the total population size is N,

$$m(T) = N - \sum_{g=1}^{N} n_g(T).$$

The infinitesimal transition probabilities associated with the process are as follows:

$$Pr\{m(T+\delta T) = m-1, n_g(T+\delta T) = n_g + 1 \,|$$
$$m(T) = m, n_{g-1}(T) = n_{g-1}, n_g(T) = n_g\}$$
$$= \beta m n_{g-1} \delta T \quad \text{if } g > 1. \tag{9.39}$$

These equations express the fact that the size of the gth generation increases by one as a result of contact between a member of the $(g-1)$th generation and an ignorant. The case $g = 1$ has to be treated separately because the size of the first generation can only increase by contact with the source so that

$$Pr\{m(T+\delta T) = m-1, n_1(T+\delta T) = n_1 + 1 \,|\, m(T) = m, n_1(T) = n_1\}$$
$$= \alpha m \delta T \tag{9.40}$$

where α is the intensity of the source. In principle, these probabilities, with the initial conditions, are sufficient to deduce the complete development of the diffusion process. Our main interest is in the limiting behaviour of the generation distribution. This can be found by taking advantage of an embedded Markov chain.

Let $P(n_1, n_2, \ldots, n_N)$ be the probability that the system is in the state $\mathbf{n} = (n_1, n_2, \ldots, n_N)$ at some time during the course of diffusion (that is that $n_h(T) = n_h, h = 1, 2, \ldots, N$ for some T). The method is to find a difference equation for the probabilities using the fact that the process is a random walk on the set of states $\mathbf{n}$. The state vector $\mathbf{n}$ is subject to certain important constraints. The sum of its elements obviously cannot exceed N and it will attain this value at the end of the epidemic. Secondly, the only vectors $\mathbf{n}$ which have a non-zero probability of occurring are those having the form $(X, X, \ldots X, 0, 0, \ldots 0)$, where X represents a non-zero entry. This property follows from the fact that there cannot be any members belonging to the gth generation unless there is first at least one member of the $(g-1)$th generation. For example, when $N = 3$ the possible $\mathbf{n}$-vectors are

$$(3, 0, 0), (2, 1, 0), (1, 2, 0), (1, 1, 1).$$

Let us call all $\mathbf{n}$-vectors satisfying the foregoing conditions *admissible* vectors; then the states of our Markov chain will be the set of admissible vectors. To simplify the exposition it will be convenient to define $P(\mathbf{n}) = 0$ for all inadmissible states so that ranges of summation can be kept as simple as possible.

Immediately prior to being in the state $(n_1, n_2, \ldots, n_N)$ the system must have been in one of the admissible states among $(n_1, n_2, \ldots n_h - 1, n_{h+1} \ldots n_N)$ $(h = 1, 2, \ldots, N)$. The probability of the transition

$$(n_1, n_2, \ldots n_h - 1, n_{h+1} \ldots n_N) \to (n_1, n_2, \ldots n_h, n_{h+1}, \ldots n_N)$$

is easily obtained from the infinitesimal transition probabilities as

$$\frac{\beta n_{h-1} m}{\beta m N - \beta m + \alpha m} = \frac{n_{h-1}}{N + \omega - 1} \qquad (h > 1)$$

and

$$\frac{\alpha m}{\beta m N - \beta m + \alpha m} = \frac{\omega}{N + \omega - 1} \qquad (h = 1)$$

where $\omega = \alpha/\beta$. Hence

$$P(n_1, n_2, \ldots, n_N) = \sum_{h=1}^{N} \frac{n_{h-1} P(n_1, n_2, \ldots, n_h - 1, \ldots n_N)}{N - 1 + \omega} \qquad (9.41)$$

if we define $n_0 \equiv \omega$. The initial condition is

$$P(1, 0, 0, \ldots, 0) = 1.$$

Equation (9.41) expresses the required probability for a population of size N in terms of a set of probabilities for a population of size $N - 1$. In building up the distribution for a given N we therefore obtain the distribution for all sizes smaller than N. The equation has been programmed for a computer, but as the number of admissible vectors increases as the square of N the demands made by the program on computer storage space limit the size of population which can be considered. We shall give some numerical results later, but first we shall pursue the analysis by means of generating functions. Let

$$\Pi_N(s) = E(s_1^{n_1} s_2^{n_2} \ldots s_N^{n_N}) \qquad (9.42)$$

where $\sum_{h=1}^{N} n_h = N$. Then, multiplying both sides of (9.41) by $s_1^{n_1} s_2^{n_2} \ldots s_N^{n_N}$ and summing over all admissible $\mathbf{n}$-vectors, we obtain

$$\left. \Pi_N(\mathbf{s}) = \frac{1}{N - 1 + \omega} \left[\omega s_1 \Pi_{N-1}(\mathbf{s}) + \sum_{h=1}^{N-1} s_h s_{h+1} \frac{\partial \Pi_{N-1}(\mathbf{s})}{\partial s_h} \right] (N > 1) \right\} \qquad (9.43)$$
$$\Pi_1(s_1) = s_1.$$

The dimension of the vector $\mathbf{s}$ in such formulae is indicated by the subscript of Π. An alternative form of (9.43) which has certain advantages can be derived by an application of Euler's theorem about homogeneous functions. Since $\Pi_N(\mathbf{s})$ has the property that all its terms are of degree N it is a homogeneous function, and the same applies to $\Pi_{N-1}(\mathbf{s})$, which is of degree $N-1$. Hence, by Euler's theorem,

$$\sum_{h=1}^{N-1} s_h \frac{\partial \Pi_{N-1}(\mathbf{s})}{\partial s_h} = (N-1)\Pi_{N-1}(\mathbf{s}). \tag{9.44}$$

Adding the right-hand side of (9.44) to that of (9.43) and subtracting the left-hand side yields

$$\Pi_N(\mathbf{s}) = \left(\frac{N-1+s_1\omega}{N-1+\omega}\right)\Pi_{N-1}(\mathbf{s}) + \frac{1}{N-1+\omega}\sum_{h=1}^{N-1} s_h(s_{h+1}-1)$$
$$\times \frac{\partial \Pi_{N-1}(\mathbf{s})}{\partial s_h}. \tag{9.45}$$

Neither of the expressions for $\Pi_N(\mathbf{s})$ is very helpful for determining the probability distribution unless N is small, and it does not appear easy to deduce a solution for $\Pi_N(\mathbf{s})$ in closed form. The value of the equations lies in their usefulness for the study of special cases and for deriving such things as moments.

The number who hear directly from the source

In the case of the first generation, the marginal distribution of n_1 can be found explicitly. The probability generating function of n_1 is obtained from (9.45) by putting $s_2 = s_3 = \ldots = s_N = 1$. The second term on the right then vanishes leaving

$$\Pi_N(s_1) = \left(\frac{N-1+s_1\omega}{N-1+\omega}\right)\Pi_{N-1}(s_1), \tag{9.46}$$

whence

$$\Pi_N(s_1) = \frac{\omega s_1(1+\omega s_1)(2+\omega s_1)\ldots(N-1+\omega s_1)}{\omega(1+\omega)(2+\omega)\ldots(N-1+\omega)}$$
$$= \frac{\Gamma(N+\omega s_1)\Gamma(\omega)}{\Gamma(N+\omega)\Gamma(\omega s_1)} \tag{9.47}$$

where $\Gamma(x)$ denotes the gamma function. Extracting the coefficient of $s_1^{n_1}$ we obtain the probability distribution as

$$P_N(n_1) = \frac{\omega^{n_1}\Gamma(\omega)|S_N^{n_1}|}{\Gamma(N+\omega)} \qquad (n_1 = 1, 2, \ldots, N) \tag{9.48}$$

where $S_N^{n_1}$ is the Stirling number of the first kind, $|S_N^{n_1}|$ being the coefficient of x^{n_1} in $x(x+1)\ldots(x+N-1)$. Miles (1959) discussed the distribution in another context when $\omega = 1$ and he tabulated the Stirling numbers for $N = 1(1)12$. A further discussion and tabulation of the distribution is given in Barlow and co-workers (1972, p. 142 ff and Table A.5, p. 363). In order to give an idea of the shape of the distribution for small N and varying ω it has been tabulated in Table 9.2.

In small groups it is clear that n_1/N is highly variable except for the extreme

Table 9.2 The distribution $P(n_1|N = 10)$ and $E(n_1|N = 10)$ for various ω

ω	n_1										$E(n_1|N = 10)$
	1	2	3	4	5	6	7	8	9	10	
0	1.0000	—	—	—	—	—	—	—	—	—	1.00
1	0.1000	0.2829	0.3232	0.1994	0.0742	0.0174	0.0026	0.0002	0.0000	0.0000	2.93
2	0.0182	0.1029	0.2350	0.2901	0.2159	0.1014	0.0303	0.0056	0.0006	0.0000	4.04
5	0.0005	0.0071	0.0404	0.1245	0.2317	0.2722	0.2032	0.0936	0.0242	0.0027	5.84
10	0.0000	0.0003	0.0035	0.0216	0.0803	0.1887	0.2819	0.2595	0.1342	0.0298	7.19
∞	—	—	—	—	—	—	—	—	—	1.0000	10.00

values of ω. As the last column shows, the mean number informed by the source increases with ω but not as rapidly as might have been expected. The only way to ensure that most people first hear the news from the source is to have a very high value of ω. This can only be achieved by increasing the strength of the source or decreasing the degree of contact between members of the population. It is a characteristic of rumours that the source is weak, consisting, perhaps, of a single person. Under these circumstances it is not surprising that distortion often occurs because almost everyone receives the rumour at second hand or worse. These conclusions apply also when N is large as we shall now show.

The exact mean and variance of n_1 given N can be found directly from the generating function. They are

$$
\left.
\begin{aligned}
E(n_1 \mid N) &= \omega \left\{ \frac{1}{\omega} + \frac{1}{\omega + 1} + \dots + \frac{1}{N + \omega - 1} \right\} \\
&= \omega \{ \phi(N + \omega - 1) - \phi(\omega - 1) \} \\
\operatorname{var}(n_1 \mid N) &= E(n_1 \mid N) - \omega^2 \sum_{i=0}^{N-1} \frac{1}{(i + \omega)^2}
\end{aligned}
\right\}
\tag{9.49}
$$

If ω is fixed and N is large

$$
E(n_1 \mid N) \sim \omega \ln N,
$$

and hence the proportion who have heard direct from the source is

$$
\frac{E(n_1 \mid N)}{N} \sim \omega \frac{\ln N}{N}.
\tag{9.50}
$$

The proportion who hear first hand from the source under these conditions thus tends to zero as the population size increases. In order for the source to communicate directly with a high proportion of the population it will obviously be necessary to make ω very large. When ω is large

and

$$
\left.
\begin{aligned}
E(n_1 \mid N) &\sim \omega \ln \left(\frac{N + \omega - 1}{\omega} \right) \\
\frac{E(n_1 \mid N)}{N} &\sim \frac{\omega}{N} \ln \left(\frac{1 + \omega/N}{\omega/N} \right)
\end{aligned}
\right\}.
\tag{9.51}
$$

If $\omega = \frac{1}{2} N$ the expected proportion of hearers is 0.55 and if $\omega = N$ it is 0.69. Thus, for example, if the source consists of N spreaders introduced into a population of size N about 70 per cent would hear the information at firsthand.

The formula for the variance in (9.49) suggests that the limiting distribution of n_1 may be Poisson in form because the mean and variance tend to equality. This is not quite true. The asymptotic form of the probability generating function is

$$
\Pi_N(s_1) \sim N^{\omega(s_1 - 1)} \Gamma(\omega) / \Gamma(\omega s).
\tag{9.52}
$$

This has the form of the generating function of the convolution of a Poisson variable with mean $\omega \ln N$ and a random variable with generating function $\Gamma(\omega)/\Gamma(\omega s)$. Since the latter does not depend on N its contribution will be negligible for very large N and the Poisson part should be a good approximation. Barton and Mallows (1961) showed that the limiting form of $P_N(n_1)$ as $N \to \infty$ with n_1 fixed and $\omega = 1$ was a Poisson probability with parameter $\ln N$. This result can be extended to cover the case of a general ω, but as it relates only to the lower tail of the distribution it adds little to the Poisson approximation based on (9.52).

The distribution of n_1 was first considered in the diffusion context by Taga and Isii (1959), who derived the probability distribution from a specialization of the argument leading to (9.41). In fact, if we sum both sides of (9.46) over admissible values of $n_2, n_3, \ldots, n_N$ we obtain the equation

$$P_N(n_1) = \frac{\omega}{N-1+\omega} P_{N-1}(n_1-1) + \frac{N-1}{N-1+\omega} P_{N-1}(n_1)$$

$$(1 \le n_1 \le N; N \ge 1) \quad (9.53)$$

with initial conditions

$$P(0|N) = 1 \quad \text{if } N = 0$$
$$= 0 \text{ otherwise,}$$

which is Taga and Isii's result.

The expected number of hearers in the gth generation

The joint generating function of the n_g's provides a convenient means of finding the expectations of the generation sizes. Thus

$$E_N(n_g) = \left. \frac{\partial \Pi_N(\mathbf{s})}{\partial s_g} \right|_{s_1 = s_2 = \ldots = s_N = 1} \quad (g = 1, 2, \ldots, N). \quad (9.54)$$

The case $g = 1$ has to be treated separately from $g > 1$. The result in the former case has already been obtained in the last section and we shall not duplicate the derivation here. Taking $g > 1$ and differentiating both sides of (9.45) with respect to s_g,

$$\frac{\partial \Pi_N(\mathbf{s})}{\partial s_g} = \left(\frac{N-1+s_1\omega}{N-1+\omega} \right) \frac{\partial \Pi_{N-1}(\mathbf{s})}{\partial s_g} + \frac{1}{N-1+\omega}$$

$$\sum_{h=1}^{N-1} s_h(s_{h+1}-1) \frac{\partial^2 \Pi_{N-1}(\mathbf{s})}{\partial s_g^2} + \frac{1}{N-1+\omega}$$

$$\left\{ s_{g-1} \frac{\partial \Pi_{N-1}(\mathbf{s})}{\partial s_{g-1}} + (s_{g+1}-1) \frac{\partial \Pi_{N-1}(\mathbf{s})}{\partial s_g} \right\}$$

$$(g = 2, 3, \ldots, N). \quad (9.55)$$

Putting $s_1 = s_2 = \ldots = s_N = 1$ and substituting in (9.54),

$$E_N(n_g) = E_{N-1}(n_g) + E_{N-1}(n_{g-1})/(N-1+\omega)$$
$$(g = 2, 3, \ldots, N). \quad (9.56)$$

The advantage of the representation of (9.45) over (9.43) will now be clear from the way terms vanish when we put the s's equal to one. Starting with $E_1(n_1) = 1$ it is easy to build up the complete set of expectations.

Equation (9.56) can also be used to deduce an approximation to the expected generation sizes, valid when N is large. We arrive at this, when ω is fixed, by treating N as a continuous variable and replacing (9.56) by

$$\frac{dE_N(n_g)}{dN} = \frac{E_N(n_{g-1})}{N} \quad (g = 2, 3, \ldots, N). \quad (9.57)$$

Starting with $E_N(n_1) \sim \omega \ln\{(N-1+\omega)/\omega\}$ we may easily deduce that

$$E_N(n_g) \sim \omega \left\{ \ln\left(\frac{N-1+\omega}{\omega}\right)\right\}^g \Big/ g! \quad (g = 1, 2, \ldots, N). \quad (9.58)$$

Notice that when these expectations are summed over g we obtain $N - 1$ which is asymptotically the same as in the exact case when the sum is N. The expected sizes of successive generations thus form an exponential series with the first term missing. Typically, this distribution will be unimodal with an upper tail dying away rapidly for large g.

The method of deriving this approximation does not depend on ω being small. The argument holds if ω is of the same order as N. In this case we shall wish to express the expected size of each grade as a proportion of the population size N. As a function of ω/N we then have

$$\frac{E_N(n_g)}{N} \sim \frac{\omega}{N} \left\{ \ln\left(1+\frac{N}{\omega}\right)\right\}^g \Big/ g! \quad (g = 1, 2, \ldots, N). \quad (9.59)$$

Some numerical values for $\omega/N = \frac{1}{2}$ and 1 are given in Table 9.3.

One way in which these results might be used is to see the effect of mixing two populations, one of which possesses information which is passed to members of the other. In this interpretation ω would represent the size of the transmitting population and N the receiving population. The table shows that most people will hear first or second hand if the influx is at least half as big as N.

Table 9.3 *The expected proportion who hear at gth-hand derived from the asymptotic approximation*

g	1	2	3	4	5
$\omega/N = \frac{1}{2}$	0.55	0.30	0.11	0.03	0.01
$\omega/N = 1$	0.69	0.24	0.06	0.01	0.00

Daley (1967a) derived the approximation of (9.59) as the solution to a deterministic version of the problem. When discussing the generation distribution for a general epidemic model in the Chapter 10 it will not be possible to make any progress with the stochastic model. We shall therefore have to rely on deterministic methods, and so to prepare the ground we shall rederive (9.59), starting with the deterministic version of the model.

Let $n_g(T)(g = 1, 2, \ldots, N)$ and $m(T)$ refer to the same quantities as before, but now we treat them as continuous functions of T and interpret them deterministically. The rates of change under the assumptions of the model are then

$$\frac{dn_1(T)}{dT} = \alpha m(T) \tag{9.60a}$$

$$\frac{dn_g(T)}{dT} = \beta n_{g-1}(T)m(T) \qquad (g = 2, 3, \ldots, N) \tag{9.60b}$$

$$\frac{dm(T)}{dT} = -\alpha m(T) - \beta\{N - m(T)\}m(T) \tag{9.60c}$$

with initial conditions

$$m(0) = N, \qquad n_g(0) = 0 \qquad (g = 1, 2, \ldots).$$

The solution of (9.60c) has already been found in (9.25) because $m(T) = N - n(T)$ so

$$m(T) = N\left(1 + \frac{\omega}{N}\right)\bigg/\left(1 + \frac{\omega}{N}e^{\beta(N+\omega)T}\right).$$

Substitution of this in (9.60a) gives $n_1(T)$ by straightforward integration as

$$n_1(T) = \omega \ln\left\{\left(1 + \frac{\omega}{N}\right)\bigg/\left(\frac{\omega}{N} + e^{-\beta(N+\omega)T}\right)\right\}. \tag{9.61}$$

The limit of $n_1(T)$ as $T \to \infty$ agrees with (9.51) and (9.59). The remaining $n_g(T)$'s are obtained from (9.60b) by putting $g = 2, 3, \ldots$ in turn and finding each $n_g(T)$ by an integration involving its predecessor. It may be verified that this procedure yields

$$n_g(T) = \frac{\omega}{g!}\left\{\ln\left(1 + \frac{\omega}{N}\right)\bigg/\left(\frac{\omega}{N} + e^{-\beta(N+\omega)T}\right)\right\}^g. \tag{9.62}$$

The deterministic approach gives not only the limiting values of the generation sizes but it also shows how these numbers change throughout the course of the diffusion. The stochastic approach does not allow us to do this without reverting to the much more difficult problem posed by (9.39) and (9.40).

9.4   DIFFUSION IN SPACE

Introduction

When we turn to the study of diffusion in space the theory available is very limited. The emphasis in this section will therefore be on the model-building aspect with relatively little analysis or testing.

Most applications of spatial diffusion theory envisage a geographical context in which hearers are located at points in a two-dimensional plane. However, the theory is often simpler for populations distributed in one dimension on a line. Many of the essential characteristics of spatial diffusion processes are preserved in passing from two dimensions to one, so we shall sometimes take advantage of the simplification which this specialization offers.

The pattern of diffusion of the simple epidemic in a randomly distributed population

Suppose that members of a population are distributed at random over a plane area. By this we mean that the number of inhabitants per unit area is a Poisson variable and that the numbers in non-overlapping areas are independent. If diffusion takes place according to the assumptions of the birth process then the chance of contact between any pair of individuals will be unrelated to their location. At any stage, therefore, the knowers will form a random subset of the original population and so will themselves be randomly distributed. The pattern revealed if the hearers are plotted on a map will thus be a purely random one. This may seem a somewhat trivial point at which to begin because the spatial aspect only enters through the original scatter of the population. However, even a random distribution of points will show a certain degree of clustering. In order to decide whether clusters arising in practice are due to special local effects or are purely random in origin it is necessary to consider whether they are compatible with the random hypothesis. Statistical methods for this purpose are well known and need not detain us here. The initial assumption of a randomly distributed population is not necessary, of course. If we had any arbitrary distribution we could test the randomness of the spread by comparing the proportion of hearers in non-overlapping regions with the expected value of $n(T)/N$. Hägerstrand (1967) proposed such a model in which the population was assumed to be spread evenly over a plane area with a density of 30 per 25 km square. He used a Monte Carlo technique to study the pattern of diffusion and concluded that such a simple model could not account for the observed patterns of his various indicators. Notice that in this, as in all other spatial spread models, the population members are assumed to have a fixed location. This is reasonable if 'hearing' corresponds to adopting an agricultural innovation which takes place on a particular farm. It may also be reasonable if the 'space' is social rather than geographical, but it will not be appropriate for the study of highly mobile populations.

The pattern of diffusion when the chance of contact depends on distance apart

Stochastic models

When potential hearers are distributed over an area it is natural to expect contact to be more common among people who are close together than those who are far apart. In such circumstances we need to study the number of hearers as a function of both time and location. Several stochastic models have been proposed in which the essential idea is to make the chance of contact a function of distance apart. Little progress has been made on the mathematical analysis of these models, but some of their basic qualitative characteristics have been determined from Monte Carlo experiments. We shall first describe two stochastic models and then pursue the analysis by means of deterministic approximations.

Hägerstrand (1967) introduced both one- and two-dimensional versions of a simple discrete model. The one-dimensional model was used simply to aid the exposition, but there is little difficulty in proceeding straight to the more realistic two-dimensional version. The population is again supposed to be evenly distributed over a plane area with a specified density. This area is divided up by a square grid. The diffusion starts with a given distribution of knowers—usually a certain number in a particular central square. Each of the original knowers then communicates with one other person selected in a manner to be described below. Next the original knowers, together with those contacted on the first round, make one further contact. The process continues in this way. At each stage every knower makes one contact; an addition to the total stock of knowers occurs each time a knower makes contact with an ignorant.

The contacts are selected in this model by a two-stage probability process. At the first stage the grid square from which the hearer is to come is chosen and at the second stage one member of that square is chosen. The second stage involves a simple random choice in which each member is equally likely to be chosen. The first stage depends on a probability distribution over the squares in the neighbourhood of the originating knower. In Hägerstrand's investigation he assumed that contacts could be made in any of the 25 squares centred on the originator's square with probabilities illustrated in the diagram that follows. The probabilities, which had an empirical basis, fall off with increasing distance from the shaded square which marks the origin.

Hagerstrand represented the realizations of this stochastic process on contour diagrams, showing the position at various stages of the development. A typical realization shows an initial cluster of knowers near the original source with several subsidiary clusters derived from a few 'long distance' contacts made in the early stages. As time advances, these clusters tend to thicken and merge to give an

0.0096	0.0140	0.0168	0.0140	0.0096
0.0140	0.0301	0.0547	0.0301	0.0140
0.0168	0.0547	0.4431	0.0547	0.0168
0.0140	0.0301	0.0547	0.0301	0.0140
0.096	0.0140	0.0168	0.0140	0.0096

irregular concentration of knowers centred near the region where the process started. Hägerstrand used the term 'central stability' to describe this phenomenon. This does not mean that the density of hearers will necessarily be greatest at the origin but that the regions of high concentration will be in its vicinity. The precise way in which the pattern develops will depend, of course, on the form of the contact probability distribution. Hagerstrand's example is symmetrical with probability falling off in all directions. We shall see later that the precise way in which probability of contact falls off with distance is a crucial factor in determining the pattern of spread.

The model described above could be represented as a Markov chain in discrete time. The states of the chain are all possible dispositions of hearers in the area. Each step in the chain involves the making of one contact as a result of which the chain either stays in the same state—if the contact already knows—or moves to a new state. The probabilities of contact are independent and depend only on the present state of the system. In any non-trivial problem the number of states will be enormous and it does not appear possible to use the theory of Markov chains to advantage except, perhaps, to arrive at the otherwise obvious result that the system will ultimately arrive at the only absorbing state—in which everyone knows.

A rather similar model has been investigated by Mollison (1972b) for diffusion in one dimension. Individuals are located on the real line with σ persons at each integer point (positive and negative). Each knower makes contacts at random intervals governed by a Poisson process of rate $\alpha\sigma$. The destination of a given contact is determined by a probability distribution, which for simplicity is assumed to be symmetrical about the source of the contact and is written in the form

$$v(u) = \alpha(|u|)/\alpha \qquad (u = 0, \pm 1, \pm 2, \ldots). \tag{9.63}$$

The main difference between this model and the one-dimensional version of Hägerstrand's model is that the former places the 'contact events' in time and so the development of the process in time can be observed. Hägerstrand only ordered his events according to the number of stages which, in practice, would only partly reflect the time scale. Since Hägerstrand was primarily interested in pattern this was adequate for his purposes. Mollison's approach takes us further by enabling us to study the rate of diffusion. The assumption that the rate at which an individual makes contacts is proportional to the number of individuals at his location is not very plausible in the rumour situation, but this only becomes important if we wish to compare processes with different σ's.

Mollison used his model to test the validity of theoretical deductions made from a deterministic model and to suggest lines for further theoretical analysis. We shall treat matters in the reverse order, using the deterministic analysis to illuminate the empirical findings from simulation of the stochastic model.

Mollison (1972b) found two main kinds of behaviour which are related to the form of the contact distribution $v(u)$—in particular to its variance and the nature of its tails. The first kind of behaviour appears to occur when $v(u)$ has exponentially bounded tails (and, hence, finite variance). In such a case contacts are usually made in the neighbourhood of the source and the 'front' of knowers tends to advance outwards from the origin at a steady rate. There is a problem about defining the front but it is clear intuitively what is meant. The profile of knowers might appear at any time in Figure 9.2. The dotted curve suggests a wave travelling outwards as the diffusion develops. At any time there will be an inner region of knowers, an outer region of ignorants, and a twilight region in which some know and some are ignorant. In any particular realization there will, of course, be stochastic variation which will blur the outlines, but the broad pattern

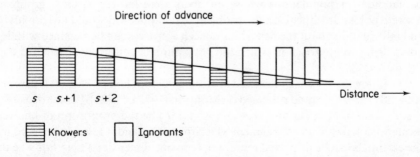

Figure 9.2   The profile of knowers for a steadily advancing epidemic.

will be clear. Mollison simulated the case when

$$v(u) = \tfrac{1}{3}(\tfrac{1}{2})^{|u|} \qquad (u = 0, \pm 1, \pm 2, \ldots)$$

and found that, on average, the wave advanced at a fairly constant rate. When $v(u)$ has infinite variance, theory suggested and simulation demonstrated a different kind of behaviour. When the position of the wave is plotted as a function of time the steady advance is broken at intervals by what Mollison calls 'great leaps forward'. This describes what happens when a contact is made at a considerable distance ahead of the advancing wave. Such a contact gives rise to a colony of knowers ahead of the wave and so accelerates the overall rate of advance. A similar kind of behaviour can also occur when the variance is finite but when the tails are not exponentially bounded.

We thus have two distinct kinds of behaviour depending on the extent to which chance of contact depends on distance apart. There will either be a steadily advancing wave or a wave preceded by local clusters; as old clusters merge with the main body of knowers, new ones will arise. Mollison suggests that a necessary and sufficient condition for the steady rate of advance may be that $v(u)$ has exponentially bounded tails, but this has not been proved. If $v(u)$ is highly dispersed so as to be almost uniform we are almost back to the case when chance of contact is independent of distance. In this case there will be no discernible wave at all—only a gradually merging set of clusters.

Deterministic models

Some progress can be made with the theory by resorting to a deterministic treatment. In fact, the model of Rushton and Mautner (1955) leading to (9.37) is very similar to Mollison's stochastic model. The k strata can be made to represent the groups of individuals located at points on the line. The between-strata contact rates can be brought into correspondence with the chance of contact for different distances apart. In principle, therefore, the spatial spread can be investigated by solving the set of differential equations (9.37) in the manner described by Rushton and Mautner (1955). In a similar way, spatial diffusion in two dimensions could be handled for populations whose members were located in large groups at discrete points. In either case the problem of computation would be formidable and information about the behaviour of such a process can be obtained with less effort using an approach in which the population is uniformly distributed over the plane.

The following approach was used in an epidemic context by Bartlett (1956) and Kendall (1957, 1965) and extended by Mollison (1972a, 1972b). Let the density of the population be σ per unit area and let $y(s, T)$ be the proportion of knowers located at s at time T. (The argument which follows extends to two dimensions but the essentials can be displayed more simply in one dimension.) Let $v(u)\delta u$ be the probability that a given contact is made at a distance in $(u, u + \delta u)$. Then if contacts are made at a rate β the rate of change of knowers at s will be proportional to the product of the number of ignorants there and the number of knowers elsewhere who succeed in making a contact at s. The probability that a contact is made at s from a knower at $(s-u, s-u+\delta u)$ is proportional to $y(s-u)v(u)\delta u$, and the total probability arising from all u may thus be expressed by

$$\bar{y}(s,T) = \int_{-\infty}^{+\infty} y(s-u,T)v(u)\mathrm{d}u. \qquad (9.64)$$

This argument thus leads to

$$\frac{\partial y(s,T)}{\partial T} = \beta\sigma\bar{y}(T)\{1-y(s,T)\}. \qquad (9.65)$$

This equation generalizes (9.24), to which it reduces if $v(u)$ is uniform. The boundary conditions will specify the distribution of knowers at $T=0$.

It does not appear to be possible to find an explicit solution to (9.65), even for densities $v(u)$ of very simple form. Attention has therefore been concentrated on approximate solutions and on the deduction of relevant properties of the solution. A method of approximation used by Kendall (1965) is to replace $\bar{y}(T)$ by the diffusion approximation

$$\bar{y}(T) \doteqdot y(s,T) + K\frac{\partial^2 y(s,T)}{\partial s^2} \qquad (K > 0). \qquad (9.66)$$

This can be thought of as resulting from a Taylor expansion of $y(s-u,T)$ as a function of u about $u=0$ as far as the third term (the second term vanishes because of the assumed symmetry of $v(u)$). The constant K is then half the variance of $v(u)$. For the approximation to be good we shall therefore require $v(u)$ to be concentrated around zero, implying that contacts are usually made in the immediate vicinity of the knower. Substituting (9.66) in (9.65) we have, dropping the arguments of y,

$$\frac{\partial y}{\partial T} = \beta\sigma\left\{y(1-y) + K(1-y)\frac{\partial^2 y}{\partial s^2}\right\}. \qquad (9.67)$$

Even after this approximation no general solution appears to be available, but progress can be made by looking for a particular sort of solution. We observed with Mollison's stochastic model that the wave of new knowers appeared to advance at a constant rate. It therefore seems reasonable to look for solutions of (9.65) having this property. That is, we search for a solution of the form $y(s,T) = y(s-cT)$. The values of y will be the same at all points (s,T) satisfying $s-cT=$ constant. If, between T_1 and T_2, a fixed point on the wave moves from s_1 to s_2 then

$$s_1 - cT_1 = s_2 - cT_2 \quad \text{or} \quad c = (s_2 - s_1)/(T_2 - T_1),$$

showing that c must be the velocity of the wave. Introducing the new variable $x = s - cT$ we have

$$\frac{\partial y}{\partial T} = \frac{\partial x}{\partial T}\frac{dy}{dx} = -c\frac{dy}{dx}$$

and

$$\frac{\partial^2 y}{\partial s^2} = \frac{\partial}{\partial s}\left\{\frac{\partial x}{\partial s}\frac{dy}{dx}\right\} = \frac{\partial^2 x}{\partial s^2}\frac{dy}{dx} + \frac{\partial x}{\partial s}\frac{d^2 y}{dx^2} = \frac{d^2 y}{dx^2}.$$

Substituting in (9.67) we obtain

$$\beta\sigma K(1-y)\frac{d^2 y}{dx^2} + c\frac{dy}{dx} + \beta\sigma y(1-y) = 0. \tag{9.68}$$

When x is large we shall be ahead of the wave and so as $x \to \infty$ we require $y \to \infty$. When x is large and negative we shall be behind the wave where $y \to 1$ as $x \to \infty$.

If there is a solution of (9.68) bounded between zero and one, then it will clearly satisfy the original equation with x replaced by $s - cT$ and will establish the possibility of a wave solution. Let us first consider the situation when y is near to zero, in which case (9.68) may be approximated by

$$K\frac{d^2 y}{dx^2} + \left(\frac{c}{\beta\sigma}\right)\frac{dy}{dx} + y = 0. \tag{9.69}$$

This equation has a general solution of the form

$$y = Ae^{r_1 x} + Be^{r_2 x}$$

where r_1 and r_2 are the roots of the quadratic equation

$$Kr^2 + \left(\frac{c}{\beta\sigma}\right)r + 1 = 0.$$

That is,

$$r_1, r_2 = -\frac{1}{2K}\left\{\frac{c}{\beta\sigma} \pm \sqrt{\frac{c^2}{\beta\sigma} - 4K}\right\}.$$

If both roots are real they will be negative, since $K > 0$, and so y will approach zero as $x \to \infty$. If both roots are complex, y will behave in an oscillatory manner and will repeatedly change sign. A solution with negative values of y cannot represent a solution of our problem and therefore we conclude that no waveform solution exists if r_1 and r_2 are complex. The condition for this is

$$c^2 \geq 4K\beta^2\sigma^2,$$
$$c \geq 2\sqrt{K}\,\beta\sigma. \tag{9.70}$$

In other words, there is a minimum velocity, given by (9.70), below which a waveform solution is impossible. At or above that minimum velocity the above analysis leaves the question open, but it may be shown that a waveform solution is possible; it further appears that if the source consists of a concentration of knowers at a point that the wave will move outwards at the minimum velocity.

The foregoing analysis is based on an approximation which assumes that contacts are made by knowers only in their immediate neighbourhood. Mollison (1972a, 1972b) was able to take the analysis further by showing that a similar result held for (9.65). Two kinds of behaviour are possible depending on the degree of dispersion of $v(u)$. If $v(u)$ has two tails which approach zero at least as fast as an exponential then all velocities are possible above a certain minimum. The results derived from the diffusion approximation then give a good indication of the behaviour of the process. On the other hand, if the tails of $v(u)$ are not bounded by an exponential the rate of advance of the epidemic tends to infinity with time.

It is interesting to relate these conclusions to those derived by simulation of the stochastic model. The critical factors governing whether the wave advanced at a steady rate or 'by leaps and bounds' was the behaviour of the tails of $v(u)$. The accelerating rate of advance in the deterministic model corresponds to the phenomenon of the 'great leap forward', and both arise when contacts can be made well beyond a spreader's immediate neighbourhood.

These results, extended into two dimensions, provide some guidance on the interpretation of patterns of diffusion of the kind studied by Hägerstrand (1967). The occurrence of central stability suggests that the chance of a pair making contact depends on their distance apart—and is generally confined to a neighbourhood. An area of stability with occasional clusters at a distance is not necessarily evidence for several sources but may indicate a contact distribution which allows occasional contact at a considerable distance from the main body of knowers.

It is clear that the information we have been able to glean from our analysis of spatial models is fragmentary and further research is needed. However, some broad and useful qualitative characteristics have emerged which will be extended in Chapter 10.

Perimeter and hierarchical models

In concluding this chapter we shall briefly mention a group of models in which the contacts made between spreaders and ignorants are constrained by a particular geographical or social structure. The stochastic features of such models have not been explored in any depth in the present context,† and the purpose in introducing the topic is to direct attention to an area in which further research is highly desirable.

---

† See, however, Hammersley's (1966) work on 'percolation processes'.

Suppose that the members of the population are immobile and are only able to communicate with their immediate neighbours. If information is passed through the population from a single source we may expect the area covered by the knowers to move outwards steadily from the source. New knowers will only be added at the boundary of the region occupied by knowers and for this reason we speak of *perimeter models*. There is an obvious similarity with waveform properties of the models discussed above. The distinction is that in the perimeter model contact is restricted to immediate neighbours; in the models discussed by Kendall and Bartlett contacts can be made with any other member of the population, albeit with small probability in the case of those a long distance apart. The pattern of diffusion in a perimeter model will follow the density of the population—the frontier will advance most rapidly where the population is most sparsely spread. For this reason the analysis cannot be expected to yield much of interest about pattern. We shall, instead, be more concerned with the rate of growth of the total numbers in a situation where contacts are constrained by geography.

A stochastic model of the perimeter variety was proposed by Bailey (1967) and analysed by simulation. According to this model individuals are located at the vertices of a square lattice. To begin with there is a single knower at the centre. This person is able to communicate with any of his eight neighbours, as illustrated in Figure 9.3, but with no one else. Spreading takes place at discrete points in time and at each such time every knower is able to make contact with his neighbours. The chance that a given ignorant hears the message from a neighbour who knows is p and all contacts are independent. At the first stage the number of new knowers

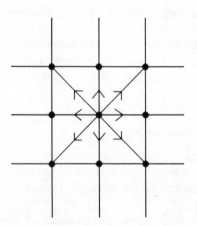

Figure 9.3 Showing the eight neighbours who can be contacted for Bailey's (1967) model.

will thus have the binomial distribution

$$\binom{8}{r} p^r (1-p)^{8-r} \qquad (r = 0, 1, 2, \ldots, 8).$$

At the next stage all ignorants who are neighbours of knowers will be exposed to the risk of contact. If a given ignorant has a knowers as neighbours the chance that he will become a knower is $1 - (1 - p)^a$. Bailey (1967) simulated such a process on grids of size 11×11 and 21×21. To begin with he found that the number of knowers rose as the number of ignorants exposed to risk increased, but as the boundaries were approached an increasing number of knowers had knowers for neighbours so the rate of growth tailed off. The resulting epidemic curve was not unlike that observed in practice with the simple epidemic—rising to a peak and then falling away rather more slowly. The manner of spreading in the perimeter model is, however, quite different from the simple epidemic model. In the one case contact is equally likely between all pairs of individuals. In the other case contacts are severely limited by the structure of the population.

A deterministic investigation of a perimeter model, very similar to Bailey's, has been made by Day (1970). He was interested in the diffusion of agricultural innovations. His model is highly simplified but it provides a pointer for the development of more realistic stochastic models. A large area is divided up into square farms on a chequerboard pattern. Initially, a farm at the centre adopts an innovation. After observing it for one time period all of the four neighbours having a common boundary with the innovator adopt it. These four adopters have eight neighbours who follow suit after one further period has elapsed. At time T there will be $4T - 2$ new adopters and the cumulative number of adopters at that point will be $1 + 2T(T + 1)$. The number of adopters will thus be a quadratic function of time until the boundary eventually stops growth. Such a growth pattern does not correspond very closely to what happens in practice, and Day extended his model by supposing that the innovation starts at farms spaced d units apart in both directions. The initial growth then follows the quadratic pattern, but as soon as $T > \frac{1}{2}(d + 1)$ the clusters begin to overlap and the growth rate slows down. In fact, for $T > \frac{1}{2}(d + 1)$

$$n(T) = \frac{1}{2}[2d^2 + 1 - \{1 - 2(d - T)\}^2].$$

The resulting growth curve has the characteristic sigmoid form and so suggests the desirability of constructing stochastic models in which diffusion spreads simultaneously from a number of centres.

A rather similar approach to spatial diffusion involves an extension of the idea on which our original deterministic treatment of the random mixing model was based. Both T and $n(T)$ are treated as continuous variables. Consider first a perimeter model in which news diffuses from a single source through a population of uniform density. Growth will take place on the perimeter of the circle enclosing those who already know. Therefore, the number of knowers who are active at T will be proportional to $\sqrt{n(T)}$. These people will be in contact with an equal number of people on the inner boundary of the ignorants. If any knower on the boundary can communicate with any ignorant on the adjacent boundary we shall have

$$\frac{dn(T)}{dT} \propto \sqrt{n(T)} \times \sqrt{n(T)} \tag{9.71}$$

and growth will be exponential. If the knowers on the perimeter can only communicate with their ignorant neighbours then

$$\frac{dn(T)}{dT} \propto \sqrt{n(T)} \tag{9.72}$$

and the growth law will be quadratic as in Day's model. We could construct stochastic versions of these models by using a pure birth model with $\lambda_n \propto \sqrt{n}$ or n as the case may be. McNeil (1972) has investigated such a model with $\lambda_n \propto \{n(N-n)\}^{\frac{1}{2}}$, but it is not easy to relate this to a geographical model of diffusion.

Hierarchical models are very similar to perimeter models as far as their mathematical structure is concerned. The 'space' of such models is usually a social or organizational structure through which the information diffuses. Such a system may be viewed as a network in which information is introduced at the highest level and then spreads downwards through the hierarchy defined by the network. We shall not pursue the discussion of these models here. A good introduction will be found in Hudson (1969), who also reviews some of the other models discussed in this chapter, drawing attention to their inadequacies in explaining diffusion.

CHAPTER 10

General Epidemic Models for the Diffusion of News, Rumours, and Ideas

10.1 INTRODUCTION

The distinguishing feature of the simple epidemic models of the last chapter was that those who had heard the news continued to spread indefinitely. Common experience and experimental work suggest that epidemics may die out before everyone has heard. People may cease to be spreaders for a variety of reasons. They may forget, lose interest, or gain the impression that 'everyone knows'. In order to achieve the greater degree of realism which these considerations suggest we shall have to consider what we call 'general' epidemic models. This term is used in epidemic theory for Kermack and McKendrick's model, to be discussed in the next section. Since its generality lies in the provision of a mechanism for the cessation of spreading it seems appropriate to extend the usage to cover other models which incorporate the same feature.

The theory of general epidemic models is difficult and it is not possible to make very much progress with the stochastic aspects of the rate of diffusion in time or space. Greater reliance has to be placed on deterministic methods. There is, however, an important new feature of general epidemics which can be treated stochastically and to which part of this chapter will be devoted. This concerns the number who will ultimately hear. Since people eventually stop being active spreaders the epidemic may cease before everyone has heard. Indeed, it is possible that spreading will die out very quickly so that hardly anyone hears. This leads us to the consideration of threshold effects concerning whether or not an epidemic, in the usual sense of the word, is likely to develop.

At time T the state of the system can be described by three random variables as follows:

$m(T)$ persons who have not heard—the ignorants;

$n(T)$ persons who have heard and are actively spreading—the spreaders;

$l(T)$ persons who, having heard the news, have ceased to spread it.

Because the size of the population is assumed to be constant it follows that

$$N = m(T) + n(T) + l(T);$$

any two of the three random variables are sufficient to describe the state of the system. In the simple epidemic $l(T)$ is zero for all T. We shall be particularly interested in the limiting behaviour of the process $T \to \infty$. It is obvious that $n(T)$ must ultimately take the value zero with probability 1; hence $m(\infty) = N - l(\infty)$

will be the random variable of particular interest. Later in the chapter we shall meet models where those who cease activity become indistinguishable from ignorants. When this happens the meaning of $m(T)$ will be extended to include both groups; $l(T)$ is then zero. When it is necessary to identify different kinds of ignorant or spreader we shall do so by means of subscripts.

10.2   KERMACK AND McKENDRICK'S MODEL

Background

The earliest contribution to mathematical epidemic theory appears to be the celebrated paper by Kermack and McKendrick (1927), in which the authors formulated a deterministic version of what has subsequently become known as the general epidemic model. Most of the basic theory is well known and may be found in Bailey (1975). Our initial formulation is slightly different from Bailey's and we shall give the topic a slant more suited to the diffusion of news and ideas.

The model is the same as the pure birth model with the addition that spreaders are only active for a random period of time. More precisely, we assume that the period of spreading for each individual is an exponential random variable with mean μ^{-1}. We further assume that cessations are independent and that once a person has ceased spreading they do not resume their activity. An essential characteristic of the model is that cessation as specified above is independent of the state of knowledge in the population. This means that an individual will spread the information with the same zeal whether many or few of his hearers have heard before. The plausibility of this assumption must be judged by the success with which the model accounts for observed diffusion. It seems most reasonable if the item of news is fairly trivial so that cessation of spreading on the part of an individual is due to forgetfulness. Obviously there will be individual variation in the time taken to forget, but the choice of the exponential distribution to describe that variability is more questionable. We shall see later that the form of this distribution may not be crucial.

The distribution of $m(T)$ and $n(T)$

We shall use $m(T)$ and $n(T)$ to describe the system at time T and shall say that the system is in state (m, n) if $m(T) = m$ and $n(T) = n$. One way of studying the development of the process in time is to consider the joint probability distribution of $m(T)$ and $n(T)$. Let us write this as

$$Pr\{m(T) = m, n(T) = n\} = P_{m, n}(T).$$

Another way would be to study the time taken for the system to reach the state (m, n). However, the system need not reach a given state (m, n) and so the simple inverse relation which existed in the analogous situation in the pure birth model is lost and there are no compensating mathematical advantages.

We shall not be able to obtain explicit formulae for the joint distribution of $m(T)$ and $n(T)$. It is nevertheless a worthwhile exercise to set up equations for them for the light which they throw on the process. From the state (m, n) two transitions are possible. They are set out below with the probabilities that they take place in $(T, T + \delta T)$.

$$\text{(a)} \quad (m, n) \rightarrow (m - 1, n + 1): m(\alpha + \beta n)\delta T$$

for $m = 1, 2, \ldots, N, n = 0, 1, \ldots, N - 1$ such that $0 \le n + m \le N$.

$$\text{(b)} \quad (m, n) \rightarrow (m, n - 1): n\mu \, \delta T$$

for $n = 1, 2, \ldots, N - 1$.

The transition (a) takes place when a spreader meets an ignorant; transition (b) occurs when a spreader forgets and ceases to spread.

Using these transition probabilities we can relate the joint probability at time $T + \delta T$ to that at T in the usual way and obtain the following bivariate differential–difference equation for $P_{m, n}(T)$:

$$\left.\begin{aligned}
P'_{m, n}(T) = &-\{m(\alpha + \beta n) + n\mu\}P_{m, n}(T) + (m + 1) \\
&(\alpha + \beta(n - 1))P_{m + 1, n - 1}(T) + (n + 1)\mu P_{m, n + 1}(T) \\
P'_{N, 0}(T) = &-N\alpha P_{N, 0}(T) \\
P_{N, 0}(0) = &1 \\
&(0 \le m \le N, 0 \le n < N, 0 \le m + n \le N)
\end{aligned}\right\} \quad (10.1)$$

where it is to be understood that probabilities with subscripts not satisfying $0 \le m + n \le N, n, m \ge 0$ are zero. It may be deduced that $P_{m, n}(T)$ can be expressed as a series of descending exponentials, but the quadratic coefficients in (10.1) make further progress difficult.

We can draw certain general conclusions about the process by noting that the pure birth model is a special case. If $\mu = 0$ there is no forgetting and it is intuitively obvious that the rate of spread must therefore be greater than when $\mu > 0$. Thus, for example, $\bar{n}(T)$ for the pure birth model provides an upper bound for the same function in our present model. Secondly, if $\beta = 0$ or $\mu = \infty$ forgetting is irrelevant since then no one is ever actively spreading the news and hence $n(T)$ is always zero. By putting $\mu = \infty$ our model thus becomes identical with the pure birth model with $\beta = 0$. The rate of diffusion will be greater if $\beta > 0$ than if $\beta = 0$ so, this time, we can obtain a lower bound for $\bar{n}(T)$. The two bounds provided by considering the extreme values of μ will usually be rather wide, but no further progress has been made with the stochastic model in this form.

The foregoing model does not allow for the possibility, which we envisaged at the beginning of the section, of the epidemic dying out. This is because the source continues to transmit indefinitely and thus, ultimately, all people will be informed. An interesting variant is obtained by supposing that the source transmits for a limited period only. One way in which this could happen is if the source consists of a group of a individuals with the same law of forgetting as the other members of the population. Under these circumstances the infinitesimal transition probabilities become

$$\text{(a)} \quad \beta mn \quad \text{and} \quad \text{(b)} \quad \mu n$$

where n now refers to the total number of spreaders whether they originate from inside or outside the population. The differential–difference equations for the probabilities $P_{m,n}(T)$ are now

$$\left.\begin{aligned}
P'_{m,n}(T) &= -(\beta mn + \mu n)P_{m,n}(T) + \beta(m+1)(n-1)P_{m+1,n-1}(T) \\
&\quad + \mu(n+1)P_{m,n+1}(T) \\
P'_{N,a}(T) &= -\{\beta aN + \mu a\}P_{N,a}(T) \\
P_{N,a}(0) &= 1 \\
&(0 \le m \le N, 0 \le n < N+a, 0 \le m+n \le N+a)
\end{aligned}\right\} . \quad (10.2)$$

We again define probabilities to be zero if their subscripts are outside the stated ranges. The equations (10.2) are those that arise in epidemic theory and a considerable body of information has accumulated about their solution. Gani (1965) and Siskind (1965) have obtained methods for finding explicit solutions using a generating function technique. Their methods are extremely unwieldy unless N is very small and so are of little immediate practical value. Two other approaches remain open. One is to concentrate on finding partial solutions, in particular, for the limiting distribution of the number of ignorants. The second is to use a deterministic approximation for the system of equations (10.2). We shall follow both of these courses but, before doing so, we point out a second way in which the present model can arise.

Suppose that the source transmits the information to exactly a people before the diffusion starts and then ceases to operate. From that point onwards the system behaves like one of size $N' = N - a$ into which a spreaders are introduced. It only requires trivial modifications of notation to make the theory cover a situation of this kind.

The terminal state of the system

Although the equations (10.2) are difficult to solve it is relatively easy to find the limiting values of the probabilities $P_{m,n}(T)$ as $T \to \infty$. After a sufficiently long period the diffusion will cease, either because everyone has heard or because the spreaders have ceased to be active. In either event $n(\infty)$, the final number of spreaders, is zero with probability 1. Consequently

$$P_{m,n}(\infty) = 0 \quad \text{if } n > 0. \quad (10.3)$$

When $n = 0$, $P_{m,0}(\infty)$ will be the probability distribution of the terminal number of ignorants. We shall determine this distribution by exploiting the existence of an embedded random walk.

Let $P_{m,n}$ denote the probability that, *at some time* during the diffusion, there are m ignorants and n spreaders. Then clearly

$$P_{m,0} = P_{m,0}(\infty). \tag{10.4}$$

If we consider the process only at those points in time when a change of state takes place we may represent it as following a random walk over the lattice points (m, n). The situation is illustrated in Figure 10.1. We imagine a particle starting at the point (N, a) and moving, at each step, either diagonally upwards or vertically downwards as shown. When the system is in state (m, n) the two transitions which it can make and their associated probabilities are

$$\text{(a)} \quad (m, n) \to (m - 1, n + 1) : \frac{m}{m + \rho}$$

$$\text{(b)} \quad (m, n) \to (m, n - 1) : \frac{\rho}{m + \rho}$$

except that every state $(m, 0)$ is an absorbing state. These probabilities are the relative values of the infinitesimal transition probabilities given in (a) and (b) earlier in this section. The random walk is Markovian because the transition probabilities depend only on the present state of the system. Here, $\rho = \mu/\beta$ and is

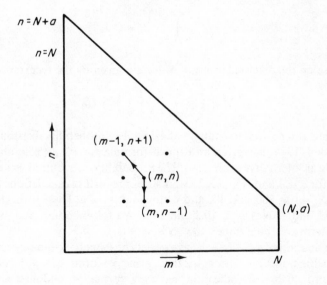

Figure 10.1 The embedded random walk of Kermack and McKendrick's model.

often called the *relative removal rate*. In words, it may be expressed as

$$\rho = \frac{\text{average time taken for a randomly chosen pair to communicate}}{\text{average length of time for which a spreader is active}}.$$

A large value of ρ indicates that forgetting takes place relatively rapidly and a small value the converse. The m-axis is an absorbing barrier corresponding to the complete elimination of spreaders from the population. If the particle reaches a point on the n-axis it descends and is absorbed at the origin.

If the particle passes through the point (m, n) it must previously have passed through either $(m + 1, n - 1)$ or $(m, n + 1)$. This enables us to set up a difference equation, with the aid of (a) and (b) above, as follows:

$$\left. \begin{aligned} P_{m,\,n} &= \left(\frac{m+1}{m+1+\rho} \right) P_{m+1,\,n-1} + \left(\frac{\rho}{m+\rho} \right) P_{m,n+1} \\ &\qquad (m \geq 0, \quad n > 1, \quad m+n < N+a) \\[2mm] P_{m,\,0} &= P_{m,\,1} \left(\frac{\rho}{m+\rho} \right) \qquad (0 < m \leq N) \\[2mm] P_{m,1} &= P_{m,2} \left(\frac{\rho}{m+\rho} \right) \qquad (0 < m \leq N) \end{aligned} \right\} . \qquad (10.5)$$

The initial condition is $P_{N,a} = 1$. For $m = N$ and $1 \leq n \leq a$ the probabilities are easily seen to be

$$P_{N,\,a-i} = \left(\frac{\rho}{N+\rho} \right)^{i} \qquad (i = 0, 1, \ldots, a), \qquad (10.6)$$

while those on the diagonal $m + n = N + a$ are given by the recurrence formula

$$P_{N-i,a+i} = P_{N-i+1,a+i-1} \left(\frac{N-i+1}{N-i+1+\rho} \right) \qquad (i = 1, 2, \ldots, N). \qquad (10.7)$$

These results can be used to compute the complete probability distribution from (10.5). Bailey (1975) has given an explicit formula for $P_{m,\,0}$ (6.59) and Siskind (1965) gave an alternative expression. The probability distribution was computed by Bailey for $a = 1$ and $N = 10, 20$, and 40. Some further calculations have been made for $N = 100, 200$, and 400 and various values of a. These form the basis of Figure 10.2 and Tables 10.1, 10.2 and 10.3. We follow Bailey and express the results in terms of the number $n_H = N - m$.

The distribution of $n_H = N - m$ has a variety of shapes depending on the values of N and ρ. Figure 10.2 illustrates the three principal forms for $a = 1$. If $N \leq \rho$ the distribution is J-shaped, indicating that the information seldom reaches more than a handful of people. If N is a little greater than ρ a mode appears in the upper tail. As N/ρ increases the mode becomes larger and moves to the end of the range until the distribution is U-shaped. Further increase in N/ρ results in a reduction in the probability concentrated near the origin. Finally, in the limit, the process degenerates into a pure birth model with all of the probability at $n_H = N$. The development of the diffusion thus depends critically on the relative sizes of N and ρ. There may be no epidemic at all, there may be an epidemic of uncertain size, or

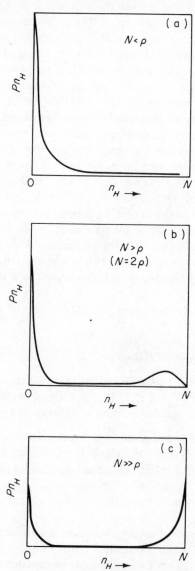

Figure 10.2   Forms of the distribution
of n_H, the ultimate number who hear.

there may be an epidemic in which everyone is almost certain to receive the information. These conclusions hold good in broad outline for any fixed a, but the position will be investigated quantitatively in more detail below.

The discussion given above is based on extensive calculations of the exact distribution and on Whittle's (1955) stochastic threshold theorem. Before giving some sample calculations and stating the theorem it is instructive to give an intuitive discussion of the threshold effect. The assumptions of our model imply that any individual communicates, on average, with $N\beta$ others per unit time irrespective of his own state. He himself is actively engaged in spreading for an average length of time μ^{-1}. Hence the expected number of tellings for each person will be $N\beta/\mu = N/\rho = d$, say. We shall see that if $d \leq 1$ the epidemic does not develop and if $d > 1$ it may do so. We might reasonably have expected the diffusion to peter out if the average number of tellings was less than one per head. However, the position can be clarified by comparing our process with a branching process such as that discussed in Chapter 2. Any member of the gth generation of hearers gives rise to a random number of *new* hearers in the $(g + 1)$th generation. The expected value of this random number must be less than or equal to 1 if $d \leq 1$ (d is the expected total number of hearers—new and old). Under these circumstances extinction of the branching process is certain. In our application this means that, if N is large and $d \leq 1$, only a small proportion will hear the news. It is very important to notice that this argument does not depend on the distribution of the period of spreading. This is why the kind of threshold effect which we have observed for this model also occurs in a much larger class of models.

The relationship with the branching process can also be used to establish the following theorem.

Theorem (*Stochastic threshold theorem*)

If N is sufficiently large the probability of the epidemic exceeding any arbitrarily chosen size tends to zero for $d \leq 1$. If $d > 1$ the probability of the epidemic exceeding any arbitrarily small proportion of N tends to $1 - d^{-a}$.

This theorem formalizes the conclusions expressed above. The extent to which it holds for moderate N can be judged from Tables 10.1–10.3. They give computations for the case $N = 200$. Other calculations for $N = 100$ and 400 have been made and lead to essentially the same conclusions. In particular, the proportions given in Tables 10.1 and 10.2 for $N = 200$ are almost identical with those for $N = 400$.

The threshold effect at $d = 1$ is very obvious in these tables. By increasing a it is possible to increase the number informed but the overall characteristics of the process remain the same. From Table 10.3 it can be seen that the expected number of hearers changes smoothly on each side of the threshold value $d = 1$.

Daniels (1966) has investigated the distribution of $N - n_H$. He showed that if d

Table 10.1 *Probabilities that the ultimate number of hearers will be small or large for various values of* d *and* $N = 200$, $a = 1$

n_H	0	$\frac{1}{2}$	$\frac{2}{3}$	1	2	4	20	∞
0–9	1.000	0.985	0.950	0.830	0.496	0.251	0.050	0.000
0–9	1.000	0.998	0.986	0.889	0.506	0.251	0.050	0.000
181–200	0.000	0.000	0.000	0.000	0.006	0.749	0.950	1.000
191–200	0.000	0.000	0.000	0.000	0.000	0.704	0.950	1.000

Table 10.2 *Probabilities that the ultimate number of hearers will be small or large for various values of* d *and* $N = 200$, $a = 5$

n_H	0	$\frac{1}{2}$	$\frac{2}{3}$	1	2	4	20	∞
0–9	1.000	0.860	0.670	0.322	0.028	0.001	0.000	0.000
0–19	1.000	0.981	0.885	0.495	0.034	0.001	0.000	0.000
181–200	0.000	0.000	0.000	0.000	0.017	0.999	1.000	1.000
191–200	0.000	0.000	0.000	0.000	0.000	0.954	1.000	1.000

Table 10.3 *Means of* n_H *for* $N = 200$

a	0	$\frac{1}{2}$	$\frac{2}{3}$	1	2	4	20	∞
1	0.0	1.0	1.8	7.0	78.2	146.8	190.0	200.0
2	0.0	1.4	3.6	13.3	118.7	183.6	199.5	200.0
5	0.0	4.8	8.7	28.8	154.4	196.0	200.0	200.0
10	0.0	9.3	16.6	47.4	162.9	196.6	200.0	200.0

and N are large then the probability that $N - n_H = x$ is approximately a Poisson probability with mean $N \, e^{-d}$ for small values of x. He made a comparison of the exact and approximate probabilities for $N = 1000, 100, d = 4$, and $a = 1$. In both cases the agreement was fair but he also showed that a much better, though more complicated, approximation could be found. The Poisson result holds regardless of the value of a. Cane (1966) arrived at a similar result.

The deterministic approximation

The distribution theory for the terminal state of the system tells us nothing about the duration or rate of growth of the epidemic. Except in degenerate cases, the mathematical problems of obtaining this information from the stochastic formulation are formidable. We shall therefore treat the process deterministically. The adequacy of this approach can be judged by comparing its terminal predictions with those already obtained for the stochastic case.

Instead of supposing that the number of transitions from (m, n) to $(m - 1, n + 1)$ in $(T, T + \delta T)$ is a random variable taking the values 0 or 1 with expectation $\beta m(T) n(T) \delta T$, we now assume that N is large enough for the expectation to be treated as the actual increase in $n(T)$ during the interval. Similarly, the decrease in $n(T)$ due to cessation of spreading will be assumed to be exactly $\mu n(T) \delta T$. Thus $n(T), m(T)$ and $l(T)$ are treated as continuous variables no longer being restricted to integer values. As $\delta T \to 0$ the change in $n(T)$ may then be represented by the differential equation

$$\frac{dn(T)}{dT} = n(T)\{\beta m(T) - \mu\}. \tag{10.8}$$

Likewise, the derivatives of $m(T)$ and $l(T) = N + a - m(T) - n(T)$ are given by

$$\frac{dm(T)}{dT} = -\beta m(T) n(T) \tag{10.9}$$

and

$$\frac{dl(T)}{dT} = \mu n(T). \tag{10.10}$$

The initial conditions are $n(0) = a$, $m(0) = N$, $l(0) = 0$ and throughout the diffusion we must have $n(T) + m(T) + l(T) = N + a$.

An important result concerning the behaviour of the system can be found without actually solving the equations. First we note that $dn(T)/dT$ is negative or zero if $m(T) \leq \rho$ for any T. Since the number of ignorants cannot increase, this condition will certainly be satisfied if $m(0) = N \leq \rho$. Thus no epidemic occurs when $N \leq \rho$, while if $N > \rho$, $dn(T)/dT > 0$ at $T = 0$, the number of spreaders rises initially and an epidemic occurs. This is essentially the same threshold result which we obtained for the stochastic version of the model. The deterministic approximation has therefore been successful in reproducing this important characteristic of the epidemic.

If $N > \rho$, we can find the ultimate size of the epidemic and compare it with the stochastic values given in Table 10.3. To do this we first establish a simple relationship between $m(T)$ and $l(T)$. Dividing each side of (10.9) by the corresponding side of (10.10) we have

$$\frac{dm(T)}{dl(T)} = -\frac{\beta}{\mu}m(T) = -\frac{m(T)}{\rho}. \qquad (10.11)$$

Integrating this equation and substituting the initial conditions we find that

$$\frac{m(T)}{N} = \exp[-l(T)/\rho]. \qquad (10.12)$$

When the diffusion has ceased $n(T)$ will be zero and hence $m(T) = N + a - l(T)$. We have earlier denoted the ultimate number of hearers by n_H, which is the value taken by $l(T)$ when $n(T) = 0$. Hence from (10.12) we have

$$1 + \frac{a}{N} - p = e^{-dp}$$

where $p = n_H/N$ and $d = N/\rho$. If a is fixed and N large, p satisfies

$$1 - p = e^{-dp} \qquad (10.13)$$

approximately. This equation can easily be solved by using the tables of Barton and co-workers (1960). A short table is given below (Table 10.4).

The last row of the table is included to facilitate the comparison of the deterministic predictions with the values of the stochastic mean given in Table 10.3. In the stochastic case the expected size depends on the value of a. The agreement between the deterministic and stochastic solutions is reasonably good when $a = 5$ or 10 but is very poor for $a = 1$ or 2. It might thus appear that the deterministic approximation is of little value. However, further investigation shows that the agreement is only poor when the distribution of n_H is bimodal or U-shaped. This happens when there is an appreciable probability that no

Table 10.4 *Deterministic approximation to the ultimate number of hearers in the general epidemic model*

		d		
	1	2	4	20
p	0	0.797	0.980	1.000
$200p$	0	159.4	196.0	200.0

epidemic will develop. According to the deterministic model an epidemic *always* occurs if $N > \rho$. We are therefore prompted to ask whether the agreement can be improved by omitting those cases where only a few people hear. There is, of course, some degree of arbitrariness about where we draw the line, but in Table 10.5 we have given the mean of n_H *given that* $n_H > 20$.

Table 10.5   Values of $E(n_H | n_H > 20)$
for $N = 200$

		d	
a	2	4	20
1	157.3	195.9	200.0
2	157.7	195.9	200.0
5	159.7	196.2	200.0
10	163.1	196.6	200.0

It is clear that the overall agreement between the stochastic and deterministic predictions as given by Tables 10.4 and 10.5 is greatly improved. The minor exception to this rule at $a = 10, d = 2$ may be accounted for by the omission of the term a/N from (10.13). It thus appears that the deterministic approach provides a satisfactory terminal description of the process *when an epidemic occurs*. The probability of an occurrence is, of course, given by the stochastic threshold theorem.

The number who have heard at time T is $N - m(T)$; the 'epidemic curve', which is defined as the rate of increase of the number of hearers, is thus obtained by plotting $(-dm(T)/dT)$ against T. (In the theory of epidemics the epidemic curve is defined by $dl(T)/dT$. This is because only the $l(T)$ members have recognizable symptoms although a further $n(T)$, who are incubating the disease, are actively spreading it.) Combining (10.9) and (10.12) we find

$$\frac{-dm(T)}{dT} = \beta m(T)\{N + a - m(T)\} + \mu m(T)\ln(m(T)/N). \qquad (10.14)$$

We may compare this with the corresponding equation for the pure birth model in Chapter 9 which is obtained by putting $\mu = 0$. Since the last term on the right-hand side of (10.14) is always negative the epidemic curve for $\mu > 0$ lies everywhere below that for $\mu = 0$. Solving the equation with the appropriate initial conditions we have

$$T = \frac{1}{\beta}\int_{m(T)}^{N} \frac{dx}{x\{N + a - x + \rho\ln x/N\}} \qquad (m(\infty) \le m(T) \le N). \quad (10.15)$$

Using numerical integration this equation may be used to plot $m(T)$, and hence

$- dm(T)/dT$, against T. Some calculations are given in Kendall (1956). It will be recalled that in the birth model the stochastic and deterministic epidemic curves approached one another as a increased. We may expect that the same will be true in the present case.

The distribution of the number of hearers by generation

We can classify the hearers at time T according to the generation to which they belong. To this end let $n_g(T)$ and $l_g(T)$ denote the numbers of spreaders and inactive knowers respectively, who first heard the news at gth hand $(g = 1, 2, \ldots, N)$. Since the $n_g(T)$'s will be zero in the limit as $T \to \infty$ we shall be primarily interested in the joint distribution of the $l_g(\infty)$'s. Daley (1967b) showed that it was possible to calculate this distribution as a set of absorption probabilities for a certain random walk on a lattice. The argument is of the same kind as that leading up to (9.41), but the greater complexity arising from the two kinds of knower makes the calculations prohibitive in this case.

Instead of pursuing this approach we shall therefore turn to the deterministic treatment—also due to Daley (1967b). For this purpose we introduce $n_0(T)$ to denote the number out of the original a source members who are still active at time T. Then, on the assumptions of the model, the $n_g(T)$'s and the $l_g(T)$'s will change at the following rates:

$$\frac{dn_0(T)}{dT} = -\mu n_0(T), \tag{10.16a}$$

$$\frac{dn_g(T)}{dT} = \beta m(T)n_{g-1}(T) - \mu n_g(T) \qquad (g = 1, 2, \ldots), \tag{10.16b}$$

$$\frac{dl_g(T)}{dT} = \mu n_g(T) \qquad (g = 0, 1, 2, \ldots). \tag{10.16c}$$

If (10.16b) is compared with (9.60b) it will be seen that it differs by the introduction of the term $-\mu n_g(T)$, which reflects the loss to $n_g(T)$ through cessation of spreading. The size of $l_g(T)$ only changes by the addition of gth generation spreaders who cease activity, and this fact is expressed in (10.16c). Since changes in $m(T)$ are determined only by the total number of spreaders we have, as before,

$$\frac{dm(T)}{dT} = -\beta m(T)n(T). \tag{10.17}$$

By solving these equations for $\{l_g(T)\}$ with initial conditions $l_g = 0$ when $m = N$, Daley (1967b) was able to show that

$$l_g = \frac{\rho}{g!} \int_m^N \left\{ \int_u^N \frac{dv}{\psi(v)} \right\}^g \exp\left[-\int_u^N \frac{dv}{\psi(v)} \frac{du}{u} \right] \qquad (g = 0, 1, 2, \ldots), \tag{10.18}$$

where $\psi(v) = N + a - v + \rho \ln(v/N)$. This equation enables us to find l_g as a function of m. It shows, incidentally, that l_g can be represented as a mixture of Poisson distributions, but it does not seem possible to turn this observation to good effect. When $T \to \infty$, $m(T) \to m(\infty)$ and so

$$\lim_{T \to \infty} l_g = \lim_{m \to m(\infty)} l_g = \frac{\rho}{g!} \int_{m(\infty)}^{N} \left\{ \int_u^N \frac{dv}{\psi(v)} \right\}^g \exp\left[-\int_u^N \frac{du}{\psi(u)} \right] \frac{dv}{v}$$

$$(g = 0, 1, 2, \ldots,). \qquad (10.19)$$

No simple approximations to this integral have been discovered, but there is no difficulty about evaluating the integrals numerically if required.

The spatial spread of a general epidemic model

We can formulate a spatial version of the general epidemic model discussed above by an extension of the arguments used for the simple epidemic. Most of the theory available is deterministic, but we shall refer to a simulation study by Mollison (1972b) which adds a note of caution to our deterministic analysis. The method followed is to formulate the partial differential equations governing the development of the process in time and then, in a heuristic fashion, to deduce some relevant properties of their solution.

As before, suppose the population to be uniformly spread over a line of unlimited extent with density σ. At time T and location s the members can be divided into three exhaustive and mutually exclusive categories, as follows:

$x(s, T)$ is the proportion of ignorants;

$y(s, T)$ is the proportion of active spreaders;

$z(s, T)$ is the proportion of spreaders who have ceased activity,
    called passive knowers.

These three proportions add up to one for all s and T. Contacts are made at a rate β and the distance at which a contact is made is governed by the probability density function $v(u)$, $s + u$ being the location of the receiver $(-\infty < u < +\infty)$. The new feature is the introduction of the rate μ at which active spreaders become passive. The rates of change of x, y, and z now become

$$\frac{\partial x}{\partial T} = -\beta \sigma x \bar{y}, \qquad (10.20a)$$

$$\frac{\partial y}{\partial T} = \beta \sigma x \bar{y} - \mu y, \qquad (10.20b)$$

$$\frac{\partial z}{\partial T} = \mu y. \qquad (10.20c)$$

The arguments of x, y and z have been suppressed for simplicity but we shall reintroduce them whenever they are needed to make the reasoning clear. As before, $\bar{y}$ is a space average of y around the point s.

We begin with the question of threshold effects. This was very easily answered in the non-spatial case by considering the sign of dy/dT at $T = 0$. The situation is now more complicated. An epidemic will certainly be impossible if $\partial y/\partial T$ is non-positive at every point s on the line. To take the easiest case first, suppose that the news is injected into the population by informing a small proportion ε uniformly distributed over the line. This means that $y(s, 0) = \bar{y}(s, 0) = \varepsilon$. Initially $x(s, 0) = 1 - \varepsilon$ and so growth will be impossible if

$$\beta\sigma(1 - \varepsilon)\varepsilon \leq \mu\varepsilon. \tag{10.21}$$

Thus we have the result that a small injection of the news will not start an epidemic if $\sigma < \rho$. In other words there must be a sufficient density of ignorants for diffusion to take place.

The symmetry of the problem in this particular case ensures that the proportions x, y, and z will be independent of s throughout the development of the process. That is, the epidemic pattern will be the same at all locations and the differential equations become identical in form to those given in (10.8), (10.9), and (10.10). This result should not surprise us because by introducing the information uniformly over the region we have effectively eliminated the spatial element from the problem.

It would be more natural to suppose that the original spreaders are concentrated around a particular location. If they are confined to a small interval then some of their initial contacts will be with ignorants outside the region and so, initially at least, there will be no circumstances under which growth is impossible at all locations. This does not, however, imply that continuing growth throughout the region is certain since growth at some points does not ensure that the total level of knowledge will rise above the initial level. The growth outside the initial area may be more than matched by a decline within it.

The position becomes clearer if we suppose those who are initially informed are distributed over the whole line according to an initial smooth density $y(s, 0)$ chosen to be small for all s. This allows us to have an initial concentration and can be made to approximate very closely to the situation in which the source is confined to a restricted part of the space. Suppose also that $v(u)$ is concentrated in a neighbourhood of $u = 0$ then

$$\bar{y}(s, T) = \int_{-\infty}^{+\infty} y(s - u, T)v(u)\mathrm{d}u \doteq y(s, T).$$

Having made this approximation we observe that $\partial y/\partial T$ at $T = 0$ is non-positive if

$$\beta\sigma \leq \mu \quad \text{or} \quad \beta \leq \rho. \tag{10.22}$$

Thus if (10.22) is satisfied there will be no growth in the number of spreaders and if it is not satisfied the number of active spreaders will rise above the initial number at all locations. These results indicate a threshold density which must be exceeded if there is to be any diffusion at all. Kendall (1965) investigated the condition under which the epidemic would be propagated as a wave using the diffusion approximation to the space average $\bar{y}$ and arrived at the same threshold condition, which he called the wave threshold theorem.

If diffusion does take place, the next thing is to consider the terminal state and to enquire whether the epidemic will die out before the whole population has been informed. The result obtained earlier for Kermack and McKendrick's model suggests that the proportion who ultimately hear will depend on the amount by which the population density exceeds the threshold. We now show that this expectation is fulfilled. From (10.20c) we have

$$\frac{\partial z(s-u,T)}{\partial T} = \mu y(s-u,T).$$

Multiplying by $v(u)$ and integrating between $-\infty$ and $+\infty$ gives

$$\frac{\partial \bar{z}}{\partial T} = \mu \bar{y},   \tag{10.23}$$

where the bar indicates the space average with respect to $v(u)$. Dividing (10.20a) by (10.20c) gives

$$\frac{\mathrm{d}x}{\mathrm{d}\bar{z}} = -x\frac{\sigma}{\rho},   \tag{10.24}$$

which may be compared with (10.11). Solving this equation with the initial condition $\bar{z} = 0$, $x = 1$ we have

$$x = \mathrm{e}^{-\sigma\bar{z}/\rho}.   \tag{10.25}$$

This equation expresses the relationship between $x(s,T)$ and $\bar{z}(s,T)$ at each point s on the line. As $T \to \infty$ the epidemic must die out at such a point s either because everyone knows or because the epidemic has 'passed on', leaving no active spreaders in the neighbourhood of s. In either event this implies $y(s, \infty) = 0$ and so, in the limit, we have, from (10.25),

$$1 - z(s, \infty) = \mathrm{e}^{-\sigma\bar{z}(s, \infty)/\rho}.   \tag{10.26}$$

If the ultimate number of hearers at each location were to be the same then we should have $\bar{z} = z$ and (10.26) would determine the terminal proportion of hearers. It is not obvious that this condition will hold, at least near to the original source, but it does seem likely that $z(s, \infty) \to z(\infty)$, say, as s becomes increasingly distant from the source. This limiting value will then satisfy

$$1 - z(\infty) = \mathrm{e}^{-\sigma z(\infty)/\rho},   \tag{10.27}$$

which is the same as (10.13) with $d = \sigma/\rho$. This equation has a non-zero root in $(0, 1)$ only if $\sigma > \rho$, which is the condition for there to be an epidemic at all. Some values of $z(\infty)$ satisfying (10.27) can be read off from Table 10.4 by entering it with $d = \sigma/\rho$. The 'interesting' values of σ/ρ lie between 1 and 4. Below this interval there will be no epidemic and above it practically everyone will hear. Kendall (1957) uses the term 'pandemic' to describe the situation when $\sigma > \rho$ since then the effects of the epidemic are felt equally at all locations however distant from the source. A fuller discussion of the theory underlying these results may also be found in Kendall's paper.

Kendall (1965) also investigated the existence of waveform solutions to (10.20a, b and c) by making the diffusion approximation used in Chapter 9. After deducing the threshold condition he showed that if there is a waveform solution it cannot have a velocity less than

$$c_{\min} = 2\sqrt{K}\,(1 - \rho/\sigma)^{\frac{1}{2}}. \tag{10.28}$$

The effect of introducing the cessation of spreading is to slow down the rate of advance by the factor $(1 - \rho/\sigma)^{\frac{1}{2}}$. In a simulation study of the stochastic version of the model Mollison (1972b) found that in an example with $\rho/\sigma = \frac{1}{2}$ the speed was reduced to about one-fifth of $2\sqrt{K}$ instead of 0.7 as the deterministic theory suggests. Subsequent investigation of deterministic and stochastic versions of epidemics in space (see Mollison, 1977) suggests that this result is not untypical. Our practice of regarding the deterministic model as an approximation to a more realistic stochastic model is therefore undermined, at least as far as matters relating to velocity are concerned. Whether or not the deterministic model has meaning or value on its own account is open to question.

Having reached this point the reader may feel that the simplifications which have had to be made in order to make the models tractable are so drastic as to make the results almost meaningless. The assumption of a population uniformly distributed over an infinite line, or plane, is far removed from the highly irregular distribution of human populations. Similarly, the assumption of a concentrated symmetrical distribution $v(u)$ supposed to be the same for all people at every location ignores the inevitable individual and topographical variation. Such arguments have considerable force if the object is to predict diffusion in some detail. Detailed prediction is not, however, a reasonable objective in the present state of the theory. The value of our highly simplified models lies in their ability to suggest broad qualitative features like threshold effects. From the manner of their derivation one might reasonably hope such effects would survive a good deal of change in the assumptions in the direction of greater realism.

10.3 RAPOPORT'S MODELS

The basic model

During the period 1948–54 a number of models for the diffusion of information were proposed by Rapoport and his associates (see Section 10.7 for details). His assumptions were the same as those of the general epidemic model except for the one governing the cessation of spreading. He supposed that each spreader told the news to exactly d other people and then ceased activity. The choice of person to be told was supposed to be random and hence independent of whether or not they had received the information. It is difficult to imagine a real-life situation in which this assumption would be true. The chain-letter in which each recipient is asked to write to d other persons is perhaps the nearest approximation. The value of studying the model is best seen by considering it as a special case of a more general model in which the number told is a random variable. If this random variable is denoted by $\tilde{d}$, Rapoport stated that his results would hold by taking $d = E(\tilde{d})$. Later in this section we shall discuss the conditions under which this statement can be justified.

The present model thus arises as an extreme case when we take the distribution of $\tilde{d}$ as concentrated at the point d. Kermack and McKendrick's model is another special case since then $\tilde{d}$ has a geometric distribution. This may be seen as follows. If a spreader is active for length of time x the number of people he communicates with, excluding the initial spreaders, will have a Poisson distribution with mean βNx. But x has an exponential distribution with parameter μ. Therefore,

$$Pr\{\tilde{d} = j\} = \int_0^\infty e^{-\mu x} \frac{(\beta Nx)^j}{j!} e^{-\beta Nx} dx$$

$$= \frac{\mu}{\mu + N\beta} \left(1 - \frac{\mu}{\mu + n\beta}\right)^j$$

$$= \frac{1}{1+d} \left(\frac{d}{1+d}\right)^j \qquad (j = 1, 2, \dots,). \qquad (10.29)$$

As another example, take the case where the individual spreads the news for a fixed length of time; $\tilde{d}$ would then have a Poisson distribution. By comparing the results obtained under a variety of fairly extreme assumptions we may hope to establish our results on a broader basis.

Except for the case $d = 1$ we shall not be able to obtain any results about the rate of diffusion. Instead we shall study the random variables $\{n_g\}$—the numbers in the different generations of hearers. Some additional notation is required as follows. Let

$$N_g = \sum_{i=1}^g n_i.$$

For some $g \le N$, N_g will attain its maximum value at which it remains constant. This limiting value is the ultimate size of the epidemic. In the last section we denoted this quantity by n_H, but here it is more natural to use N_∞ because it is obtained by allowing g rather than T to tend to infinity. The equivalence between

n_H and N_∞ is established by noting that

$$N_\infty = \lim_{g \to \infty} \sum_{i=1}^{g} n_i = \lim_{T \to \infty} \{N - m(T)\} = n_H. \qquad (10.30)$$

Finally, let

$$p_g = E(n_g)/N.$$

This may be interpreted as the probability that a randomly chosen member of the population belongs to the gth generation.

Exact theory when $a = d = 1$

In this case it is possible to derive a simple explicit expression for the distribution of N_∞. Since each spreader tells only one individual who is either an ignorant or former spreader, it follows that $n_g = 0$ or 1 for all g. The process will continue until a spreader tells a former spreader. Suppose that this happens at the ith generation. Then

$$n_g = 1, \qquad i \leq g$$
$$= 0, \qquad i > g.$$

The process can be represented as a random walk on the integers $1, 2, \ldots, N$ as follows:

$$\frac{i-1}{N-1} \longleftarrow \qquad \longrightarrow \frac{N-i}{N-1}$$

$$\underset{1 \qquad\quad 2 \qquad\quad 3 \;\ldots\; i-1 \qquad\quad i \qquad\quad i+1 \;\ldots\; N}{\vert \quad\quad\quad \vert \quad\quad\quad \vert \quad\quad\quad \vert \quad\quad\quad \vert \quad\quad\quad \vert \quad\quad\quad \vert}$$

A particle starts at the left-hand end. It moves to the right whenever a spreader tells an ignorant and to the left if a spreader tells a former spreader. The probabilities of each kind of transition are $(N-i)/(N-1)$ and $(i-1)/(N-1)$ respectively, and the process terminates as soon as the first move to the left is about to occur. The point of termination is equal to N_∞. A direct argument thus gives

$$Pr\{N_\infty = i\} = \left(\frac{i-1}{N-1}\right) \prod_{j=1}^{i-1} \left(\frac{N-j}{N-1}\right)$$

$$= \frac{(N-2)!\,(i-1)}{(N-i)!\,(N-1)^{i-1}} \qquad (i = 2, 3, \ldots, N). \qquad (10.31)$$

For small N the distribution is easily tabulated; the result for $N = 11$ is given in Table 10.6.

The distribution is positively skewed with mean value 4.660. If communication with the source is allowed the Table 10.6 would apply to the case $N = 10$ with i reduced by one.

Table 10.6 Distribution of N_∞ for $N = 11$, $d = 1$, $a = 1$

i	$Pr\{N_\infty = i\}$	i	$Pr\{N_\infty = i\}$
2	0.1000	7	0.0907
3	0.1800	8	0.0423
4	0.2160	9	0.0145
5	0.2016	10	0.0033
6	0.1512	11	0.0004

For large N an approximation to $Pr\{N_\infty = i\}$ can be found as follows. Let K be a constant such that $i = K\sqrt{N}$ and let N be large. Then

$$\prod_{j=1}^{i-1}\left(1 - \frac{j-1}{N-1}\right) = \exp\left[\sum_{j=1}^{i-1}\ln\left(1 - \frac{j-1}{N-1}\right)\right]$$
$$\sim e^{-K^2/2}.$$

Therefore

$$Pr\{N_\infty = K\sqrt{N}\} \sim \frac{K}{\sqrt{N}}e^{-K^2/2}. \tag{10.32}$$

Omitting the factor $1/\sqrt{N}$, (10.32) gives the probability density function of the continuous approximation to the distribution of N_∞ at $K\sqrt{N}$. As a check, we may note that the density integrates to one. It may be shown that the median of the distribution of N_∞ is at $\sqrt{(2\ln 2)N} = 1.177\sqrt{N}$. The asymptotic moments may be deduced by observing that $K = N_\infty / \sqrt{N}$ is distributed like χ^2 with 4 degrees of freedom. Hence

$$E(N_\infty) \sim \sqrt{\frac{\pi N}{2}} = 1.253\sqrt{N}$$

and

$$\left.\vphantom{\begin{array}{c}1\\1\\1\\1\end{array}}\right\} \tag{10.33}$$

$$\text{var}(N_\infty) \sim \frac{4-\pi}{2}N = 0.4292N$$

It thus follows that the expected proportion of the population who eventually hear tends to zero as N increases. When $N = 11$ the asymptotic formula for the mean gives $E(N_\infty) = 4.1557$ as compared with the exact value of 4.6604.

In the case $a = d = 1$ it is also possible to find the distribution of the duration of the diffusion. On the assumption of random mixing the length of time taken for a spreader to communicate with another person has an exponential distribution with mean value $1/\beta(N-1)$. Given that $N_\infty = i$ the distribution of the duration will then be like that of $\chi^2_{2(i-1)}/2\beta(N-1)$. We have already found the distribution of N_∞ so that an expression for the unconditional distribution of the duration can be written down at once. It is simpler to work with the moments, for which we find

$$E(T_{N_\infty}) = \sum_{i=2}^{N} Pr\{N_\infty = i\}(i-1)/\beta(N-1)$$

$$= \frac{E(N_\infty)-1}{\beta(N-1)} \sim \sqrt{\frac{\pi}{2\beta^2 N}} = \frac{1.253}{\beta\sqrt{N}}. \qquad (10.34)$$

Similarly,

$$\text{var}(T_{N_\infty}) = \sum_{i=2}^{N} Pr\{N_\infty = i\}E(T_i^2|i) - \{E(T_{N_\infty})\}^2$$

$$\sim \left(\frac{2}{\beta^2} - \frac{\pi}{2\beta^2}\right)\Big/ N = \frac{0.4292}{\beta^2 N}. \qquad (10.35)$$

The duration of the epidemic thus decreases as the population size increases, as we found with the pure birth model.

An approximation to $E(N_\infty)$ in the general case

It is possible to find the exact joint distribution of the random variables $\{n_g\}$ and hence of N_∞ by the methods of classical occupancy theory (see, for example, Barton and David, 1962). However, an approximation to the expectation of N_∞ can be found directly by the following argument. The probability that a randomly chosen individual does not receive *any* of the $n_g d$ transmissions made by members of the gth generation is

$$\left(1 - \frac{1}{N-1}\right)^{n_g d}.$$

The probability that he receives at least one such transmission is thus

$$1 - \left(1 - \frac{1}{N-1}\right)^{n_g d}.$$

A randomly chosen hearer is an ignorant with probability

$$\left(1 - \frac{N_g - 1}{N-1}\right).$$

Therefore the expected number of new hearers in the gth generation is

$$E(n_{g+1}|n_g, N_g) = (N-1)\left(1 - \frac{N_g-1}{N-1}\right)\left\{1 - \left(1 - \frac{1}{N-1}\right)^{n_g d}\right\}. \qquad (10.36)$$

A simple rearrangement gives

$$E\left\{\frac{N - N_{g+1}}{N - N_g}\bigg|n_g, N_g\right\} = \left(1 - \frac{1}{N-1}\right)^{n_g d} \qquad (g \geq 1). \qquad (10.37)$$

Suppose that we now attempt to take the expectations of both sides of this equation with respect to all the random variables appearing. There is no simple expression for the result, but an approximation to the answer can be obtained by replacing all the random variables by their expectations. For this operation we need

$$E(n_g) = Np_g \quad \text{and} \quad E(N_g) = N\sum_{i=1}^{g} p_i.$$

Substitution now yields

$$\frac{N - N\sum_{i=1}^{g+1} p_i}{N - N\sum_{i=1}^{g} p_i} = \left(1 - \frac{1}{N-1}\right)^{Np_g d}. \qquad (10.38)$$

Let g^* be the smallest value of g for which $N_{g+1} = N_g$. We now take the product of both sides of this equation for $g = 1$ to $g = g^*$. On the left-hand side we have

$$N - \sum_{i=1}^{g^*} p_i = N - E(N_\infty)$$

and on the right-hand side

$$\left(1 - \frac{1}{N-1}\right)^{dE(N_\infty)}.$$

Setting $p = E(N_\infty)/N$ the equation thus becomes

$$p = 1 - \left(1 - \frac{1}{N-1}\right)^{Ndp} \sim 1 - e^{-dp}. \qquad (10.39)$$

This result was first obtained by Solomanoff and Rapoport (1951) but it is identical with that found for Kermack and McKendrick's model in (10.13). This should occasion no surprise. The two models differ only in the stochastic aspects of the cessation of spreading and both (10.13) and (10.39) were found by deterministic methods involving only averages. Table 10.4 can thus also be used for Rapoport's model.

A check on the accuracy of the approximation can be made when $a = d = 1$. In the limit as $N \to \infty$ we have $p = 0$, but we can investigate the situation when N is finite. To do this we rewrite (10.39) in the form

$$p_{g+1} = \left\{1 - \left(1 - \frac{1}{N-1}\right)^{Ndp_g}\right\}\left\{1 - \sum_{i=1}^{g} p_i\right\} \qquad (g \geq 1) \qquad (10.40)$$

and add

$$p_1 = 1 - \left(1 - \frac{1}{N-1}\right)^{ad}. \tag{10.41}$$

These equations may be solved recursively to give the p_g's and hence p. A comparison of exact and approximate values is given in Table 10.7.

Table 10.7 *Comparison of Rapoport's recursive approximation to n_H/N with the exact value for $a = d = 1$*

	N					
	10	20	50	100	200	300
Exact	0.366	0.265	0.171	0.122	0.087	0.071
Approximate	0.292	0.237	0.168	0.125	0.092	0.076

The agreement is reasonably good and suggests that the approximation may be used with confidence when N is moderate or large. Further evidence confirming this conclusion for $a = 1$ and $d = 2$ and 4 has been obtained from the simulation studies summarized in Table 10.8.

Table 10.8 *Average values and standard deviation of n_H/N in 100 simulations of Rapoport's model*

		N				Rapoport's asymptotic approximation
		20	100	300	3000	
$d = 2$	Average	0.784	0.799	0.794	0.796	0.797
	SD	0.138	0.049	0.031	0.010	—
$d = 4$	Average	0.978	0.981	0.981	0.980	0.980
	SD	0.035	0.015	0.010	0.0077	—

Some further results are summarized in Table 10.9 which gives the most common value of g and a deterministic approximation to the duration of the epidemic. The latter is obtained as the largest value of g for which $Np_g \geq 1$. Perhaps the most surprising feature of this table is the small number of generations taken for the news to spread. Nevertheless, even with d as large as 10, most people receive the news at third or fourth hand.

Table 10.9 *The most common value of g and estimated duration of the epidemic for N = 300 and 3000*

		d		
		2	4	10
$N = 300$	Mode	7	4	3
	Duration	13	7	4
$N = 3000$	Mode	11	6	4
	Duration	19	9	5

It is convenient at this point to consider the more realistic general model in which each person tells a random number $\tilde{d}$ of persons, where $E(\tilde{d}) = d$. We have already seen that (10.39) still holds in one such case when $\tilde{d}$ has a geometric distribution (see equation 10.29). That it holds in general may be demonstrated as follows. Let n_{gi} be the number of people informed by the ith member of the gth generation. Assume that the n_{gi}'s are independent. The argument leading to (10.36) depended only on the total number of contacts made by members of the gth generation. Hence $n_g d$ can be replaced by $\Sigma_i n_{gi}$. At the point where we approximate by replacing random variables by their expectations, $\Sigma_i n_{gi}$ is replaced by $n_g d$, where d now denotes the expectations of $\tilde{d}$.

This model, together with the simpler version of Rapoport's model and that of Kermack and McKendrick, can be regarded as a special case of the chain-letter model discussed by Daley (1967b). According to this model spreaders act independently and are active for a random length of time. During their active period (or at the end of it) they pass on the news to a random number of other persons. Those among the hearers who were previously ignorant become spreaders. The same deterministic approximation serves for all the special cases of this model, at least so far as the ultimate state of the population is concerned. In particular, the equations which determine the ultimate number of hearers by their generation will be valid for this wider class of models. The results given in the last part of this section therefore also provide an approximation for the models here. Since Rapoport's models are expressed in terms of generations rather than in terms of time it may not be immediately clear how the connection can be made. The difficulty is overcome by observing that the terminal state in Rapoport's models would be unaffected by the introduction into the model of a random variable representing the time for which each spreader was active. The essential parameter on which the terminal state depends is the average number of contacts made by an active spreader.

10.4 DALEY AND KENDALL'S MODEL

The basic model and some exact theory

One important feature of the diffusion of information in human populations has been disregarded when constructing the models given earlier in the chapter. On telling the news the spreader is likely to discover whether his hearer has already heard it. This knowledge may very well influence his enthusiasm for continuing to spread the news. In other words, the cessation of spreading may depend on the state of knowledge in the population. This interaction between spreader and hearer is peculiar to this application and has no counterpart in the theory of epidemics. It is thus desirable to develop new models to assess the importance of this characteristic. A class of models having the property that cessation of spreading depends on the state of knowledge in the population was proposed by Daley and Kendall (1965). Their work forms the basis of the present section.

We retain all the assumptions of Kermack and McKendrick's model except the one which governs the cessation of spreading. In the simplest version of the model the diffusion proceeds as follows. If a spreader meets an ignorant the news is transmitted and the ignorant becomes a spreader. If a spreader meets someone who has previously been informed he ceases to spread the news. Those who have heard but are no longer spreading are called 'stiflers' because, on contact with a spreader, they cause him to cease spreading. As before, $m(T)$ and $n(T)$ denote the number of ignorants and spreaders respectively, at time T. The theory for this model has only been worked out for the case when the a persons who initiate the diffusion process are indistinguishable from the other members of the population. Thus they may communicate with one another and they remain present as stiflers throughout the process. The number of stiflers at time T, denoted by $l(T)$, is thus $N + a - m(T) - n(T)$.

When the system is in state (m, n) three transitions are possible as follows:

(a) $(m, n) \to (m - 1, n + 1)$. This transition occurs whenever a spreader meets an ignorant. The probability that such a meeting occurs in $(T, T + \delta T)$ is $\beta mn \delta T$.

(b) $(m, n) \to (m, n - 1)$. The number of spreaders is reduced by one whenever a spreader meets a stifler. This happens with probability $\beta nl \delta T$.

(c) $(m, n) \to (m, n - 2)$. This transition results from contact between two spreaders when both become stiflers. There are $\frac{1}{2}n(n - 1)$ pairs of spreaders so the total probability of contact is $\beta \frac{1}{2} n(n - 1) \delta T$.

It is possible to set up differential–difference equations for the probabilities $P_{m,n}(T)$ in exactly the same way as for Kermack and McKendrick's model. As in that case they are difficult to solve, except in degenerate cases. Some information can be obtained by comparison with the pure birth process model because the expected number of hearers will always be greater in the absence of stifling. Thus the curve for $N - \bar{m}(T)$ for the birth process will provide an upper bound for the corresponding function of the stifling model. However, we shall see later that the deterministic approximation provides sufficient information.

The distribution of the ultimate number of ignorants, or knowers, can be obtained by the method of the embedded random walk. The situation closely parallels the development in Section 10.2 so we shall give a more condensed treatment here. We construct difference equations for the probabilities $\{P_{m,n}\}$, using the transition probabilities which are as follows:

$$(m, n) \to (m - 1, n + 1): \frac{m}{N + a - \frac{1}{2}(n + 1)}$$

$$(m, n) \to (m, n - 1): \frac{N + a - n - m}{N + a - \frac{1}{2}(n + 1)}$$

$$(m, n) \to (m, n - 2): \frac{\frac{1}{2}(n - 1)}{N + a - \frac{1}{2}(n + 1)}.$$

The equations are:

$$\left.\begin{aligned}
P_{m,n} &= \frac{m+1}{N + a - \frac{1}{2}n} P_{m+1, n-1} + \frac{N + a - n - m - 1}{N + a - \frac{1}{2}(n+2)} P_{m, n+1} \\
&\qquad + \frac{\frac{1}{2}(n+1)}{N + a - \frac{1}{2}(n+3)} P_{m, n+2} \\
&\qquad\qquad (m \geq 0, n > 1) \\
P_{m,1} &= \frac{N + a - m - 2}{N + a - 3/2} P_{m,2} + \frac{1}{N + a - 2} P_{m,3} \qquad (m \geq 0) \\
P_{m,0} &= \frac{N + a - m - 1}{N + a - 1} P_{m,1} + \frac{\frac{1}{2}}{N + a - 3/2} P_{m,2} \qquad (m \geq 0)
\end{aligned}\right\} \qquad (10.42)$$

If we define $P_{m,n} = 0$ whenever $m + n > N + a$ these equations hold for all m and n satisfying $0 \leq m \leq N$ and $0 \leq m + n \leq N + a$. The initial condition is $P_{N,a} = 1$. The m-axis is an absorbing barrier so $\{P_{m,0}\}$ is the probability distribution of ignorants remaining when diffusion ceases. It has not proved possible to obtain an explicit solution of (10.42) for this distribution but numerical values can easily be obtained by using a computer.

The shape of the distribution of n_H is illustrated in Figure 10.3 for $N = 50$, $a = 1$. This shows that the number reached will usually be in the neighbourhood of 40 but on a few occasions the diffusion will fail to develop. If N is increased the hump near the origin diminishes in size while the major hump becomes more nearly symmetrical with decreasing variance. Provided that N is large the value of a has hardly any effect on the main part of the distribution of n_H.

The results obtained for the distribution of the ultimate size of the epidemic are in marked contrast to those for Kermack and McKendrick's model. There is no threshold effect, by which we mean that the development of the process does not

ⁿ A ᴸᴬᵀᵀᴷᴱ

295

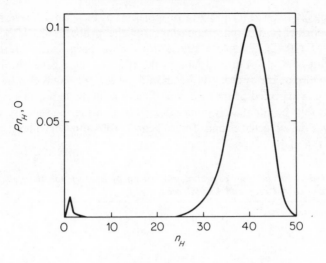

Figure 10.3

depend critically on the population size. Daley and Kendall's model is like Rapoport's in that a fixed proportion, which is strictly less than one, ultimately receive the information.

A more general version of the model

In the simple model described above the spreaders are very easily rebuffed. After meeting only one person who has already heard the news they assume that 'everyone knows' and cease to be spreaders. Various generalizations of this model are possible. We shall describe one of the two proposed by Daley and Kendall. Suppose that when a spreader communicates with another spreader or a stifler he becomes a stifler himself with probability ζ. As a further generalization let η be the probability that communication of the news actually takes place when a spreader meets another person. In our previous model both ζ and η were unity. (We could have introduced the second generalization in any of the earlier models but it would have been superfluous. Its only effect would have been to reduce the effective rate of contact to $\eta\beta$. In the stifling model things are a little more complicated.)

The infinitesimal transition probabilities are as follows:

(a) $(m, n) \to (m - 1, n + 1): \eta\beta mn\, \delta T;$

(b) $(m, n) \to (m, n - 1): \beta[\eta\zeta nl + \zeta(1 - \zeta)\{1 - (1 - \eta)^2\}n(n - 1)]\, \delta T;$

(c) $(m, n) \to (m, n - 2): \beta\{1 - (1 - \eta)^2\}\zeta^2 \frac{1}{2}n(n - 1)\, \delta T.$

The transition probabilities for the associated random walk are obtained by expressing (a), (b), and (c) above as proportions of their total. These may then be used to construct difference equations for the probabilities $\{P_{m,n}\}$ after the manner of (10.42). Extensive calculations have been made of the terminal probabilities $\{P_{m,0}\}$ in order to assess the effect of these generalizations on the ultimate state of the system. As we should expect, the value of a has almost no effect on the expected proportion who finally hear the news. The influence of ζ and η can thus be adequately demonstrated by considering the case $a = 1$. Table 10.10 gives the expected proportion of hearers for values of ζ and η intermediate between 0 and 1.

Table 10.10   The expected proportion of hearers for the general model of Daley and Kendall with $a = 1$

	$N = 50$			$N = 200$		
			η			
	0.25	0.50	0.75	0.25	0.50	0.75
0.25	0.979	0.983	0.987	0.981	0.985	0.989
0.50	0.892	0.905	0.920	0.899	0.911	0.925
0.75	0.793	0.813	0.834	0.807	0.825	0.844

The general effect of relaxing the assumption that $\zeta = \eta = 1$ is to increase the proportion who ultimately hear. It is clear from the table that η plays a relatively minor role compared with ζ. Inspection of the actual distribution of n_H/N shows that the basic form is preserved under this generalization. That is, it remains true that the diffusion reaches a fixed proportion of the population on average, but the actual values of that proportion will depend on the extent to which stifling occurs.

An alternative way of generalizing the simple model is to suppose that a spreader becomes a stifler after k unsuccessful attempts at telling the news. The quantity k could be fixed or it could be a random variable. Using deterministic methods Daley and Kendall obtained the following asymptotic values for the $E(n_H/N)$ when k is fixed.

k	1	2	3
$E(n_H/N)$	0.797	0.940	0.980

It is thus true to say that 'most people' will hear the news in a homogeneously mixing population unless there is a high degree of stifling. By supplementing our analysis by deterministic methods we shall be able to show that $E(n_H/N)$ cannot be less than 0.63 under any of the models we have considered in this section. The deterministic approach also enables us to obtain information about the rate of diffusion.

A deterministic approximation to the model can be obtained in the usual way.

If we treat m and n as continuous variables with rates of change determined by the transition probabilities given above, we obtain

$$\left. \begin{aligned} \frac{dm(T)}{dT} &= -\eta\beta m(T)n(T) \\ \frac{dn(T)}{dT} &= \eta\beta n(T)\{(1+\zeta)m(T) - \zeta(1-\eta)n(T) + \zeta(2-\eta-N-a)\} \end{aligned} \right\} \quad (10.43)$$

The initial conditions are $m = N$, $n = a$. A direct consequence of (10.43) is that

$$\frac{dn}{dm} = -(1+\zeta) + \zeta(1-\zeta)\frac{n}{m} - \frac{\zeta(2-\eta-N-a)}{m} \quad (10.44)$$

where we are now treating n as a function of m. This differential equation may be solved by standard methods to give

$$n = -\frac{(N+a+\eta-2)}{(1-\eta)} - \frac{(1+\zeta)n}{1-\zeta(1-\eta)} + \left(\frac{m}{n}\right)^{\zeta(1-\eta)}\left(\frac{2-\eta}{1-\eta}\right)$$

$$\times \left[\frac{N}{\{1-\zeta(1-\eta)\}} + (a-1)\right] \quad (\eta \neq 1). \quad (10.45)$$

When $\eta = 1$ the appropriate relationship between $m(T)$ and $n(T)$ can be obtained either directly from (10.44) or by a limiting operation on (10.45). Equation (10.45) enables us to find $n(T)$ in terms of $m(T)$ at any stage of the process. Hence we may eliminate $n(T)$ from (10.43) to obtain a differential equation for $m(T)$.

The rate at which the news spreads can be investigated by solving the first member of (10.43) for $m(T)$. A solution can be obtained numerically after using (10.45) to express $-\eta\beta m(T)n(T)$ as a function of $m(T)$. The method is exactly the same as the one used for the model discussed in Section 10.2. We shall merely point out an interesting connection between the two models in this respect. Let $a = 1$, $\zeta = \eta = 1$; then $m(T)$ satisfies

$$\frac{dm(T)}{dT} = -\beta m(T)\left\{2N+1-2m(T) + N\ln\frac{m(T)}{N}\right\}, \quad (10.46)$$

with initial condition $m(0) = N$. If we refer to (10.14), the corresponding equation for Kermack and McKendrick's model, we see that the two equations are very similar. In fact, the rate of spread, $dm(T)/dT$, in the present model is the same as it is in Kermack and McKendrick's model with a population size of $2N$ and $d = 2$. This fact was also noted by Cane (1966).

10.5 A MODEL FOR THE SURVIVAL OF A SOCIAL GROUP

All of the foregoing models assume that once a spreader ceases activity he plays no further part in the process. This is reasonable enough with infectious diseases where infection may cause death or lifelong immunity, but in social applications the assumption is less satisfactory. Consider, for example, a new political party or interest group of some kind. Its members may recruit others by personal contact or advertising and at the same time there may also be a falling away of membership. In many such processes it is perfectly possible that lapsed members might have their interest rekindled and hence become active members again. A simple way of incorporating this feature into a model is to suppose that when spreaders cease activity they return to the pool of susceptibles. From a model of such a process we might hope to discover the conditions for growth to occur and to learn something about the viability and the ultimate size of the active group. Unlike the processes considered hitherto, this one can continue indefinitely.

The Model

We propose a rather simple version of a model for the growth and survival of a social group, but one which exhibits some novel features. This will also pave the way for the more interesting case where there are two competing groups, to be discussed in the next section. The stochastic assumptions are exactly the same as those of Kermack and McKendrick's (KM's) model except for the fact that a person becomes susceptible again as soon as he ceases to be active as a spreader. This represents a simplification in that the population is now divided into only two classes instead of three. With the additional assumption we have a stochastic process whose state at time T can be described by a single variable, $n(T)$, which is the number of spreaders. The state space is $n = 0, 1, 2, \ldots N$ and the possible transitions and their associated rates are as follows.

$$n \to n + 1 : \lambda_n = \alpha(N - n) + \beta n(N - n), \qquad n = 0, 1, 2, \ldots N$$

$$n \to n - 1 : \mu_n = \mu n, \qquad n = 1, 2, \ldots N. \tag{10.47}$$

These rates define a birth and death process whose state probabilities satisfy

$$\frac{\mathrm{d}}{\mathrm{d}T} P_n(T) = -(\lambda_n + \mu_n) P_n(T) + \lambda_{n-1} P_{n-1}(T) + \mu_{n+1} P_{n+1}(T)$$

$$n = 0, 1, 2, \ldots N, \tag{10.48}$$

with $\lambda_0 = \mu_{N+1} = 0$. The quadratic birth rate, λ_n, makes it very difficult to find an exact solution to these equations. It is nevertheless possible to obtain a good deal of information about the process by a combination of techniques which we now describe.

The deterministic approximation

Proceeding in the usual way, the transition rates of (10.47) lead to the differential equation

$$\frac{\mathrm{d}n(T)}{\mathrm{d}T} = \beta(N - n(T))(n(T) + \omega) - \mu n(T), \tag{10.49}$$

where $\omega = \alpha/\beta$ as before. The solution may be written in the form

$$n(T) = n(\infty) - A\left[1 + \left\{\frac{a}{n(\infty) - n(0)} - 1\right\}\exp\frac{\beta AT}{N}\right]^{-1}, \tag{10.50}$$

where $A = \{4\omega N + (N - \rho - \omega)^2\}^{\frac{1}{2}}$ and $\rho = \mu/\beta$. This is a logistic curve rising from the initial value of $n(0)$ to a limiting value of

$$n(\infty) = \tfrac{1}{2}(N - \rho - \omega + A). \tag{10.51}$$

The ultimate size of the active group relative to the whole thus depends on the two ratios N/ρ and ω/N. The former quantity is also the expected number of contacts made by a spreader in the active phase. Increasing the mean number of contacts or the strength of the source will have the effect of increasing the eventual size of the active group. If $\rho = 0$ (no cessation of activity) the formulae reduce to those for the pure birth model of Chapter 9.

The case $\omega = 0$ requires special consideration. Diffusion then depends solely on interpersonal contact and so it is essential to have $n(0) > 0$. Even then we cannot guarantee that the group will grow because a threshold effect operates in essentially the same way as with the KM model. Growth can only take place if

$$\left.\frac{dn(T)}{dT}\right|_{T=0} > 0,$$

which implies that $N - n(0) > \rho$. If N is large and $n(0)$ small this is virtually the same as the condition $d > 1$. If the threshold condition is met, growth will take place according to (10.50) which now simplifies somewhat because $A = N - \rho$. One consequence of this is that $n(\infty) = N - \rho$. It is interesting to note that the growth curve is then identical with that for a simple epidemic in a population of size $N - \rho$.

Stochastic analysis

We shall follow two approaches to the stochastic analysis of the process. The first, due to Karmeshu and Pathria (1980a), yields an approximate solution for the whole process when N is large. The second, provides the exact limiting distribution $\{P_n(\infty)\}$. Taken together these two methods provide a fairly complete picture of the process.

Karmeshu and Pathria's (1980a) method is based on what they call the 'system-size expansion' technique which is used in statistical mechanics. It starts from the fact that we expect $\bar{n}(T)$ to be of order N and variations about this mean path to be of order $\sqrt{N}$ (this assumption breaks down in certain critical phases of the process which we ignore for the moment). Accordingly, we introduce a new process $x(T)$ related to $n(T)$ as follows:

$$n(T) = \bar{n}(\mathrm{T}) + \sqrt{N}x(T). \tag{10.52}$$

If $P_x(T)$ denotes the probability that $x(T) = x$ we can use (10.52) and (10.48) to deduce equations for $P_x(T)$. These, of course, are no easier to solve than the original equations, but the terms can be expanded as power series in $N^{-\frac{1}{2}}$. Equating the coefficients of leading powers of N we obtain first an equation for $\bar{n}(T)$ which is identical with (10.49). The deterministic approximation already obtained thus emerges as the first approximation by the new method. Proceeding to the next approximation Karmeshu and Pathria (1980a) find that $P_x(T)$ satisfies

$$\frac{\partial P}{\partial T} = -\phi_1(\bar{n})\frac{\partial(xP)}{\partial x} + \frac{1}{2}\phi_2(\bar{n})\frac{\partial^2 P}{\partial x^2}, \qquad (10.53)$$

where

$$\phi_1(\bar{n}) = \beta(1 - \omega - 2\bar{n}/N)$$
$$\phi_2(\bar{n}) = \beta(1 - \bar{n}/N)(\bar{n}/N + \omega) + \mu\bar{n}/N.$$

In physical applications (10.53) is known as the Fokker–Planck equation and its solution has the form of a normal density. Converting back from $x(T)$ to $n(T)$ it may be shown that, for fixed T,

$$n(T) \sim N(\bar{n}(T), N\sigma^2(T)). \qquad (10.54)$$

An exact expression for the variance in (10.53) is given by Karmeshu and Pathria (1980a, equation 25) but in the limit as $T \to \infty$ it reduces to

$$\sigma^2(\infty) = \bar{n}(\infty)/Ad. \qquad (10.55)$$

This approximate analysis thus gives us some idea of how much the actual size of the group will differ from what the deterministic approximation predicts. The argument depends on the initial assumption about the orders of magnitude of the mean and variance of $n(T)$ and these break down in the neighbourhood of $\omega = 0$, $d = 1$. The position will become clearer from our examination of the exact stochastic theory as $T \to \infty$.

It is well known that the steady-state probabilities of the birth and death process satisfy the equations

$$\mu_{n+1}P_{n+1} = \lambda_n P_n, \qquad n = 0, 1, 2, \dots$$

In our case this implies that

$$P_{n+1} = \frac{(N-n)(n+\omega)}{\rho(n+1)}P_n, \qquad 0 \le n < N, \qquad (10.56)$$

whence

$$P_n = \binom{N}{n}\rho^{N-n}\Gamma(n+\omega) \Big/ \sum_{n=0}^{N}\binom{N}{n}\rho^{N-n}\Gamma(n+\omega) \qquad 0 \le n \le N \quad (10.57)$$

if $\omega > 0$. If $\omega = 0$, (10.57) shows that $P_n = 0$ if $n > 0$ and so we must have $P_0 = 1$.

In other words, when there is no source, the active group will eventually die out. However, this is not the whole story and we return to this special case below.

Using the exact distribution of (10.57) it is easy to check the accuracy of Karmeshu and Pathria's (1980a) approximation of (10.54) when $T = \infty$. From calculations given in Bartholomew (1976a) it appears that it is excellent when $N = 100$ and fair if N is as small as 10. Although explicit formulae for the exact mean and variance of $\{P_n\}$ are lacking there is a simple relationship between them. Starting from (10.56) we find

$$\rho \sum_{n=0}^{N-1} (n+1)P_{n+1} = \sum_{n=0}^{N} (N-n)(n+\omega)P_n$$

or

$$\rho E(n) = NE(n) + N\omega - E(n^2) - \omega E(n). \tag{10.58}$$

Hence

$$\text{var}(n) = E(n)(N - \rho - \omega) + N\omega - E^2(n). \tag{10.59}$$

Since $\text{var}(n) \geq 0$ we can use (10.59) to place an upper bound on $E(n)$ which turns out to be

$$E(n) \leq \tfrac{1}{2}(N - \rho - \omega) + \{N\omega + \tfrac{1}{4}(N - \rho - \omega)^2\}^{\frac{1}{2}} = n(\infty) \tag{10.60}$$

from (10.51). The deterministic approximation thus provides an upper bound to the exact expectation of n at $T = \infty$. Karmeshu and Pathria (1980a) show that if the approximate variance given by (10.55) is substituted in (10.59) and the resulting equation solved for $E(n)$ the approximation obtained is excellent even when $N = 10$. In Karmeshu and Pathria (1980c) it is shown that the same approach can be used to yield an improved approximation for the variance.

The approximate expressions for the mean and variance of the size of the active group can be used to assess the relative importance of the source and interpersonal contact under various assumptions about the parameters of the process. We look mainly at the behaviour of the mean, but it is useful to do the same thing with the variance. As with the pure birth model, it needs a very strong source to match the effects of interpersonal contact. For example, if $(\omega/N) \gg 1$, which means that the source is much more powerful than the combined strength of the population, we find

$$En(\infty)/N = 1 - 1/d\{\omega/N + (1 + 1/d)\} \tag{10.61}$$

which shows that d, the mean number of contacts, still has a significant effect on the size of the group. If, on the other hand, $(\omega/N) \ll 1$, meaning that the source is comparable in strength to a single individual, its effect is negligible since then

$$\frac{En(\infty)}{N} = 1 - \frac{1}{d}, \qquad d > 1. \tag{10.62}$$

If d is much less than one, $En(\infty)/N = O(N^{-1})$ so that the active group never makes a significant impact on the population. When d is close to 1 with ω/N small the approximation on which the formulae are based breaks down.

Another special case of interest arises when d is large. Then we find

$$En(\infty)/N = 1 - 1/d\,(1 + \omega/N) \tag{10.63}$$

for any value of ω/N. In this case the mean and variance of $N - n(\infty)$ are approximately equal, which suggests that the limiting distribution of the number *outside* the active group has a Poisson distribution.

We left the special case $\omega = 0$ on one side in our general treatment because there appeared to be a contradiction between the deterministic and stochastic analyses. According to the latter the active group eventually dies out, whereas the deterministic analysis gives a limiting size of $(N - \rho)$ if $d > 1$. This apparent paradox has been examined in detail by Bartholomew (1976a) and Karmeshu and Pathria (1979). Their analysis shows that the deterministic approximation corresponds to what may be termed a quasi-stationary distribution. Although the limiting size will ultimately be zero the expected number of steps before this happens increases exponentially with N. Since the chance of becoming extinct is very small in any time interval of practical interest, it is reasonable to enquire about the form of the distribution *given that absorption does not occur*. If we envisage the process as a random walk on the integers $0, 1, 2, \ldots N$ this is achieved by making zero a reflecting barrier. This ensures that the group is kept in existence by, in effect, introducing a new member as soon as n becomes zero. As one would expect, the system then behaves very much like one with $\omega = 1$. The limiting distribution of n with this provision for regeneration is called the 'quasi-stationary distribution' and it is the mean of this distribution which corresponds to the result given by the deterministic analysis.

10.6 A MODEL FOR TWO COMPETING GROUPS

Many of the foregoing models can be regarded as special cases of the final model now to be considered. It is common in human societies for groups to form and compete with one another for the allegiance of the uncommitted. Political parties, pressure groups, sects, and cults all provide familiar examples. Although the model proposed concerns only two groups and makes many simplifying assumptions it represents a first step to understanding the dynamics of one of the most widespread and important of social phenomena. The model of the last section gives some insight into the conditions needed for a new movement to establish itself. We now consider how it will fare in competition with others of like kind. The model to be described is due to Karmeshu and Pathria (1980b); it includes, as a special case, Osei and Thompson's (1977) model for the supression of one rumour by another and, also, the model of the previous section.

Although groups may make use of the mass media we shall not include a source, or sources, in the model but examine only the effects of interpersonal activity. We have already seen that this is likely to be the most important factor in growth but the inclusion of competing sources is clearly an area for the future research. We assume a homogeneously mixing population whose members may be classified at any time in three ways, as follows:

Party Y of size $n_1(T)$ at T.
Party Z of size $n_2(T)$ at T.
Neutrals of size $m(T) = N - n_1(T) - n_2(T)$ at T.

Members of either party may lapse and become neutrals. While active they may convert neutrals to their party. If a Y meets a Z either may win over the other. A neutral person who becomes a Y or a Z will be called a convert; a person who changes from one party to the other will be called a proselyte. We do not allow transfer from one party to the other in the absence of contact though this could easily be included. With such a system we can conceive of various possibilities. Either or both parties may die out; either may convert virtually the whole population or an equilibrium may be reached in which the two parties each have a share of the population. How these outcomes depend on the initial state and on the vigour with which parties convert and proselytize is something which will emerge from our analysis of the model. We introduce the infinitesimal transition probabilities and associated rates by an obvious extension of the arguments used previously as follows:

$$
\left.
\begin{array}{ll}
Z \text{ converts a neutral:} (n_1, n_2) \to (n_1 + 1, n_2) & : \beta_1 n_1(T) m(T) \delta T \\
Y \text{ converts a neutral:} (n_1, n_2) \to (n_1, n_2 + 1) & : \beta_2 n_2(T) m(T) \delta T \\
Z \text{ becomes a neutral:} (n_1, n_2) \to (n_1 - 1, n_2) & : \mu_1 n_1(T) \delta T \\
Y \text{ becomes a neutral:} (n_1, n_2) \to (n_1, n_2 - 1) & : \mu_2 n_2(T) \delta T \\
Z \text{ proselytizes } Y : (n_1, n_2) \to (n_1 - 1, n_2 + 1) : \gamma_{12} n_1(T) n_2(T) \delta T \\
Y \text{ proselytizes } Z : (n_1, n_2) \to (n_1 + 1, n_2 - 1) : \gamma_{21} n_2(T) n_1(T) \delta T
\end{array}
\right\} . \quad (10.64)
$$

In the same manner as in previous sections we can regard the sequence of changes of state as a Markov chain with state space consisting of all points (n_1, n_2) satisfying $n_1 + n_2 \le N$. The point $(0, 0)$ is clearly the only absorbing state, all the others being transient. Ultimately, therefore, absorption will occur and the process will be at an end. It is still possible, however, to investigate quasi-stationary distributions as for the model of the last section. In this connection we note that prior to absorption in $(0, 0)$ the system must pass through $(0, 1)$ or $(1, 0)$. Before final absorption the process will thus become identical with the model for survival of a single group since once either group is eliminated it can never be regenerated. The final stage of the process is therefore covered by the theory of the last section, and the quasi-stationary distribution given there applies here. We can also find a quasi-stationary distribution by conditioning on $(n_1, n_2) > (0, 0)$ which relates to the situation when neither group is eliminated. The method of 'system size expansion' also works here and the equations of the deterministic approximation emerge by taking the leading terms. We restrict attention to this case.

The deterministic approximation

The differential equations derived from (10.64) in the usual way are

$$\frac{dn_1(T)}{dT} = \beta_1 n_1(T)m(T) - (\gamma_{12} - \gamma_{21})n_1(T)n_2(T) - n_1(T)\mu_1$$

$$\frac{dn_2(T)}{dT} = \beta_2 n_2(T)m(T) - (\gamma_{21} - \gamma_{12})n_2(T)n_1(T) - n_2(T)\mu_2. \tag{10.65}$$

Solving these equations gives the mean path of the process as $N \to \infty$. We first examine the various equilibria which arise from these equations by setting $n_1(T) = n_1, n_2(T) = n_2$. Any such values must then satisfy

$$\beta_1 n_1 m - (\gamma_{12} - \gamma_{21})n_1 n_2 - n_1 \mu_1 = 0$$

$$\beta_2 n_2 m - (\gamma_{21} - \gamma_{12})n_1 n_2 - n_2 \mu_2 = 0. \tag{10.66}$$

These equations have the following solutions:

I $n_1 = n_2 = 0, \qquad m = N;$

II $n_1 = N(1 - 1/d_1), \qquad n_2 = 0, \qquad m = N/d_1;$

III $n_1 = 0, \qquad n_2 = N(1 - 1/d_2), \qquad m = N/d_2;$

IV $n_1 = N\beta_2\{(\gamma_{12} - \gamma_{21}) + \beta_1(1/d_1 - 1/d_2) +$
$\qquad (\gamma_{21} - \gamma_{12})/d_2\}/\{(\gamma_{21} - \gamma_{12})^2 + (\gamma_{21} - \gamma_{12})(\beta_1 - \beta_2)\};$

$\quad n_2 = N\beta_1\{(\gamma_{21} - \gamma_{12}) + \beta_2(1/d_2 - 1/d_1) + (\gamma_{12} - \gamma_{21})/d_1\}/$
$\qquad \{(\gamma_{12} - \gamma_{21})^2 + (\gamma_{12} - \gamma_{21})(\beta_2 - \beta_1)\};$

$\quad m = N\{(\gamma_{12} - \gamma_{21}) + (\beta_1/d_1 - \beta_2/d_2)\}/\{(\gamma_{12} - \gamma_{21}) + (\beta_1 - \beta_2)\},$

where $d_i = N\beta_i/\mu_i$, $(i = 1, 2)$, the mean number of converts made by an active member of group i. We have already deduced the existence of case I which marks the end of the process. Cases II and III arise if one group eliminates the other; case IV represents the continuing existence of both groups. Cases II, III, and IV can obviously occur only if the values of (n_1, n_2) lie in the region $(n_1 \geq 0, n_2 \geq 0, n_1 + n_2 \leq N)$ and this means that $d_1 > 1$ and $d_2 > 1$ are necessary conditions in cases II and III respectively. Which of the four equilibrium points will apply for a given set of parameter values can be established by determining the effect of a small displacement from the point. If the tendency is to return, the equilibrium is stable; if not the point cannot represent an equilibrium of the kind we are looking for. For example if we *increase* n_1 slightly we require $dn_1(T)/dT$ to become negative implying that the effect will be to decrease n_1 towards its original value. For any set of parameter values it turns out that just one of the equilibria is stable. We can identify the required point according to the value of $(\gamma_{12} - \gamma_{21})$. Let us suppose, without loss of generality, that $d_1 > d_2$ then the following diagram identifies the various equilibria:

$$
\text{III} \longrightarrow \quad \longleftarrow \text{IV} \longrightarrow \quad \longleftarrow \text{II} \qquad \qquad scale\ of\ (\gamma_{12} - \gamma_{21}).
$$

$$
\underset{\displaystyle \frac{\mu_2(d_1 - d_2)}{d_1 - 1}}{\uparrow} \qquad \underset{\displaystyle \frac{\mu_1(d_1 - d_2)}{(d_2 - 1)}}{\uparrow}
$$

The equilibrium I is stable if $d_1, d_2 < 1$.

Some interesting qualitative conclusions can be drawn from this analysis. Suppose first that $d_1 = d_2$ so that both groups are equally good at converting neutrals. The two critical values in the diagram coincide and the group which is most effective at proselytizing will eventually eliminate its rival. Next, if the two groups are equally good at proselytizing $(\gamma_{12} = \gamma_{21})$, the group that is most effective at conversion will be the only one to survive. For both groups to persist there has to be a balance between their relative advantages. Thus if the Y-party is better at conversion $(d_1 > d_2)$ the Z-party will have to counterbalance that advantage by a more effective proselytizing. Finally if $\mu_1 = \mu_2$, so that both groups lose members at an equal rate, the region IV vanishes and one group will eventually triumph.

As with all the models in this book, it is necessary to look in detail at many more special cases than it is possible to illustrate here if real insight is to be achieved. This applies especially to the development in time of the process. Here we consider only one with $\beta_1 = \beta_2 = \beta$ and $\mu_1 = \mu_2 = \mu$. By adding the two equations in (10.65) we arrive at (10.49) with $\omega = 0$ and $n(T) = n_1(T) + n_2(T)$. As is otherwise obvious, in this case, the total size of the two groups now grows in exactly the same way as in the model of the last section. We also know that one group will eventually dominate the other. What remains to be discovered is how the relative sizes of the groups change over time. To do this consider

$$
\frac{d}{dT}\left(\frac{n_1(T)}{n_2(T)}\right) = \frac{1}{n_2(T)}\frac{dn_1(T)}{dT} - \frac{1}{n_2^2(T)}\frac{dn_2(T)}{dT}
$$

$$
= -(\gamma_{12} - \gamma_{21})\left(\frac{n_1(T)}{n_2(T)}\right)n(T) \tag{10.67}
$$

using (10.65). Solving this equation,

$$
\frac{n_1(T)}{n_2(T)} = \frac{n_1(0)}{n_2(0)}\exp\left[(\gamma_{21} - \gamma_{12})\int_0^T n(x)dx\right]. \tag{10.68}
$$

Since the integral in square brackets diverges the ratio tends to zero or infinity with T according as $\gamma_{12} < \gamma_{21}$ or $\gamma_{12} > \gamma_{21}$. If $\gamma_{12} = \gamma_{21}$ the two groups behave identically in every respect and the result is stalemate— nothing changes at all. In order to calculate $n_1(T)$, $n_2(T)$, and $m(T)$ we first find $n_1(T) + n_2(T)$ from (10.5) then $n_1(T)/n_2(T)$ from (10.68). If $n_1(0)$ and $n_2(0)$ are both small, $dn_1(T)/dT$ and $dn_2(T)/dT$ will both be positive near $T = 0$ and so both groups will grow as there are enough neutrals for both to recruit.

10.7   COMPLEMENTS TO CHAPTERS 9 AND 10

Most of the models described in these chapters are based on Markov processes and thus they depend on the same body of theory as those in Chapters 4 and 5. The feature which distinguishes them from most earlier models is that they have transition rates which are quadratic functions of the state numbers. This makes the analysis more difficult and gives rise to distinctive properties such as the threshold effects. On the mathematical side the main techniques used in solving epidemic models are those appropriate to ordinary and partial differential equations. These are treated in many standard textbooks and in Volume IV of the *Handbook of Applicable Mathematics*. A good approximation to the solution of the pure birth model given in (9.6) has been provided by Daniels (1981) using the saddle point technique applied to the Laplace transform of the distribution. He indicates that there are prospects of being able to find similar approximations for birth and death processes such as that in Section 10.5. Generalizations of the simple epidemic model have been made by Billard and co-workers (1980) using the representation of (9.9).

A wide-ranging and comprehensive account of the theory of epidemics from the viewpoint of infectious diseases is given in Bailey (1975). This is an updated and enlarged version of the book by the same author first published under the title of *The Mathematical Theory of Epidemics* in 1957. It gives more detail on the KM model including, for example, the duration of the epidemic. There are also many other models such as those for diseases such as malaria, which are transmitted from host to host via a vector, and these may find social applications. In fact Goffman and Newill (1964) have already suggested just such a model for the diffusion of new knowledge. According to their model knowledge is transmitted from one scientist to another by papers published in journals. Bailey's book contains an extensive bibliography covering the field of infectious diseases. Apart from giving a wealth of technical detail, Bailey's exposition lays great stress on the proper use of mathematics in applied research and is well worth reading for this alone.

Early empirical work on diffusion was largely concerned with observing the number of hearers as a function of time. Some of the data resulting from these studies will be found in Pemberton (1936), Ryan and Gross (1943), Dodd (1955), Griliches (1957, 1960), and Hagerstrand (1967). Further material will be found in Chapter 17 of Coleman (1964b). Mansfield (1968) gives many examples of the diffusion of technological innovations; Dixon (1980) returns to Griliches' earlier work on hybrid corn; and Teece (1980) considers the spread of an administrative innovation. The data given in these publications, mainly relating to the adoption of innovations, were obtained by observing processes occurring naturally. On the whole, such data are not sufficiently refined to estimate the models discussed here or to discriminate effectively between them. For such purposes we need more detailed data which can often be obtained only under conditions of controlled experimentation. A study of this kind was carried out by Coleman and co-workers (1957) on the diffusion of knowledge about new drugs among physicians. This work permitted a comparison to be made between those who shared an office and those who did not. The most extensive collections of experimental data on diffusion of news appear to be those of the Washington Public Opinion Laboratory in Seattle obtained by Dodd and his collaborators. Some of the experiments used small populations of school children or college students; others used the populations of small towns. In some cases information was introduced into the community by leaflets dropped from the air. A good deal of effort was devoted to fitting logistic growth curves. The fit could often be improved by allowing β, the contact rate, to depend on T. This assumption means either that the frequency of meeting changes with time or that the frequency changes with the state of knowledge in the population. Although generalizations of this kind may provide better fits they have limited explanatory value. The fit of almost any model can be improved by introducing new parameters, but unless they have substantive meaning little is gained.

The formulation of epidemic models in discrete time has been advocated for social applications by Gray and von Broembsen (1974) and for infections by Gani (1978). The differential equations then become difference equations and these can be solved recursively in a straightforward manner. A disadvantage is that some of the more subtle aspects, like threshold effects, can be lost in the discrete treatment. Bartholomew (1976a) shows this to be the case for the model of Section 10.5 which is a continuous version of one first proposed by Gray and von Broembsen (1974).

The early work on spatial diffusion was carried out mainly by Kolmogoroff and co-workers (1937) for the spread of an advantageous gene and by Bartlett (1956) and Kendall (1957) for epidemic diseases. A useful survey of the field up to the mid-1970s is given by Mollison (1977). He shows that much of what appeared to be established knowledge was less securely founded than had been generally believed. We have already noted that the approximate methods of handling the equations of the deterministic general epidemic fail to provide adequate approximations to velocities of propagation. Another discrepancy concerns the possible extinction of the general epidemic. In the discussion of Mollison's paper Kelly reported that if the initial number of infectives was finite and if the contact distribution had a finite mean then ultimate extinction was certain. In reply, Mollison showed that the converse was true in two or more dimensions under certain conditions. Mollison's paper, the discussion of it, and the references it contains expose the complexity and subtlety of the subject. Much more needs to be done before the novice can find his way with confidence. In the meantime, simulation is, perhaps, the best safeguard against the grosser errors into which uncritical acceptance of approximations may lead.

The special structure of many spatial populations makes it rather difficult to give general mathematical treatments. A recent attempt to make progress is that of Cairns (1979). She supposes the diffusion to originate at the centre of a lattice formed by the intersection of a set of rays with concentric circles in the form of a spider's web. Individuals or communities are located at the intersections and contacts depend on distance apart. Other attempts to cope with the complexities of spatial diffusion have been made by Sernadas (1980) and by Faddy and Slorach (1980). The latter authors consider an ordered sequence of colonies through which infection is spread from one colony to the next by migration. On the applied side there has been a rapid growth of interest among geographers, much of it stemming from the pioneering work of Hagerstrand (1967). Although this work was largely descriptive, he proposed a number of models one of which we have described.

During the period 1948–54 several models for the diffusion of information were prepared by Rapoport and his co-workers (see, for example Rapoport, 1948, 1951, 1953a, 1953b, 1954; Solomanoff and Rapoport, 1951; Rapoport and Rebhun, 1952). Their work was originally concerned with the random net which arises in neuro-physiological problems, but its relevance to the spread of infection and disease was soon apparent. This work predates much of that described by Bailey (1975), but the close relationship seems to have passed unnoticed. Rapoport's main model is, in fact, a simple variant of the KM model and both can be regarded as special cases of the more general model discussed in Section 10.3. In the previous editions of this book there was a much fuller discussion of the random net model. In particular we investigated the effects of imperfect mixing and showed that this was unlikely to affect the proportion ultimately receiving the news. It would however, slow down the rate of spread in the sense that the news would be liable to have passed through more hands before reaching a given individual. We also showed how the exact theory of the random net could be derived by expressing it as a Markov chain, but with such a large state space as to make the method impracticable for all but very small systems. It was, however, possible to show by this means that the approximate method used here was adequate.

The first example of a model with two competing groups is in Osei and Thompson (1977) who consider the situation where one rumour suppresses another in a closed population. Their model is a special case of Karmeshu and Pathria's (1980b) discussed in Section 10.6. It is obtained by putting $\beta_1 = \beta_2 = \gamma_{12}, \gamma_{21} = 0, \mu_1 = \mu_2 = 0$. In this case the Y-group will be eliminated and Osei and Thompson's analysis is mainly concerned with the distribution of the maximum size of the Y-group before its decline begins. All of our models assume that the environment is stable in the sense that it has no effect on the process. This is unlikely to be the case in practice. Randomness in the environment may cause the parameters to·vary or introduce other 'errors' into the equations. Karmeshu and Pathria (1980d) have begun an exploration of the consequences of this kind of generalization.

We have followed the literature of the subject in placing the main emphasis on the diffusion of information and rumours where the analogy with epidemics is close. In recent research on social applications, increasing attention has been given to those aspects—such as the generation distribution—which have no counterpart in the spread of infectious diseases. In our view, the model for the survival of a social group and that for competing groups treated in Sections 10.5 and 10.6 represent a further step towards a general stochastic theory of social dynamics. The very simple models discussed here will have to be extended by, for example, allowing for immigration and emigration, but the effort should be well repaid. It should not be too difficult to obtain data on the rise and fall of social groups because clubs, societies, pressure groups, and suchlike often keep the detailed records of membership that would be required. There is a rich field here awaiting both theoretical and empirical researchers who are interested in the dynamics of modern society.

Bibliography

The bibliography contains all the books, papers, and reports referred to in the text. It also contains references to other work relevant to the theme of the book. Some of this is to general discussions on the use of mathematics and statistics in the social sciences, some relates directly to the topics treated in the book, and the remainder has to do with stochastic modelling in other fields of application. No claim is made to completeness, especially in respect of the last group, but I think the list includes most published work on stochastic models for social processes.

Aaker, D. A. (1971a). 'The new-trier stochastic model of brand choice'. *Man. Sci.*, Application Series **17**, B435–B450.

Aaker, D. A. (1971b). 'A new method for evaluating stochastic models of brand choice'. *J. Marketing Res.*, **7**, 300–306.

Abernathy, W. J., N. Baloff, J. C. Hershey, and S. Wandel (1973). 'A three-stage manpower planning and scheduling model – a service sector example'. *Operat. Res.*, **21**, 693–711.

Abodunde, T. T., and S. I. McClean, (1980). 'Production planning for a manpower system with a constant level of recruitment'. *Appl. Statist*, **29**, 43–49.

Adelman, I. G. (1958). 'A stochastic analysis of the size distribution of firms'. *J. Amer. Statist. Ass.*, **53**, 893–904.

Agnew, R. A. (1971). 'Counter examples to an assertion concerning the normal distribution and a new stochastic price fluctuation model'. *Rev. Econ. Studies*, **38**, 381–383.

Aitchison, J. (1955). Contribution to the discussion on Lane and Andrew (1955).

Aitchison, J., and J. A. C. Brown (1957). *The Lognormal Distribution*, Cambridge University Press.

Alker, H. R., K. W. Deutsch, and A. Stoetzel (1973). *Mathematical Approaches to Politics*. Elsevier, Amsterdam, London, New York.

Allan, G. J. B., and W. Bytheway (1973). 'The effects of differential fertility on sampling studies of inter-generational social mobility'. *Sociology*, **7**, 273–276.

Alling, D. W. (1958). 'The after history of pulmonary tuberculosis: a stochastic model'. *Biometrics*, **14**, 527–547.

Altman, S. A. (1965). 'Sociology of rhesus monkeys, II: Social communication'. *Theoretical Biology*, **8**, 490–522.

Anderson, D., and R. Watson (1980). 'On the spread of a disease with gamma distributed latent and infectious periods'. *Biometrika*, **67**, 191–198.

Anderson, R. B. W. (1974). 'A Markov chain model of medical speciality choice'. *J. Math. Sociology*, **3**, 259–274.

Anderson, T. W. (1954). 'Probability models for analyzing time changes in attitudes'. In P. F. Lazarsfeld (1954), 17–66.

Andley, R. J. (1960). 'A stochastic model for individual choice behaviour'. *Psychological Review*, **67**, 1–15.

Archer, S. H. and J. McGuire (1965). 'Firm size and probabilities of growth'. *Western Econ. J.*, **3**, 233–246.

324

Armitage, P. H., C. M. Phillips, and J. Davies (1970). 'Towards a model of the upper secondary school system (with discussion)'. *J. R. Statist. Soc.*, **A133**, 166–205.

Armitage, P. H., and C. S. Smith (1972). 'Controllability: an example'. *Higher Education Review*, **5**, 55–66.

Armitage, P. H., C. S. Smith, and P. Alper (1969). *Decision Models for Educational Planning*, Allen Lane, The Penguin Press, London.

Arrow, K. J., S. Karlin, and P. Suppes (Eds.), (1961). *Mathematical Methods in the Social Sciences*, Stanford University Press.

Audley, R. J. (1960). 'A stochastic model for individual choice behaviour'. *Psychological Review*, **67**, 1–15.

Bailey, N. T. J. (1957). *The Mathematical Theory of Epidemics*, Griffin, London.

Bailey, N. T. J. (1963). 'The simple stochastic epidemic: a complete solution in terms of known functions'. *Biometrika*, **50**, 235–240.

Bailey, N. T. J. (1964). *The Elements of Stochastic Processes with Applications to the Natural Sciences*, John Wiley, New York.

Bailey, N. T. J. (1967). 'The simulation of stochastic epidemics in two dimensions'. *Proc. Fifth Berkeley Symp. Math. Statist. Prob.*, **4**, 237–257.

Bailey, N. T. J. (1968a). 'Stochastic birth, death and migration processes for spatially distributed populations'. *Biometrika*, **55**, 189–198.

Bailey, N. T. J. (1968b). 'A perturbation approximation to the simple stochastic epidemic in a large population'. *Biometrika*, **55**, 199–209.

Bailey, N. T. J. (1975). *The Mathematical Theory of Infectious Diseases and its Applications* (2nd Edn.) Griffin, London and High Wycombe.

Balachandran, V., and S. D. Deshmukh (1976). 'A stochastic model of persuasive communication'. *Man. Sci.* **22**, 829–840.

Balinsky, W., and A. Reisman (1972). 'Some manpower planning models based on levels of educational attainment'. *Man. Sci.*, **18**, B691–B705.

Balinsky, W., and A. Reisman (1973). 'A taxonomy of manpower–educational planning models'. *Socio-econ. Plan. Sci.*, **7**, 13–18.

Balleer, M. (1968). 'The exclusion of sets of persons in sickness insurance considered as a Markov process'. *Bla. Dtsch. Ges. Versichmath*, **8**, 611–632.

Barbour, A. D. (1975). 'The duration of the closed stochastic epidemic'. *Biometrika*, **62**, 477–482.

Barlow, R. E., D. J. Bartholomew, J. M. Bremner, and H. D. Brunk (1972). *Statistical Inference under Order Restrictions*, John Wiley, Chichester.

Bartholomew, D. J. (1959). 'Note on the measurement and prediction of labour turnover'. *J. R. Statist. Soc.*, **A122**, 232–239.

Bartholomew, D. J. (1963a). 'A multistage renewal process'. *J. R. Statist. Soc.*, **B25**, 150–168.

Bartholomew, D. J. (1963b). 'An approximate solution of the integral equation of renewal theory'. *J. R. Statist. Soc.*, **B25**, 432–441.

Bartholomew, D. J. (1963c). 'Two-stage replacement strategies'. *Operat. Res. Quart.*, **14**, 71–87.

Bartholomew, D. J. (1969). 'Renewal theory models for manpower systems'. In N. A. B. Wilson (1969), 120–128.

Bartholomew, D. J. (1971). 'The statistical approach to manpower planning'. *Statistician*, **20**, 3–26.

Bartholomew, D. J. (1972). 'The effect of changes in quits and hires on the length of service composition of employed workers: a comment on Stoikov's paper'. *Brit. J. Indust. Rel.*, **10**, 130–133.

Bartholomew, D. J. (1975). 'A stochastic control problem in the social sciences'. *Bull. Int. Statist. Inst.* **46**, 670–680.

Bartholomew, D. J. (1976a). 'Continuous time diffusion models with random duration of interest'. *J. Math. Sociology*, **4**, 187–199.

Bartholomew, D. J. (1976b). *Manpower Planning, Penguin Modern Management Readings*, Penguin Books, Harmondsworth, Middlesex, England.

Bartholomew, D. J. (1976c). 'Statistical problems of prediction and control in manpower planning'. *Math. Scientist*, **1**, 133–144.

Bartholomew, D. J. (1977a). 'The analysis of data arising from stochastic processes'. Chapter 5 in C. A. O'Muircheartaigh and C. Payne (Eds.). *The Analysis of Survey Data*, Vol. II *Model Fitting*, John Wiley, Chichester, 145–174.

Bartholomew, D. J. (1977b). 'Maintaining a grade or age structure in a stochastic environment'. *Adv. Appl. Prob.*, **9**, 1–17.

Bartholomew, D. J. (1979). 'The control of a grade structure in a stochastic environment using promotion control'. *Adv. Appl. Prob.*, **11**, 603–615.

Bartholomew, D. J. (1981) *Guidebook for Social Scientists*, John Wiley, Chichester.

Bartholomew, D. J., and A. D. Butler (1971). 'The distribution of the number of leavers for an organization of fixed size'. In A. R. Smith (1971), 417–426.

Bartholomew, D. J., and A. F. Forbes (1979). *Statistical Techniques for Manpower Planning*, John Wiley, Chichester.

Bartholomew, D. J., and A. R. Smith (Eds.) (1971). *Manpower and Management Science*, English Universities Press, London, and D. C. Heath and Co., Lexington, Mass.

Bartlett, M. S. (1955). *An Introduction to Stochastic Processes*. Cambridge University Press, London.

Bartlett, M. S. (1956). 'Deterministic and stochastic models for recurrent epidemics'. *Proc. Third Berkeley Symp. Math. Statist. Prob.*, **4**, 81–109.

Barton, D. E., and F. N. David (1962). *Combinatorial Chance*, Griffin, London.

Barton, D. E., F. N. David and M. Merrington (1960). 'Tables for the solution of the exponential equation, $\exp(-a) + ka = 1$'. *Biometrika*, **47**, 439–445.

Barton, D. E., and C. L. Mallows (1961). 'The randomization bases of the problem of the amalgamation of weighted means'. *J. R. Statist. Soc.*, **B23**, 423–433.

Barton, D. E., and C. L. Mallows (1965). 'Some aspects of the random sequence'. *Ann. Math. Statist.*, **36**, 236–260.

Bartos, O. J. (1967). *Simple Models of Group Behaviour*, Columbia University Press, New York and London.

Becker, N. G. (1968). 'The spread of an epidemic to fixed groups within the population'. *Biometrics*, **24**, 1007–1014.

Bedall, F. K. (1974). 'The analysis of aggregated time series data under the assumption of a simple Markov chain'. *Biom. Zeit.* **16**, 451–458.

Bell, E. J. (1974). 'Markov analysis of land use change – an application of stochastic processes to remotely sensed data'. *Socio-Econ. Plan. Sci.*, **8**, 311–316.

Benjamin, B. (1972). 'Stochastic processes as applied to life tables'. *Bull. Inst. Math. Appl.*, **8**, 12–16.

Benjamin, B., and J. Maitland (1958). 'Operational research and advertising: some experiments on the use of analogies'. *Operat. Res. Quart.*, **6**, 207–217.

Berman, A., and R. J. Plemmons (1979). *Nonnegative Matrices in the Mathematical Sciences*, Academic Press, New York.

Bernhardt, I. (1970). 'Diffusion of catalytic techniques through a population of medium size petroleum refining firms'. *J. Indust. Econ.*, **19**, 50–64.

Bernhardt, I., and K. D. Mackenzie (1970). 'Some problems in using diffusion models for new products'. In G. Fisk (Ed.), *Essays in Marketing Theory*, Allyn and Bacon, Boston, Mass.

Bernhardt, I., and K. D. Mackenzie (1972). 'Some problems in using diffusion models for new products'. *Man. Sci.*, **19**, 187–200.

Berry, B. J. L. (1971). 'Monitoring trends forecasting change and evaluating goal achievement in the urban environment'. In M. Chisholm, A. E. Frey, and P. Haggett (1971), 93–117.

Beshers, J. M., and E. O. Laumann (1967). 'Social distance: a network approach'. *Amer. Soc. Rev.*, **32**, 225–236.

Bhargava, T. N., and L. Katz (1964). 'A stochastic model for a binary dyadic relation with applications to social and biological science'. *Bull. Int. Statist. Inst.*, **40**, II, 1055–1057.

Bharucha-Reid, A. T. (1960). *Elements of the Theory of Markov Processes and Their Applications*, McGraw-Hill, New York.

Bibby, J. (1970). 'A model to control for the biasing effects of differential wastage'. *Brit. J. Indust. Rel.*, **8**, 418–420.

Bibby, J. (1975). 'Methods of measuring mobility'. *Quality and Quantity*, **9**, 107–136.

Billard, L. (1973). 'The distribution of a stochastic epidemic model'. *Bull. Int. Statist. Inst.*, **45**, Book 1, 194–198.

Billard, L. (1975). 'General stochastic epidemic with recovery'. *J. Appl. Prob.*, **12**, 29–38.

Billard, L. (1976). 'A stochastic general epidemic in m sub-populations'. *J. Appl. Prob.*, **13**, 567–572.

Billard, L. (1977). 'Mean duration time for a general epidemic process'. *J. Appl. Prob.*, **14**, 232–240.

Billard, L., H. Lacayo, and N. A. Landberg (1980). 'Generalizations of the simple epidemic process'. *J. Appl. Prob.*, **17**, 1072–1078.

Birnbaum, Z. W., and S. C. Saunders (1969). 'A new family of life distributions'. *J. Appl. Prob.*, **6**, 319–327.

Blom, A. J. and A. J. Knights (1976). 'Long term manpower forecasting in the Zambian mining industry'. *Personnel Review*, **5**, 18–28.

Blumen, I., M. Kogan, and P. J. McCarthy (1955). *The Industrial Mobility of Labour as a Probability Process*, Cornell University Press, Ithaca, New York.

Blumstein, A., and R. Larson (1969). 'Model of a total criminal justice system'. *Operat. Res.*, **17**, 199–232.

Boag, J. (1949). 'Maximum likelihood estimates of the proportion of patients cured by cancer therapy'. *J. R. Statist. Soc.*, **B11**, 15–53.

Boudon, R. (1973). *Mathematical Structures of Social Mobility*, Elsevier, Amsterdam, London, and New York.

Bowers, R. V. (1937). 'The direction of the intra-societal diffusion'. *Amer. Sociol. Rev.*, **2**, 826–836.

Bowey, A. M. (1969). 'Labour stability curves and labour stability'. *Brit. J. Indust. Rel.*, **7**, 69–84.

Box, G. E. P., and G. M. Jenkins (1971). *Times Series Analysis Forecasting and Control*, Holden-Day, San Francisco, Cambridge, London, and Amsterdam.

Breslow, N. E. (1975). 'Analysis of survival data under the proportional hazards model'. *Internat. Stat. Rev.*, **43**, 45–57.

Breslow, N. E., and J. Crowley (1974). 'A large sample study of the life table and product limit estimates under random censorship'. *Ann. Statist.*, **2**, 437–455.

Brissenden, P. F., and E. Frankel (1922). *Labour Turnover in Industry: a Statistical Analysis*, Macmillan, New York.

British Association Mathematical Tables: Vol. I, Circular and Hyperbolic Functions (1951). Cambridge University Press.

Britney, R. R. (1975). 'Forecasting educational enrolments: comparison of a Markov chain and circuitless flow network model'. *Socio-Econ. Plan. Sci.*, **9**, 53–60.

Brock, D. B., and A. H. Kshirsagar (1973). 'A χ^2 goodness-of-fit test for Markov renewal processes'. *Ann. Inst. Statist. Math.*, **25**, 643–654.

Brown, L. A. (1970). 'On the use of Markov chains in movement research'. *Econ. Geog.*, **46**, 393–403.

Brummelle, S. L., and Y. Gerchak, (1980). 'A stochastic model allowing interaction among individuals and its behavior for large populations'. *J. Math. Sociology*, **7**, 73–90.

Bryant, D. T. (1965). 'A survey of the development of manpower planning policies'. *Brit. J. Indust. Rel.*, **3**, 279–290.

Bryant, D. T. (1972). 'Recent developments in manpower research'. *Personnel Review*, **1**, 14–31.

Burack, E. H. and J. W. Walker (1972). *Manpower Planning and Programming*, Allyn and Bacon, Boston.

Burbeck, S. L., W. J. R. Lane, and M. J. Abudu Stark (1978). 'The dynamics of riot growth: an epidemiological approach'. *J. Math. Sociology*, **6**, 1–22.

Bush, R. R., and C. F. Mosteller (1955). *Stochastic Models for Learning*, John Wiley, New York.

Butler, A. D. (1970). 'Renewal theory applied to manpower systems'. Ph.D. Thesis, University of Kent.

Butler, A. D. (1971). 'An analysis of flows in a manpower system'. *The Statistician*, **20**, 69–84.

Cairns, V. E. (1979). 'Stochastic models for spatial diffusion in homogenous populations and for diffusion in certain homogeneous populations'. Ph.D. Thesis, University of London.

Cane, V. R. (1966). 'A note on the size of epidemics and the number of people hearing a rumour'. *J. R. Statist. Soc.*, **B28**, 487–490.

Canon, M. D., C. D. Cullum, and E. Polack (1970). *Theory of Optimal Control and Mathematical Programming*, McGraw-Hill, New York.

Carlsson, G. (1958). *Social Mobility and Class Structure*, CWK Gleerup, Lund, Sweden.

Carman, J. M. (1966). 'Brand switching and linear learning models'. *J. Advertising. Res.*, **6**, 23–31.

Casstevens, T. W., and W. Morris (1972). 'The cube law and the decomposed system'. *Canadian J. Pol. Sci.*, **5**, 521–532.

Champernowne, D. G. (1953). 'A model of income distribution'. *Econ. J.*, **63**, 318–351.

Champernowne, D. G. (1973). *The Distribution of Income Between Persons*, Cambridge University Press.

Charnes, A., W. W. Cooper, and R. J. Niehaus (1971). 'A generalised network model for training and recruiting decisions in manpower planning'. In D. J. Bartholomew and A. R. Smith (1971), 115–130.

Charnes, A., W. W. Cooper, R. J. Niehaus, and D. Sholtz (1970). 'A model for civilian manpower management and planning in the U.S. Navy'. In A. R. Smith (1971), 247–263.

Chatfield, C., and G. J. Goodhart (1970). 'The beta-binomial model for consumer purchasing behaviour'. *Appl. Statist.*, **19**, 240–250.

Chen, W-C. (1980). 'On the weak form of Zipf's law'. *J. Appl. Prob.*, **17**, 611–622.

Chhikara, R. S., and J. L. Folks (1974). 'Estimation of the inverse Gaussian distribution function'. *J. Amer. Statist. Ass.*, **69**, 250–255.

Chhikara, R. S., and J. L. Folks (1978). 'The inverse Gaussian distribution and its statistical application—a review'. *J. R. Statist. Soc.*, **40**, 263–289.

Chiang, C. L. (1968). *Introduction to Stochastic Processes in Biostatistics*, John Wiley, New York.

Chiang, C. L., and J. P. Hsu (1976). 'On multiple transition time in a single illness death process—a Fix–Neyman model'. *J. Math. Biosciences*, **31**, 55–72.

Chisholm, M., A. E. Frey, and P. Haggett (Eds.) (1971). *Regional Forecasting*, Butterworths, London.

Chung, K. L. (1967). *Markov Chains with Stationary Transition Probabilities*, (2nd edn.) Springer-Verlag, Berlin.

Churchill, N. C., and J. K. Shank (1975). 'Accounting for affirmative action programs: a stochastic flow approach'. *The Accounting Review*, **50**, 643–656.

Çinlar, E. (1975). 'Markov renewal theory: a survey'. *Man. Sci.*, **21**, 727–752.

Clark, W. A. V. (1965). 'Markov chain analysis in geography: an application to the movement of rental housing areas'. *Ann. Assoc. of Amer. Geog.*, **55**, 351–359.

Cliff, A. D., P. Haggett, J. K. Ord, K. Bassett, and R. Davies (1975). *Elements of Spatial Structure: A Qualitative Approach*, Cambridge University Press.

Clough, D. J., C. G. Lewis, and A. L. Oliver (1974). *Manpower Planning Models*, English Universities Press, London; Crane, Russach, and Co. Inc., New York.

Clough, D. J., and W. P. McReynolds (1966). 'State transition model of an educational system incorporating a constraint theory of supply and demand'. *Ontario J. Educational Research*, **9**, 1–18.

Clowes, G. A. (1972). 'A dynamic model for the analysis of labour turnover'. *J. R. Statist. Soc.*, **A135**, 242–256.

Coale, A. J. (1972). *The Growth and Structure of Human Populations: A Mathematical Investigation.* Princeton University Press, Princeton, New Jersey.

Cohen, B. (1963). *Conflict and Conformity: A Probability Model and its Application*, The M.I.T. Press, Cambridge, Mass.

Cohen, J. E., and B. Singer (1979). 'Malaria in Nigeria: constrained continuous-time Markov models for discrete time longitudinal data on human mixed-species infections'. *Lectures on Mathematics in the Life Sciences*, **12**, 69–133 (American Mathematical Society).

Coleman, J. S. (1961). *The Adolescent Society.* The Free Press of Glencoe, New York.

Coleman, J. S. (1964a). *Models of Change and Response Uncertainty*, Prentice-Hall, Englewood Cliffs.

Coleman, J. S. (1964b). *Introduction to Mathematical Sociology*, The Free Press of Glencoe and Collier-Macmillan, London.

Coleman, J. S., E. Katz, and H. Menzel (1957). 'The diffusion of an innovation among physicians'. *Sociometry*, **20**, 253–270.

Coleman, J. S., E. Katz, and H. Menzel (1966). *Medical Innovation: A Diffusion Study*, Bobbs-Merrill, New York.

Collins, L. (1972). *Industrial Migration in Ontario: Forecasting Aspects of Industrial Activity through Markov Chain Analysis*, Statistics Canada, Ottowa.

Collins, L. (1973). 'Industrial size distribution and stochastic processes'. *Progress in Geography*, **5**, 121–165.

Collins, L. (1974). 'Estimating Markov transition probabilities from micro-unit data'. *Appl. Statist.*, **23**, 355–370.

Collins, L. (1976). *The Use of Models in the Social Sciences*, Tavistock Publications, London.

Collins, N. R., and L. E. Preston (1961). 'The size structure of the largest industrial firms, 1907–1958'. *Amer. Econ. Rev.*, **51**, 986–1011.

Conlisk, J. (1976). 'Interactive Markov chains'. *J. Math. Sociology*, **4**, 157–185.

Conlisk, J. (1978). 'A stability theorem for an interactive Markov chain'. *J. Math. Sociology*, **6**, 163–168.

Cootner, P. H. (1964). *The Random Character of Stock Market Prices*, M.I.T. Press, Cambridge, Mass.

Cox, D. R. (1962). *Renewal Theory*, Methuen, London.

Cox, D. R., and H. D. Miller (1965). *The Theory of Stochastic Processes*, Methuen, London.

Cozzolino, J. H. (1973). 'The maximum-entropy distribution of the future market price of a stock'. *Operat. Res.*, **21**, 1200–1211.

Crain, R. L. (1966). 'Fluoridation: the diffusion of an innovation among cities'. *Social Forces*, **44**, 467–476.

Creedy, J. (1979). 'The analysis of labour market flows using a continuous time model'. Private communication.

Cripps, T. F., and R. J. Tarling (1974). 'An analysis of the duration of male unemployment in Great Britain 1932–73'. *Econ. J.*, **84**, 289–316.

Criswell, J. H., H. Soloman, and P. Suppes (Eds.) (1972). *Mathematical Methods in Small Group Processes*, Stanford University Press.

Crussard, C. (1973). 'A new thermodynamics of growth'. *Int. J. Research Management*, **16**, 13–16.

Cyert, R. M., H. J. Davidson, and G. L. Thompson (1962). 'Estimation of the allowance for doubtful accounts by Markov chains'. *Man. Sci.*, **8**, 287–303.

Dacey, M. F. (1971). 'Regularity in spatial distributions: a stochastic model of the imperfect central place plane'. *Statistical Ecology*, **1**, 287–309.

Daellenbach, H. G. (1976). 'Note on a stochastic manpower smoothing and production model'. *Operat. Res. Quart.*, **27**, 573–579.

Daley, D. J. (1967a). 'Concerning the spread of news in a population of individuals who never forget'. *Bull. Math. Biophysics*, **29**, 373–376.

Daley, D. J. (1967b). 'Some aspects of Markov chains in queueing theory'. Ph.D. Thesis, University of Cambridge.

Daley, D. J., and D. G. Kendall (1965). 'Stochastic rumours'. *J. Inst. Math. Appl.*, **1**, 42–55.

Daniels, H. E. (1966). 'The distribution of the total size of an epidemic'. *Proc. Fifth Berkeley Symp. Math. Statist. Prob.*, **4**, 281–293.

Daniels, H. E. (1981). 'The saddle point approximation for a general birth process'. *Biometrika*, **68**, to appear.

David, F. N., M. G. Kendall, and D. E. Barton (1966). *Symmetric Functions and Allied Tables*, Cambridge University Press.

David, H. A. (1970). 'On Chiang's proportionality assumption in the theory of competing risks'. *Biometrics*, **26**, 336–339.

Davies, G. S. (1973). 'Structural control in a graded manpower system'. *Man. Sci.*, **20**, 76–84.

Davies, G. S. (1975). 'Maintainability of structures in Markov chains in models under recruitment control'. *J. Appl. Prob.*, **12**, 376–382.

Davies, R., D. Johnson, and S. Farrow (1975). 'Planning patient care with a Markov model'. *Operat. Res. Quart.*, **26**, 599–607.

Dawson, D. A., and F. T. Denton (1974). 'Some models for simulating Canadian manpower flows and related systems'. *Socio-Econ. Plan. Sci.*, **8**, 233–248.

Day, R. H. (1970). 'A theoretical note on the spatial diffusion of something new'. *Geographical Analysis*, **2**, 68–76.

De Cani, J. S. (1961). 'On the construction of stochastic models of population growth and migration'. *J. Reg. Sci.*, **3**, 1–13.

Deegan, J. (1979). 'Constructing statistical models of social processes'. *Quality and Quantity*, **13**, 97–119.

Dent, W. T. (1967). 'Application of Markov analysis to international wool flows'. *Rev. of Economics and Statistics*, **49**, 613–616.

Dent, W. T. (1972). 'A note on transition probability estimation'. *Aust. J. Statist.*, **14**, 217–221.

Dent, W. T. (1973). 'State substitution and the estimation of transition probabilities'. *Man. Sci.*, **19**, 1082–1086.

Diekman, A. (1979). 'A dynamic stochastic version of the Pitcher–Hamblin–Miller model of "collective violence"'. *J. Math. Sociology*, **6**, 277–282.

Dietz, K. (1966). 'On the model of Weiss for the spread of epidemics by carriers'. *J. Appl. Prob.*, **3**, 375–382.

Dietz, K. (1967). 'Epidemics and rumours: a survey'. *J. R. Statist. Soc.*, **A130**, 505–528.

Dill, W. R., D. P. Gaver, and W. L. Weber (1966). 'Models and modelling for manpower planning'. *Man. Sci.*, **13**, B142–B166.

Dixon, R. (1980). 'Hybrid corn re-visited'. *Econometrica*, **48**, 1451–1461.

Dodd, S. C. (1955). 'Diffusion is predictable: testing probability models for laws of interaction'. *Amer. Sociol. Rev.*, **20**, 392–401.

Dodd, S. C., and S. C. Christopher (1968). 'The reactant model'. In *Essays in Honour of George. A. Lunenberg*, Behavioral Research Council, Great Barrington, Mass.

Dodd, S. C., and M. McCurtain (1965). 'The logistic diffusion of information through randomly overlapped cliques'. *Operat. Res. Quart.*, **16**, 51–63.

Doreian, P., and N. P. Hummon (1976). *Modeling Social Processes*, Elsevier, Amsterdam and New York.

Downton, F. (1967a). 'A note on the ultimate size of the general stochastic epidemic'. *Biometrika*, **54**, 314–316.

Downton, F. (1967b). 'Epidemics with carriers: a note on a paper of Dietz'. *J. Appl. Prob.*, **4**, 264–270.

Drewett, J. R. (1969). 'A stochastic model of the land conversion process'. *Reg. Studies*, **3**, 269–280.

Drinkwater, R. W., and O. P. Kane (1971). ' "Rolling up" a number of Civil Service classes'. In D. J. Bartholomew and A. R. Smith (1971), 293–302.

Dryden, M. (1968). 'Short-term forecasting of share prices: an information theory approach'. *Scottish J. Political Economy*, **15**, 227–249.

Dryden, M. (1969). 'Share price movements: a Markovian approach'. *J. Finance*, **24**, 49–60.

Duncan, G. T., and L. G. Lin (1972). 'Inference for Markov chains having stochastic entry and exit'. *J. Amer. Statist. Ass.*, **67**, 761–767.

Duncan, O. D. (1966). 'Methodological issues in the analysis of social mobility'. In N. J. Smelsner and S. M. Lipset (Eds.), *Social Structure and Mobility in Economic Development*, Aldine, Chicago, 51–97.

Dyke, B., and J. W. Macluer (Eds.) (1973). *Computer Simulation in Human Population Studies*, Academic Press, New York.

Eaton, W. W., and G. A. Whitmore (1977). 'Length of stay as a stochastic process: a general approach and application to hospitalization for schizophrenia'. *J. Math. Sociology*, **5**, 273–292.

Ehrenberg, A. S. C. (1965). 'An appraisal of Markov brand-switching models'. *J. Marketing Res.*, **2**, 347–363.

Ehrenberg, A. S. C. (1970). 'Models of fact: examples from marketing'. *Man. Sci.*, **16**, 435–445.

Eidem, R. J. (1968). 'Innovation diffusion through the urban structures of North Dakota'. Master's Thesis, University of North Dakota.

El Agizy, M. (1971). 'A stochastic programming model for manpower planning'. In D. J. Bartholomew and A. R. Smith (1971), 131–146.

Engwall, L. (1968). 'Size distributions of firms, a stochastic model'. *Swedish J. Econ.*, **70**, 138–157.

Engwall, L. (1973). *Models of Industrial Structure*, D.C. Heath and Co., Lyingbon, Mass.

Ericson, P. M., and L. E. Rogers (1973). 'New procedures for analyzing relational communication'. *Family Process*, **12**, 245–267.

Eymard, J. (1977). 'A Markovian cross-impact model'. *Futures*, **9**, 216–228.

Ezzati, A. (1974). 'Forecasting market shares of alternative home-heating units by Markov process using transition probabilities estimated by aggregate time series data'.

Man. Sci., **21**, 462–473.

Faddy, M. J., and I. H. Slorach (1980). 'Bounds on the velocity of spread of infection for a spatially connected epidemic process'. *J. Appl. Prob.*, **17**, 839–845.

Fararo, T. J. (1973). *Mathematical Sociology – An Introduction to Fundamentals.* Wiley-Interscience, New York.

Farris, P. L., and D. I. Padberg (1964). 'Measures of market structure change in the Florida fresh citrus fruit packing industry'. *Agric. Econ. Res.*, **16**, 93–102.

Farrow, S. C., D. J. H. Fisher, and D. B. Johnson (1971). 'Statistical approach to planning an integrated haemodialysis/transplantation programme'. *British Medical Journal*, **2**, 671–676.

Feeney, G. (1973). 'Two models for multi-regional population dynamics'. *Environment and Planning*, **5**, 31–43.

Feichtinger, G. (1971). *Stochastiche Modelle Demographischer Prozesse*, Springer-Verlag, Berlin.

Feichtinger, G. (1972). 'Stochastic decrement models of demography'. *Biom. Zeit.*, **14**, 106–125.

Feichtinger, G. (1973). 'Markovian models for some demographic processes'. *Statistiche Hefte*, **14**, 310–334.

Feichtinger, G. (1976). 'On the generalization of stable age distributions to Gani-type person-flow models'. *Adv. Appl. Prob.*, **8**, 433–445.

Feichtinger, G. and M. Deistler (1973) 'Notes on linear models for population dynamics' (in German). *Operat. Res. Verfahren*, **15**, 40–62.

Feichtinger, G., and A. Mehlmann (1976). 'The recruitment trajectory corresponding to particular stock sequences in Markovian person-flow models'. *Mathematics of Operat. Res.*, **1**, 175–184.

Fein, E. (1970). 'Demography and thermodynamics'. *Amer. J. Phys.*, **38**, 1373–1379.

Feller, W. (1966). *An Introduction to Probability Theory and its Applications*, Vol. II, John Wiley, New York.

Feller, W. (1968). *An Introduction to Probability Theory and its Applications*, Vol. I (3rd edn.), John Wiley, New York.

Fielitz, B. D., and T. H. Bhargava (1973). 'The behaviour of stock-price relatives – a Markovian analysis'. *Operat. Res.*, **21**, 1183–1199.

Fienberg, S. E. (1971). 'Randomization and social affairs: the 1970 draft lottery'. *Science*, **171**, 255–261.

Firey, W. (1950). 'Mathematics and social theory'. *Social Forces*, **29**, 20–25.

Fisher, J. W., M. R. Bristow, and L. Henderson (1964). 'An actuarial procedure for assessing the experience of mental hospital patients'. *Bull. Int. Statist. Inst.*, **40**, II, 1102–1120.

Fix, E., and J. Neyman (1951). 'A simple stochastic model of recovery, relapse, death and loss of patients'. *Human Biology*, **23**, 205–241.

Fleur, Melvin L. de (1956). 'A mass communication model of stimulus response relationships: an experiment in leaflet message diffusion'. *Sociometry*, **19**, 12–36.

Foley, J. D. (1967). 'A Markovian model of the University of Michigan executive system'. *Communications of the A.C.M.*, **10**, 584–588.

Forbes, A. F. (1971a). 'Markov chain models for manpower systems'. In D. J. Bartholomew and A. R. Smith (1971), 93–113.

Forbes, A. F. (1971b). 'Non-parametric methods of estimating the survivor function'. *Statistician*, **20**, 27–52.

Forbes, A. F. (1971c). 'Promotions and recruitment policies for the control of quasi-stationary hierarchical systems'. In A. R. Smith (1971), 401–414.

Frank, R. E. (1962). 'Brand choice as a probability process'. *J. Business*, **35**, 43–56.

Frazer, R. A., W. J. Duncan, and A. R. Collar (1946). *Elementary Matrices*, Cambridge

University Press.

Frydman, H., and B. Singer (1979). 'Total positivity and the embedding problem'. *Math. Proc. Camb. Phil. Soc.*, **86**, 339–344.

Fuguitt, G. V. (1965). 'The growth and decline of small towns as a probability process'. *Amer. Sociol. Rev.*, **30**, 403–411.

Funkhauser, G. R., and M. E. McCoombs (1972). 'Predicting the diffusion of information to mass audiences'. *J. Math. Sociology*, **2**, 121–130.

Gani, J. (1963). 'Formulae for projecting enrolments and degrees awarded in universities'. *J. R. Statist. Soc.*, **A126**, 400–409.

Gani, J. (1965). 'On a partial differential equation of epidemic theory, I'. *Biometrika*, **52**, 617–622.

Gani, J. (1973). 'On the age structure of some stochastic processes'. *Bull. Int. Statist. Inst.*, **45**, Book 1, 434–436.

Gani, J. (1977). 'A spatial distribution for the spread of infection'. *Bull. Int. Statist. Inst.*, **48**, 544–550.

Gani, J. (1978). 'Some problems in epidemic theory' (with discussion). *J. R. Statist. Soc.*, **A141**, 323–347.

Gantmacher, F. R. (1964). *The Theory of Matrices* (2 vols.). Chelsea, New York.

Gart, J. J. (1968). 'The mathematical analysis of an epidemic with two kinds of susceptibles'. *Biometrics*, **24**, 557–566.

Gear, A. E., J. S. Gillespie, A. G. Lockett, and A. W. Pearson (1971). 'Manpower modelling: a study in research and development'. In D. J. Bartholomew and A. R. Smith (1971), 147–162.

Geary, R. C. (1980). 'Prais on strikes'. *J. R. Statist. Soc.*, **A143**, 76–77.

Gelfand, A. E., and H. Solomon (1973). 'A study of Poisson's models for jury verdicts in criminal and civil trials'. *J. Amer. Statist. Ass.*, **68**, 271–278.

Gilbert, G. (1973). 'Semi-Markov processes and mobility: a note'. *J. Math. Sociology*, **3**, 139–145.

Ginsberg, R. B. (1971). 'Semi-Markov processes and mobility'. *J. Math. Sociology*, **1**, 233–262.

Ginsberg, R. B. (1972a). 'Critique of probabilistic models: application of the semi-Markov model to migration'. *J. Math. Sociology*, **2**, 63–82.

Ginsberg, R. B. (1972b). 'Incorporating causal structure and exogenous information with probabilistic models: with special reference to gravity, migration and Markov chains'. *J. Math. Sociology*, **2**, 83–103.

Ginsberg, R. B. (1973). 'Stochastic models of residential and geographical mobility for heterogeneous populations'. *Environment and Planning*, **5**, 113–124.

Ginsberg, R. B. (1978a). 'Probability models of residence histories: analysis of times between moves', in W. A. V. Clark and E. G. Moore (Eds.), *Population Mobility and Residential Change*, Northwestern studies in Geography, No. 24, Evanston, Illinois.

Ginsberg, R. B. (1978b). 'The relationship between timing of moves and choice of destination in stochastic models of migration'. *Environment and Planning*, **A10**, 667–679.

Ginsberg, R. B. (1978c). 'Timing and duration effects in residential histories and other longitudinal data, II: studies of duration effects in Norway 1965–1971'. *Fels Discussion Paper No. 121*, School of Public and Urban Policy, University of Pennsylvania.

Ginsberg, R. B. (1979a). *Stochastic Models of Migration: Sweden 1961–1975*, North-Holland, New York and Amsterdam.

Ginsberg, R. B. (1979b). 'Tests of stochastic models of timing in mobility histories: comparison of information derived from different observation plans'. *Environment and Planning*, **A11**, 1387–1404.

Ginsberg, R. B. (1979c). 'Timing and duration effects in residential histories and other longitudinal data, I: stochastic and statistical models'. *Regional Science and Urban Economics*, **9**, 311–331.

Girmes, D. H., and A. E. Benjamin (1975). 'Random walk hypothesis for 543 stocks and shares registered on the London stock exchange'. *J. Business, Finance and Accounting*, **2**, 135–145.

Glass, D. V. (Ed.) (1954). *Social Mobility in Britain* Routledge and Kegan Paul, London.

Glen, J. J. (1977). 'Length of service distributions in Markov manpower models'. *Operat. Res. Quart.*, **28**, 975–982.

Goffman, W. (1965). 'An epidemic process in an open population'. *Nature*, **205**, 831–832.

Goffman, W., and V. A. Newill (1964). 'Generalization of epidemic theory—an application to the transmission of ideas'. *Nature*, **204**, 225–228.

Goffman, W., and V. A. Newill (1967). 'Communication and epidemic processes'. *Proc. Roy. Soc.*, **A298**, 316–334.

Goodman, L. A. (1961). 'Statistical methods for the "mover–stayer" model'. *J. Amer. Statist. Ass.*, **56**, 841–868.

Goodman, L. A. (1963). 'Statistical methods for the preliminary analysis of transaction flows'. *Econometrica*, **31**, 197–208.

Goodman, L. A. (1965). 'On the statistical analysis of mobility tables'. *Amer. J. Sociol.*, **70**, 564–585.

Goodman, L. A. (1968). 'Stochastic models for the population growth of the sexes'. *Biometrika*, **55**, 469–487.

Gray, L. N., and M. H. von Broembsen (1974). 'On simple stochastic diffusion models'. *J. Math. Sociology*, **3**, 231–241.

Green, J. R., and D. N. Martin (1973). 'Absconding from approved schools as learned behaviour: a statistical study'. *J. Res. Crime and Delinquency*, **10**, 73–86.

Greville, T. N. E. (Ed.) (1973). *Population Dynamics*, Academic Press, New York and London.

Gribbens, W. D., S. Halperin, and P. R. Loynes (1966). 'Application of stochastic models in research and career development'. *J. Counseling Psychology*, **13**, 403–408.

Griffiths, D. A. (1974). 'A catalytic model of infection for measles'. *Appl. Statist.*, **23**, 330–339.

Grigelionis, B. I. (1964). 'Limit theorems for series of renewal processes'. *Cybernetics in the Service of Communism*, **2**, 246–266.

Griliches, Z. (1957). 'Hybrid corn: an exploration in the economics of technical change'. *Econometrics*, **25**, 501–522.

Griliches, Z. (1960). 'Hybrid corn and the economics of innovation'. *Science*, **132**, 257–280.

Griliches, Z. (1980). 'Hybrid corn re-visited: a reply'. *Econometrica*, **48**, 1463–1465.

Grinold, R. C. (1976). 'Input policies for a longitudinal manpower flow model'. *Man. Sci.*, **22**, 570–575.

Grinold, R. C., and K. T. Marshall (1977). *Manpower Planning Models*, North-Holland, New York.

Grofman, B. (1974). 'Helping behaviour and group size: some exploratory stochastic models'. *Behavioural Science*, **19**, 219–224.

Gumbel, E. J. (1958). *The Statistics of Extremes*, Columbia University Press, New York.

Gunderon, M. (1974). 'Retention of trainees: a study of dichotomous dependent variables'. *J. Econometrics*, **2**, 79–94.

Gupta, I., J. Zareda and N. Kramer (1971). 'Hospital manpower planning by use of queueing theory'. *Health Serv. Res.*, **6**, 76–82.

Gurevitch, M., and Z. Loevy (1972). 'The diffusion of television as an innovation: the case of the Kibbutz'. *Human Relations*, **25**, 181–197.

Hagerstrand, T. (1952). *The Propagation of Innovation Waves*, Lund Studies in

Geography, Ser. B, No. 4.

Hagerstrand, T. (1967). *Innovation Diffusion as a Spatial Process*, The University of Chicago Press, Chicago and London.

Haines, G. H. (1964). 'A theory of market behaviour after innovation'. *Man. Sci.*, **10**, 634–666.

Hajnal, J. (1956). 'The ergodic properties of non-homogeneous finite Markov chains'. *Proc. Camb. Phil. Soc.*, **52**, 67–77.

Hajnal, J. (1976). 'On products of non-negative matrices'. *Math. Proc. Camb. Phil. Soc.*, **79**, 521–530.

Hamblin, R. L., R. B. Jacobson, and J. L. L. Miller (1973). *A Mathematical Theory of Social Change*, John Wiley, New York.

Hammersley, J. M. (1966). 'First-passage percolation'. *J. R. Statist. Soc.*, **B28**, 491–496.

Harary, F., and B. Lipstein (1962). 'The dynamics of brand loyalty: a Markovian approach'. *Operat. Res.*, **10**, 19–40.

Harden, W. R., and M. T. Tcheng (1971). 'Projection of enrolment distributions with enrolment ceilings by Markov processes'. *Socio-econ. Plan. Sci.*, **5**, 467–473.

Hart, P. E., and S. J. Prais (1956). 'The analysis of business concentration'. *J. R. Statist. Soc.*, **A119**, 150–191.

Haskey, H. W. (1954). 'Stochastic cross-infection between two otherwise isolated groups'. *Biometrika*, **44**, 193–204.

Hassani, H. (1980). 'Markov renewal models for manpower systems'. Ph.D. Thesis, University of London.

Hawes, L. C. and J. M. Foley (1973). 'A Markov analysis of interview communication'. *Speech Monographs*, **40**, 208–219.

Hawkes, A. G. (1969). 'An approach to the analysis of electoral swing'. *J. R. Statist. Soc.*, **A132**, 68–79.

Hedberg, M. (1961). 'The turnover of labour in industry, an actuarial study'. *Acta Sociologica*, **5**, 129–143.

Hedge, B. J., B. S. Everitt, and C. Frith, (1978). 'The role of gaze in dialogue'. *Acta Psychologica*, **42**, 453–475.

Henderson, L. F. (1971). 'The statistics of crowd fluids'. *Nature*, **229**, 381–383.

Henry, N. W. (1971). 'The retention model: a Markov chain with variable transition probabilities'. *J. Amer. Statist. Ass.*, **66**, 264–267.

Henry, N. W., R. McGinnis, and H. W. Tegtmeyer (1971). 'A finite model of mobility'. *J. Math. Sociology*, **1**, 107–116.

Herbst, P. G. (1954). 'Analysis of social flow systems'. *Human Relations*, **7**, 327–336.

Herbst P. G. (1963). 'Organizational commitment: a decision model'. *Acta Sociologica*, **7**, 34–45.

Herniter, J. (1971). 'A probabilistic market model of purchasing timing and brand selection'. *Man. Sci.*, **18**, B102–B113.

Hill, B. M. (1970). 'Zipf's law and prior distributions for the composition of a population' *J. Amer. Statist. Ass.*, **65**, 1220–1232.

Hill, B. M. (1974). 'The rank-frequency form of Zipf's law'. *J. Amer. Statist. Ass.*, **69**, 1017–1026.

Hill, B. M. and M. Woodroofe, (1975). 'Stronger forms of Zipf's law'. *J. Amer. Statist. Ass.*, **70**, 212–219.

Hill, J. M. M. (1951). 'A consideration of labour turnover as the resultant of a quasi-stationary process'. *Human Relations*, **4**, 255–264.

Hill, R. J. (1957). 'An experimental investigation of the logistic model of message diffusion'. *Social Forces*, **36**, 21–26.

Hill, R. T., and N. C. Severo (1969). 'The simple epidemic for small populations with one or more initial infectives'. *Biometrika*, **56**, 183–196.

Hiorns, R. W., G. A. J. Harrison, and C. F. Kuchemann (1970). 'Social class relatedness in some Oxfordshire parishes'. *J. Bio. Soc. Sci.*, **2**, 71–80.

Hodge, R. W. (1966). 'Occupational mobility as a probability process'. *Demography*, **3**, 19–34.

Hoem, J. M. (1969). 'Purged and partial Markov chains'. *Skand. Aktuarietidskrift*, **52**, 147–155.

Hoem, J. M. (1971). 'Point estimation of forces of transition in demographic models'. *J. R. Statist. Soc.*, **33**, 275–289.

Hoem, J. M. (1976). 'The statistical theory of demographic rates: a review of current developments'. *Scandanavian. J. Statistics*, **3**, 169–178.

Hoem, J. M. (1977). 'A Markov chain model of working life tables'. *Scand. Actuarial. J.*, **4**, 1–20.

Holland, P. W., and S. Leinhardt (1977). 'A dynamic model for social networks'. *J. Math. Sociology*, **5**, 5–20.

Hopkins, D. S. P. (1974). 'Faculty early-retirement programs'. *Operat. Res.*, **22**, 455–467.

Horden, W. R., and M. T. Tcheng (1971). 'Projection of enrolment distributions with enrolment ceilings by Markov processes'. *Socio-econ. Plan. Sci.*, **5**, 467–473.

Horowitz, A. and I. (1968). 'Entropy, Markov processes and competition in the brewing industry'. *J. Indust. Econ.*, **16**, 196–211.

Horvath, W. J. (1966). 'Stochastic models of behaviour'. *Man. Sci.*, **12**, B513–B518.

Horvath, W. J. (1968). 'A statistical model for the duration of wars and strikes'. *Behavioural Science*, **13**, 18–28.

Howard, R. A. (1963). 'Stochastic process models of consumer behaviour'. *J. Advertising Res.*, **3**, 35–42.

Huang, C. J., and B. W. Bolch (1976). 'Some stochastic aspects of trial by jury'. *Metron*, **34**, 73–80.

Hudson, J. C. (1969). 'Diffusion in a central place system'. *Geographical Analysis*, **1**, 45–58.

Hyman, R. (1970). 'Economic motivation and labour stability'. *Brit. J. Indust. Rel.*, **8**, 159–178.

Hymer, S., and P. Pashigan (1962). 'Firm size and rate of growth'. *J. Pol. Econ.*, **70**, 556–569.

Ichino, Shozo (1973). 'Rodo ido bunseki' (An analysis of labour mobility). *Operations Research as a Management Science*, **18**, 9–17.

Ijiri, Y., and H. A. Simon (1964). 'Business firm growth and size'. *Amer. Econ. Rev.*, **54**, 77–89.

Ijiri, Y., and H. A. Simon (1967). 'A model of business firm growth'. *Econometrica*, **35**, 348–355.

Ijiri, Y., and H. A. Simon (1977). *Skew Distributions and the Sizes of Business Firms*, Elsevier, Amsterdam.

Isaacson, D. L., and R. W. Madsen (1976). *Markov Chains Theory and Applications*, John Wiley, New York.

Johnstone, J. N., and H. Philp (1973). 'The application of a Markov chain in educational planning'. *Socio-econ. Plan. Sci.*, **7**, 283–294.

Jones, E. (1946). 'An actuarial problem concerning the Royal Marines'. *J. Inst. Actu. Students Soc.*, **6**, 38–42.

Jones, E. (1948). 'An application of the service-table technique to staffing problems'. *J. Inst. Actu. Students Soc.*, **8**, 49–55.

Jones, J. M. (1971). 'A stochastic model for adaptive behaviour in a dynamic situation'. *Man. Sci.*, **17**, 484–509.

Jones, R. C., S. R. Morrison, and R. P. Whiteman (1973). 'Helping to plan a bank's manpower resources'. *Operat. Res. Quart.*, **24**, 365–374.

Joseph, G. (1974). 'Inter-regional population distribution and growth in Britain—a projection exercise'. *Scottish J. Political Economy*, **21**, 159–170.

Kadane, J. B., and G. Lewis (1969). 'The distribution of participation in group discussion: an empirical and theoretical appraisal'. *Amer. Sociol. Rev.*, **34**, 710–723.

Kalescar, P. (1970). 'A Markovian model for hospital admissions scheduling'. *Man. Sci.*, **16**, B384–B396.

Kamat, A. R. (1968a). 'Estimating wastage in a course of education'. *Sankhya*, **B30**, 5–12.

Kamat, A. R. (1968b). 'Mathematical schemes for describing progress in a course of education'. *Sankhya*, **B30**, 13–24.

Kamat, A. R. (1968c). 'A stochastic model for progress in a course of education'. *Sankhya*, **B30**, 25–32.

Kao, E. P. C. (1972). 'A semi-Markov model to predict recovery progress of coronary patients'. *Health Serv. Res.*, **7**, 191–208.

Kao, E. P. C. (1973). 'A semi-Markovian population model with applications to hospital planning'. *I.E.E.E. Transactions on Systems, Man and Cybernetics*, **SMC-3**, 327–336.

Kao, E. P. C. (1974). 'Modelling the movement of coronary patients within a hospital by semi-Markov processes'. *Operat. Res.*, **22**, 683–698.

Kao, R. C. (1953). 'Note on Miller's finite Markov processes in psychology'. *Psychometrika*, **18**, 241–243.

Kaplan, R. S. (1972). 'Stochastic growth models'. *Man. Sci.*, **18**, 249–264.

Karlin, S., and H. Taylor (1975). *A First Course in Stochastic Processes*, Academic Press, New York.

Karmeshu and R. K. Pathria, (1979). 'Co-operative behaviour in a non-linear diffusion of information'. *Canadian J. Physics*, **57**, 1572–1578.

Karmeshu and R. K. Pathria, (1980a). 'Stochastic evolution of a non-linear model of diffusion of information'. *J. Math. Sociology*, **7**, 59–71.

Karmeshu and R. K. Pathria, (1980b). 'Stochastic evolution of competing social groups'. *J. Math. Sociology*, **7**, 47–58.

Karmeshu and R. K. Pathria, (1980c). 'Time development of a Markov process in a finite population: application to diffusion of information'. *J. Math. Sociology*, **7**, 229–240.

Karmeshu and R. K. Pathria, (1980d). 'Diffusion of information in a random environment'. *J. Math. Sociology*, **7**, 215–227.

Katz, L., and C. H. Procter (1959). 'The concepts of configurations of inter-personal relations in a group as a time-dependent stochastic process'. *Psychometrika*, **24**, 317–327.

Kayser, B. D. (1976). 'Educational aspirations as a Markovian process. A panel study of college and non-college plans during high school'. *J. Math. Sociology.*, **4**, 295–306.

Kelker, D. (1973). 'A random walk epidemic simulation'. *J. Amer. Statist. Ass.*, **68**, 821–823.

Kelley, A. C., and L. W. Weiss (1969). 'Markov processes and economic analysis: the case of migration'. *Econometrica*, **37**, 280–297.

Kemeny, J. G., and L. Snell (1962). *Mathematical Models in the Social Sciences*, Ginn and Co., Boston.

Kemeny, J. G., and L. Snell (1976). *Finite Markov Chains*, Springer-Verlag, Berlin (1st Edn., van Nostrand, 1960, Princeton, N.J.).

Kemp, A. G., and G. C. Reid (1971). 'The random walk hypothesis and the recent behaviour of equity prices'. *Economica*, **38**, 28–51.

Kendall, D. G. (1956). 'Deterministic and stochastic epidemics in closed populations'. *Proc. Third Berkeley Symp. Math. Statist. Prob.*, **4**, 149–165.

Kendall, D. G. (1957). 'La propogation d'une epidemie au d'un bruit une population limitée'. *Publ. De. L'Inst. de l'Université de Paris*, **6**, 307–311.

Kendall, D. G. (1965). 'Mathematical models in the spread of infection'. *In Mathematics and Computer Science in Biology and Medicine*, H.M.S.O., London, 213–225.

Kendall, M. G. (1961). 'Natural law in the social sciences'. *J. R. Statist. Soc.*, **A124**, 1–19.

Kermack, W. O., and A. G. McKendrick (1927). 'Contributions to the mathematical theory of epidemics'. *Proc. Roy. Soc.*, **A115**, 700–721.

Keyfitz, N. (1968). *Introduction to the Mathematics of Population*, Addison-Wesley, Reading, Mass.

Khintchine, A. J. (1960). *Mathematical Methods in the Theory of Queueing*, Griffin, London and High Wycombe.

Kingman, J. F. C. (1962). 'The imbedding problem for finite Markov chains'. *Zeit. Wahrscheinlichkeitstheorie*, **1**, 14–24.

Kolmogoroff, A. N., I. Petrovsky, and N. Piscounoff (1937). 'Etude de l'équation de la diffusion avec croissance de la quantité de matière et son application à un problème biologique'. *Bull. de l'Univ. d'Etat à Moscou*, **A1**, fasc 6, 1–25.

Komulainen, E. (1971). 'Investigations into the instructional process: IV. Teaching as a stochastic process'. *Institute of Education University of Helsinki Research Bulletin*, No. 29.

Koskenniemi, M., and E. Komulainen (1969). 'Investigations into the instructional process: I. Some methodological problems'. *Institute of Education University of Helsinki Research Bulletin*, No. 26.

Kotler, P. (1968). 'Mathematical models of individual buyer behaviour'. *Behavioural Science*, **13**, 274–287.

Kottas, J. F., A. Hing-Ling Lau, Hong-Shiang Lau (1978). 'A general approach to stochastic management planning models: an overview'. *The Accounting Review*, **53**, 389–401.

Krantz, D. H., D. R. Luce, P. Suppes, and A. Tversky (1971). *Additive and Polynomial Representations Foundations of Measurement*, 1, Academic Press, New York and London.

Krenz, R. D. (1964). 'Projection of farm numbers for North Dakota with Markov chains'. *Agric. Econ. Res.*, **16**, 77–83.

Krishnan, P. (1971). 'A stochastic process model of conjugal history'. *Proc. of the Social Statistics Section of the American Statist. Ass.*, 324–327.

Krishnan, P. (1974). 'Preliminary report on an epidemic model approach to the propagation of family planning ideas'. *Socio-econ. Plan. Sci.*, **8**, 225–228.

Krishnan, P. (Ed.) (1977a). *Mathematical Models of Sociology*, Sociological Review Monograph 24, University of Keele, Keele, Staffs, U.K.

Krishnan, P. (1977b). 'A stochastic process model of conjugal history', in *Mathematical Models of Sociology* (P. Krishnan Ed.) Sociological Review Monograph 24, University of Keele, Keele, Staffs, U.K., 117–125.

Krishnan, P. (1977c). 'A Markov chain approximation of conjugal history', in *Mathematical Models of Sociology* (P. Krishnan Ed.) Sociological Review Monograph 24, University of Keele, Keele, Staffs, U.K., 127–133.

Kruskal, W. (Ed.) (1970). *Mathematical Sciences and Social Sciences*, Prentice-Hall, Englewood Cliffs, N.J.

Kryscio, R. J. (1972). 'The transition probabilities of the extended simple stochastic epidemic model and the Haskey model'. *J. Appl. Prob.*, **9**, 471–485.

Kryscio, R. J. (1974). 'On the extended simple stochastic epidemic model'. *Biometrika*, **61**, 200–202.

Kryscio, R. J. (1975). 'The transition probabilities of the general stochastic epidemic'. *J. Appl. Prob.*, **12**, 415–424.

Kuehn, A. A. (1961). 'A model of budget advertising'. In Bass and co-workers (Eds.), *Mathematical Models and Methods in Marketing*, Irwin, Homewood, Illinois, 302–356.

Kuehn, A. A. (1962). 'Consumer brand choice as a learning process'. *J. Advertising Res.*, **2**, 10–17.

Kuhn, A., A. Poole, P. Sales, and H. P. Wynn (1973). 'An analysis of graduate job mobility'. *Brit. J. Indust. Rel.*, **11**, 124–142.

Kushner, H. (1967). *Stochastic Stability and Control*, Academic Press, New York.

Kushner, H. (1971). *Introduction to Stochastic Control*, Holt, Rinehart and Winston, New York.

Kushner, H. J., and A. J. Kleinman (1971). 'Mathematical programing and the control of Markov chains'. *Int. J. Control.*, **13**, 801–820.

Kwak, N. K., W. A. Garrett, and S. Barone (1977). 'A stochastic model of demand forecasting for technical manpower planning'. *Man. Sci.*, **23**, 1089–1098.

Lancaster, A. (1972). 'A stochastic model for the duration of a strike'. *J. R. Statist. Soc.*, **A135**, 257–271.

Lancaster, A., and S. J. Nickell (1980). 'The analysis of re-employment probabilities for the unemployed'. *J. R. Statist. Soc.*, **A143**, 141–165.

Lancaster, P. (1969). *Theory of Matrices*. Academic Press, New York.

Land, K. C. (1969). 'Duration of residence and prospective migration'. *Demography*, **6**, 133–140.

Land, K. C. (1971). 'Some exhaustible Poisson process models of divorce by marriage cohort'. *J. Math. Sociology*, **2**, 213–232.

Landau, H. G. (1952). 'On some problems of random nets'. *Bull. Math. Biophysics.*, **14**, 203–212.

Landau, H. G., and A. Rapoport (1953). 'Contribution to the mathematical theory of contagion and spread of information I: through a thoroughly mixed population'. *Bull. Math. Biophysics.*, **15**, 173–183.

Lane, K. F., and J. E. Andrew (1955). 'A method of labour turnover analysis'. *J. R. Statist. Soc.*, **A118**, 296–323.

LaSalle, J. P. (1972). 'Appointments, promotion and tenure under steady-state staffing'. *Notices of the American Mathematical Society*, **19**, 69–73.

Lawrence, J. R. (Ed.) (1966). *Operational Research and the Social Sciences*, Tavistock Publications, London.

Lazarsfeld, P. F. (Ed.) (1954). *Mathematical Thinking in the Social Sciences*, Free Press of Glencoe, New York.

Ledermann, W. (Editor-in-Chief) (1980). *Handbook of Applicable Mathematics*, (6 vols.) John Wiley, Chichester.

Lee, T. C., G. C. Judge and A. Zellner (1970). *Estimating the Parameters of the Markov Probability Model from Aggregate Time Data*, North-Holland, New York and Amsterdam.

Leeson, G. W. (1979). 'Wastage in a hierarchical manpower system'. *J. Operat. Res. Soc.*, **30**, 341–348.

Leeson, G. W. (1980). 'A projection model for hierarchical manpower systems'. *J. Operat. Res. Soc.*, **31**, 247–256.

Lehoczky, J. P. (1980). 'Approximations for interactive Markov chains in discrete and continuous time', *J. Math. Sociology*, **7**, 139–157.

Lever, W. F. (1973). 'A Markov approach to the optimal size of cities in England and Wales'. *Urban Studies*, **10**, 353–365.

Lieberson, S., and G. V. Fuguitt (1967). 'Negro–white occupational differences in the absence of discrimination'. *Amer. J. Sociol.*, **73**, 188–200.

Lillard, L. A., and R. J. Willis (1978). 'Dynamic aspects of earning mobility'. *Econometrica*, **46**, 985–1012.

Lipstein, B. (1965). 'A mathematical model of consumer behaviour'. *J. Marketing. Res.*, **2**, 259–265.

Long, L. H. (1970). 'On measuring geographic mobility'. *J. Amer. Statist. Ass.*, **65**, 1195–1203.

Lu, K. H. (1968). 'A Markov chain analysis of caries process with consideration for the effect of restoration'. *Archs. Oral. Biol.*, **13**, 1119–1132.

Ludwig, D. (1973). 'Stochastic approximation for the general epidemic'. *J. Appl. Prob.*, **10**, 263–276.

Lui, P. T., and L. P. Chow (1971). 'A stochastic approach to the prevalence of IUD: example of Taiwan, Republic of China'. *Demography*, **8**, 341–353.

Lundberg, O. (1964). *On Random Processes and their Applications to Sickness and Accident Statistics*, Almquist and Wiksells, Uppsala, Sweden.

McCall, J. J. (1971). 'A Markovian model of income dynamics'. *J. Amer. Statist. Ass.*, **66**, 439–447.

McCarthy, C., and T. M. Ryan (1977). 'Estimates of voter transition probabilities for the British General Elections of 1974'. *J. R. Statist. Soc.*, **A140**, 78–85.

McClean, S. I. (1976a). 'The two-stage model of personnel behaviour'. *J. R. Statist. Soc.*, **A139**, 205–217.

McClean, S. I. (1976b). 'A continuous time population model with Poisson recruitment'. *J. Appl. Prob.*, **13**, 348–354.

McClean, S. I. (1976c). 'Some models for company growth'. *J. R. Statist. Soc.*, **A139**, 501–507.

McClean, S. I. (1978). 'Continuous-time stochastic models of a multi-grade population'. *J. Appl. Prob.*, **15**, 26–37.

McClean, S. I. (1980). 'A semi-Markov model for a multi-grade population with Poisson recruitment'. *J. Appl. Prob.*, **17**, 846–852.

McClean, S. I., and T. Abodunde (1978). 'Entropy as a measure of stability in a manpower system'. *J. Operat. Res. Soc.*, **29**, 885–890.

McClean, S. I., and L. Karageorgos (1979). 'An age stratified manpower model applied to the educational system'. *Statistician*, **28**, 9–18.

MacCrea, E. C. (1977). 'Estimation of time-varying Markov processes with aggregate data'. *Econometrica*, **45**, 183–198.

McFarland, D. (1970). 'Inter-generational social mobility as a Markov process: including a time-stationary Markovian model that explains observed declines in mobility rates over time'. *Amer. Sociol. Rev.*, **35**, 463–476.

McGinnis, R. (1964). *Mathematical Foundations for Social Analysis*, Bobbs-Merrill, Chicago.

McGinnis, R. (1968). 'A stochastic model of social mobility'. *Amer. Sociol. Rev.*, **33**, 712–721.

McGuire, T. W. (1969). 'More on least squares estimation of the transition matrix in a stationary first-order Markov process from sample population data'. *Psychometrika*, **34**, 335–345.

McNamara, J. F. (1971). 'Mathematical programming models in educational planning'. *Review of Educational Research*, **41**, 419–446.

McNamara, J. F. (1973). 'Mathematical programming applications in educational planning'. *Socio-econ. Plan. Sci.*, **7**, 19–36.

McNeil, D. R. (1972). 'On the simple stochastic epidemic'. *Biometrika*, **59**, 494–497.

McPhee, W. N. (1963). *Formal Theories of Mass Behaviour*, Free Press, New York.

Maffei, R. B. (1960). 'Brand preferences and simple Markov processes'. *Operat. Res.*, **8**, 210–218.

Maffei, R. B. (1961). 'Brand preferences and simple Markov processes'. In Bass and co-workers (Eds.), *Mathematical Models and Methods of Marketing*, Irwin, Homewood, Illinois, 103–120.

Mahoney, T. A., and G. T. Milkovich (1971). 'The internal labor market as a stochastic process'. In D. J. Bartholomew and A. R. Smith (1971), 75–91.

Mandl, P. (1973). 'Some applications of martingales in controlled Markov processes'. *Bull. Int. Statist. Inst.*, **45**, Book 2, 170–175.

Mandl, P. (1974). 'Estimation and control in Markov chains'. *Adv. Appl. Prob.*, **6**, 40–60.

Manpower Planning, Proceedings of NATO Conference, Brussels, 1965, English Universities Press, London.

Mansfield, E. (1961). 'Technical change and the rate of imitation'. *Econometrica*, **29**, 741–766.

Mansfield, E. (1962). 'Entry, Gibrat's law, innovation and the growth of firms'. *Amer. Econ. Rev.*, **52**, 1023–1051.

Mansfield, E. (1968). *Industrial Research and Technological Innovation*, W. W. Norton, New York.

Mansfield, E., and C. Hensley (1960). 'The logistic process: tables of the stochastic epidemic curve and applications'. *J. R. Statist. Soc.*, **B22**, 332–337.

Mapes, R. (1968). 'Promotion in static hierarchies'. *J. Man. Studies*, **5**, 365–379.

Marinelli, G. (1972). 'Alcuni Contributi della R. O. alla Planificazione del Personale'. *Ricerca Operativa*, **2**, 41–49.

Marma, V. J., and K. W. Deutsch (1973). 'Survival in unfair conflict: odds, resources and random walk models'. *Behavioural Science*, **18**, 313–334.

Marris, R. (1970). *The Economic Theory of 'Managerial' Capitalism*, Macmillan, New York and London.

Marshall, A. W., and H. Goldhamer (1955). 'An application of Markov-processes to the study of the epidemiology of mental disease'. *J. Amer. Statist. Ass.*, **50**, 99–129.

Marshall, K. T. (1973). 'A comparison of two personnel prediction models'. *Operat. Res.*, **21**, 810–822.

Marshall, K. T., and R. M. Oliver (1970). 'A constant-work model for student attendance and enrolment'. *Operat. Res.*, **18**, 193–206.

Marshall, M. L. (1971a). 'Some statistical methods for forecasting wastage'. *Statistician*, **20**, 53–68.

Marshall, M. L. (1971b). 'The use of probability distributions for comparing the turnover of families in a residential area'. *London Papers in Regional Science*, **2**, (Urban Regional Planning) A. G. Wilson (Ed.), Pion, London, 171–193.

Martel, A., and A. Al-Nuaimi (1973). 'Tactical manpower planning via programming under uncertainty'. *Operat. Res. Quart.*, **24**, 571–585.

Martin, J. J. (1967). *Bayesian Decision Problems and Markov Chains*, John Wiley, New York.

Maslov, P. P. (1962). 'Model building in sociological research'. *Soviet Sociology*, **1**, 11–23.

Massy, W. F. (1966). 'Order and homogeneity of family specific brand-switching processes'. *J. Marketing Res.*, **3**, 48–54.

Massy, W. F. (1968). 'Stochastic models for monitoring new product introductions'. In F. M. Bass, C. W. King, and E. A. Pessimier (Eds.), *Applications of the Sciences in Marketing Management*, John Wiley, New York.

Massy, W. F., D. B. Montgomery, and D. G. Morrison (1970). *Stochastic Models of Buying Behaviour*, M.I.T. Press, Cambridge, Mass.

Matras, J. (1960a). 'Comparison of intergenerational occupational mobility patterns: an application of the formal theory of social mobility'. *Population Studies*, **14**, 163–169.

Matras, J. (1960b). 'Differential fertility, intergenerational occupational mobility and change in the occupational distribution: some elementary interrelationships'. *Population Studies*, **15**, 187–197.

Matras, J. (1967). 'Social mobility and social structure: some insights from the linear model'. *Amer. Sociol. Rev.*, **32**, 608–614.

Mayer, T. (1972). 'Models of intra-generational mobility'. In J. Berger, M. Zelditch, Jr., and B. Anderson (Eds.), *Sociological Theories in Progress*, **2**, Houghton Mifflin, New York.

Mayhew, B. H. (1972). 'Growth and decay of structure in interaction'. *Comparative Group Studies*, May, 131–160.

Mayhew, B. H., and L. N. Gray, (1971). 'The structure of dominance relations in triadic interaction systems'. *Comparative Group Studies*, May, 161–190.

Mehlmann, A. (1977a). 'A note on the limiting behaviour of discrete-time Markovian manpower models with inhomogeneous independent Poisson input'. *J. Appl. Prob.*, **14**, 611–613.

Mehlmann, A. (1977b). 'Markovian manpower models in continuous time'. *J. Appl. Prob.*, **14**, 249–259.

Mehlmann, A. (1979). 'Semi-Markovian manpower models in continuous time'. *J. Appl. Prob.*, **16**, 416–422.

Mehlmann, A. (1980). 'An approach to optimal recruitment and transition strategies for manpower systems using dynamic programming'. *J. Operat. Res. Soc.*, **31**, 1009–1015.

Meier, P. (1955). 'Note on estimation in a Markov process with constant transition rates'. *Human Biology*, **27**, 121–124.

Menges, G., and G. Elstermann (1971). 'Capacity models in university management'. In D. J. Bartholomew and A. R. Smith (1971), 207–221.

Merck, J. W. (1961). 'A mathematical model of personnel structure of large scale organizations based on Markov chains'. Ph.D. Thesis, Duke University, North Carolina.

Meredith, J. (1973). 'A Markovian analysis of a geriatric ward'. *Man. Sci.*, **19**, 604–612.

Midlarsky, M. (1970). 'Mathematical models of instability and a theory of diffusion'. *International Studies Quart.*, **14**, 60–84.

Miles, R. E. (1959). 'The complete amalgamation into blocks, by weighted means, of a finite set of real numbers'. *Biometrika*, **46**, 317–327.

Miller, G. A. (1952). 'Finite Markov processes in psychology'. *Psychometrika*, **17**, 149–167.

Miller, W. L. (1972). 'Measures of electoral change using aggregate data'. *J. R. Statist. Soc.*, **A135**, 122–142.

Milton, R. C. (1972). 'Computer evaluation of the multivariate normal integral'. *Technometrics*, **14**, 881–889.

Mode, C. J. (1972). 'A study of a Malthusian parameter in relation to some stochastic models of human reproduction'. *Theoret. Pop. Biology*, **3**, 300–323.

Mollison, D. (1972a). 'Possible velocities for a simple epidemic'. *Adv. Appl. Prob.*, **4**, 233–257.

Mollison, D. (1972b). 'The rate of spatial propagation of simple epidemics'. *Proc. Sixth Berkeley Symp. Math. Statist. Prob.*, **3**, 579–614.

Mollison, D. (1977). 'Spatial contact models for ecological and epidemic spread' (with discussion). *J. R. Statist. Soc.*, **B39**, 283–326.

Mollison, D. (1978). 'Markovian contact processes'. *Adv. Appl. Prob.*, **10**, 85–108.

Montgomery, D. B. (1967). 'Stochastic modelling of the consumer'. *Industrial Management Review*, Spring 1967, 31–42.

Montgomery, D. B. (1969). 'A stochastic response model with application to brand choice'. *Man. Sci.*, **15**, 323–337.

Montgomery, D. B. (1972). 'Note on a limit distribution arising in certain stochastic response models'. *J. R. Statist. Soc.*, **C21**, 204–207.

Moore, S. G. (1975). 'On Johnstone and Philp's Markov chain educational planning model'. *Socio-econ. Plan. Sci.*, **9**, 89–91.

Moran, P. A. P. (1968). *An Introduction to Probability Theory*, Clarendon Press, Oxford.

Morgan, R. W. (1970). 'Manpower planning in the Royal Air Force: an exercise in linear programming'. In A. R. Smith (1971), 317–325.

Morgan, R. W. (1971). 'The use of a steady state model to obtain the recruitment, retirement and promotion policies of an expanding organization'. In D. J. Bartholomew and A. R. Smith (1971), 283–291.

Morgan, R. W. (1979). 'Some models for hierarchical manpower system'. *J. Operat. Res. Soc.*, **30**, 727–736.

Morgan, R. W., G. A. Keenay, and K. Ray (1974). 'A steady state model for career planning'. In D. J. Clough and co-workers (Eds.) (1974).

Morrill, R. L. (1965). 'The negro ghetto: problems and alternatives'. *The Geographical Review*, **55**, 349–361.

Morrill, R. L. (1968). 'Waves of spatial diffusion'. *J. Reg. Sci.*, **8**, 1–18.

Morrill, R. L. (1970). 'The shape of diffusion in space and time'. *Econ. Geog.*, **46** (1970 supplement), 259–268.

Morrison, D. G. (1966). 'Testing brand-switching models'. *J. Marketing Res.*, **3**, 401–409.

Morrison, D. G., W. F. Massy, and F. N. Silverman (1971). 'The effect of non-homogeneous populations on Markov steady state probabilities'. *J. Amer. Statist. Ass.*, **66**, 268–274.

Morrison, P. (1967). 'Duration of residence and prospective migration: the evaluation of a stochastic model'. *Demography*, **4**, 533–561.

Morse, W. J. (1975). 'Estimating the human capital associated with an organization'. *Accounting and Business Research*, No. 21, 48–56.

Moya-Angeler, J. (1976). 'A model with shortage of places for educational and manpower systems'. *Omega*, **4**, 719–730.

Myers, G. C., R. McGinnis, and G. Masnick (1967). 'The duration of residence approach to a dynamic stochastic model of internal migration: a test of the axiom of cumulative inertia'. *Eugenics Quarterly*, **14**, 121–126.

Nagaev, A. V., and A. N. Startsev (1970). 'The asymptotic analysis of a stochastic model of an epidemic'. *Theor. Prob. Appl.*, **15**, 98–107.

Nakamura, M. (1973). 'Some programming problems in population projection'. *Operat. Res.*, **21**, 1048–1062.

Nash, J. C. (1977). 'A discrete alternative to the logistic growth function'. *Appl. Statist.*, **26**, 9–14.

Naslund, B. (1970). 'Size distribution and the optimal size of firms'. *Zeit. Nat. Okon.*, **30**, 271–282.

Navarro, V. (1969). 'Planning personal health services: a Markovian model'. *Medical Care*, **7**, 242–249.

Navarro, V., R. Parker and K. L. White (1970). 'A stochastic and deterministic model of medical care utilization'. *Health Serv. Res.*, **5**, 342–357.

Nemhauser, G. L., and H. L. W. Nuttle (1965). 'A quantitative approach to employment planning'. *Man. Sci.*, **11(B)**, 155–165.

Newman, E. B. (1951). 'The pattern of vowels and consonants in various languages'. *Amer. J. Psych.*, **64**, 369–379.

Newman, P., and J. N. Wolfe (1961). 'A model for the long-run theory of value'. *Rev. Econ. Studies*, **29**, 51–61.

Neyman, J., and E. L. Scott (1964). 'A stochastic model of epidemics'. In J. Gurland (Ed.), *Stochastic Models in Medicine and Biology*, University of Wisconsin Press, Madison, 45–83.

Nicholson, M. B. (1968). 'Mathematical models in the study of international relations'. *The Year Book of World Affairs*, **22**, 47–63.

Nickell, S. J. (1979). 'Estimating the probability of leaving unemployment'. *Econometrica*, **47**, 1249–1266.

Nicosia, F. M. (1967). 'New developments in advertising research: stochastic models'.

Proc. Amer. Ass. Public Opinion Research.

Nielson, G. L., and A. R. Young (1973). 'Manpower planning: a Markov chain application'. *Public Personnel Management,* **2,** 133–143.

Nordhaus, W. D. (1973). 'World dynamics: measurement without data'. *Econ. J.,* **83,** 1156–1183.

Oliver, R. M. (1972). 'Operations research in university planning'. In A. Drake, R. Keeney and P. M. Morse (Eds.), *Analysis of Public Systems,* M.I.T. Press, Cambridge, Mass.

Oliver, R. M., D. S. P. Hopkins, and R. L. Armacost (1972). 'An equilibrium flow model of a university campus'. *Operat. Res.,* **20,** 249–264.

Osborne, M. F. (1959). 'Brownian motion in the stock market'. *Operat. Res.,* **7,** 145–173.

Osei, G. K., and J. W. Thompson (1977). 'The supersession of one rumour by another'. *J. Appl. Prob.,* **14,** 127–134.

Padberg, D. I. (1962). 'The use of Markov processes in measuring changes in market structure'. *J. Farm. Economics,* **44,** 189–199.

Paige, C. C., C. P. H. Styan, and P. G. Wachter (1975). 'Computation of the stationary distribution of a Markov chain'. *J. of Statist. Computations and Simulations,* **4,** 173–186.

Parvin, M. (1974). 'Planning human capital structure: a study of economic, demographic and educational determinants'. *Man. Sci.,* **20,** 1543–1553.

Parzen, E. (1962). *Stochastic Processes,* Holden-Day, San Francisco.

Peltier, R. (Ed.) (1966). *Model Building in the Social Sciences,* Union Européene d'Editions, Monaco.

Pemberton, H. E. (1936). 'The curve of culture diffusion rate'. *Amer. Sociol. Rev.,* **1,** 547–556.

Pemberton, H. E. (1938). 'The spatial order of culture diffusion'. *Sociology and Social Research,* **22,** 246–251.

Peria, D. H. (1968). 'The comparison of sequences from the Flanders interaction category system'. *Classroom Interaction Newsletter,* **4,** 27–33.

Perrin, E. B., and M. C. Sheps (1964). 'Human reproduction: a stochastic process'. *Biometrics,* **20,** 28–45.

Pestieau, P. and U. M. Possen (1979). 'A model of wealth distribution'. *Econometrica,* **47,** 761–772.

Pike, M. C. (1970). 'A note on Kimball's paper "Models for the estimation of competing risks from grouped data"'. *Biometrics,* **26,** 579–581.

Pitcher, B. L., R. L. Hamblin, and J. L. Miller (1978). 'Diffusion of collective violence'. *Amer. Sociol. Rev.,* **43,** 23–35.

Pitts, F. R. (1963). 'Problems in computer simulation of diffusion'. *Papers and Proc. Regional Science Ass.,* **11,** 111–119.

Plewis, I. (1980). 'Using longitudinal data to model teachers' ratings of classroom behaviour as a dynamic process'. Private communication.

Pollard, A. H. (1970a). 'Minimum rate of progress formulae at universities'. *Aust. J. Statist.,* **12,** 92–103.

Pollard, A. H. (1970b). 'Some hypothetical models in systems of tertiary education'. *Aust. J. Statist.,* **12,** 78–91.

Pollard, J. H. (1966). 'On the use of the direct matrix product in analysing certain stochastic population models'. *Biometrika,* **53,** 397–415.

Pollard, J. H. (1967). 'A note on certain discrete time stochastic population models with Poisson immigration'. *J. Appl. Prob.,* **4,** 209–213.

Pollard, J. H. (1969). 'Continuous-time and discrete-time models of population growth'. *J. R. Statist. Soc.,* **A132,** 80–88.

Pollard, J. H. (1973). *Mathematical Models for the Growth of Human Populations,* Cambridge University Press.

Praetz, P. D. (1969). 'Australian share prices and the random walk hypothesis'. *Aust. J. Statist.*, **11**, 123–139.

Prais, S. J. (1955a). 'Measuring social mobility'. *J. R. Statist. Soc.*, **A118**, 56–66.

Prais, S. J. (1955b). 'The formal theory of social mobility'. *Population Studies*, **9**, 72–81.

Prais, S. J. (1978). 'The strike-proneness of large plants in Britain'. *J. R. Statist. Soc.*, **A141**, 368–384.

Preston, L. E., and E. J. Bell (1961). 'The statistical analysis of industry structure: an application to food industries'. *J. Amer. Statist. Ass.*, **56**, 925–932.

Pullum, T. W. (1970). 'What can mathematical models tell us about occupational mobility?'. *Sociological Inquiry*, **40**, 258–280.

Pyke, R. (1961a). 'Markov renewal processes: definitions and preliminary properties'. *Ann. Math. Statist.*, **32**, 1231–1242.

Pyke, R. (1961b). 'Markov renewal processes with finitely many states'. *Ann. Math. Statist.*, **32**, 1243–1259.

Quandt, R. E. (1968). 'On the size distribution of firms'. *Amer. Econ. Rev.*, **56**, 416–432.

Quetelet, A. (1835). *Essai de Physique Sociale*, (2 vols), Bachalier, Paris (2nd Edn. 1869, Brussels).

Quetelet, A. (1849). *Letters on the Theory of Probabilities as Applied to the Moral and Political Sciences* (translated from the French by O. G. Downes), C. and E. Layton, London.

Radcliffe, J. (1976). 'The convergence of a position-dependent branching process as an approximation to a model describing the spread of an epidemic'. *J. Appl. Prob.*, **13**, 338–344.

Rainio, K. (1961). 'Stochastic processes of social contacts'. *Scandinavian J. Psychology*, **2**, 113–128.

Rao, B. L. S. (1972). 'Maximum likelihood estimation for Markov processes'. *Ann. Inst. Statist. Maths.*, **24**, 333–345.

Rapoport, A. (1948). 'Cycle distributions in random nets'. *Bull. Math. Biophysics*, **10**, 145–157.

Rapoport, A. (1951). 'Nets with distance bias'. *Bull. Math. Biophysics*, **13**, 85–91.

Rapoport, A. (1953a). 'Spread of information through a population with socio-structural bias: I. Assumption of transitivity'. *Bull. Math. Biophysics*, **15**, 523–533.

Rapoport, A. (1953b). 'Spread of information through a population with socio-structural bias: II. various models with partial transitivity'. *Bull. Math. Biophysics*, **15**, 535–546.

Rapoport, A. (1954). 'Spread of information through a population with socio-structural bias: III. Suggested experimental procedures'. *Bull. Math. Biophysics*, **16**, 75–81.

Rapoport, A., and L. I. Rebhun (1952). 'On the mathematical theory of rumour spread'. *Bull. Math. Biophysics*, **14**, 375–383.

Rashevsky, N. (1959). *Mathematical Biology of Social Behaviour* (revised edn.), University of Chicago Press, Chicago.

Ray, W. D., and F. Margo (1976). 'The inverse problem in reducible Markov chains'. *J. Appl. Prob.*, **13**, 49–56.

Rees, P. H. (1973). 'A revised notation for spatial demographic accounts and models'. *Environment and Planning*, **5**, 147–155.

Rees, P. H., and A. G. Wilson (1973). 'Accounts and models for spatial demographic analysis: I. aggregate population'. *Environment and Planning*, **5**, 61–90.

Reiger, M. H. (1968). 'A two-state Markov model for behavioural chance'. *J. Amer. Statist. Ass.*, **63**, 993–999.

Reisman, A. (1966). 'A population flow feedback model'. *Science*, **153**, 89–91.

Renshaw, E. (1977). 'Velocities of propagation for stepping-stone models of population growth'. *J. Appl. Prob.*, **14**, 591–597.

Renshaw, E. (1979). 'Wave-forms and velocities for non-nearest neighbour contact distributions'. *J. Appl. Prob.*, **16**, 1–11.

Revelle, C., F. Feldman, and W. Lynne (1969). 'An optimization of tuberculosis epidemiology gives a Markov process model for tuberculosis'. *Man. Sci.*, **16**, B190–B211.

Rice, A. K., J. M. M. Hill, and E. L. Trist (1950). 'The representation of labour turnover as a social process'. *Human Relations*, **3**, 349–381.

Ridler-Rowe, C. J. (1967). 'On a stochastic model of an epidemic'. *J. Appl. Prob.*, **4**, 19–33.

Roberts, H. V. (1959). 'Stock market "patterns" and financial analysis: methodological suggestions'. *J. Finance*, **14**, 1–10.

Robertson, T. S. (1967). 'The process of innovation and the diffusion of innovation'. *J. Marketing*, **31**, 14–19.

Robinson, D. (1974). 'Two stage replacement strategies and their application to manpower planning'. *Man. Sci.*, **21**, 199–208.

Rogers, A. (1966). 'A Markovian policy model of inter-regional migration'. *Papers of the Regional Science Association*, **17**, 205–224.

Rogers, A. (1968). *Matrix Analysis of Inter-regional Population Growth and Distribution*, University of California Press, Berkeley, California.

Rogers, A. (1973a). 'The mathematics of multi-regional demographic growth'. *Environment and Planning*, **5**, 3–29.

Rogers, A. (1973b). 'The multi-regional life table'. *J. Math. Sociology*, **3**, 127–137.

Rogers, A. (1975). *Introduction to Multi-regional Mathematical Demography*, Wiley, New York.

Rogers, E. (1962). *Diffusion of Innovations*, Free Press of Glencoe, New York.

Rogers, E. M., and F. F. Shoemaker, (1971). *Communication of Innovations: A Cross-cultural Approach*, 2nd Edn., Free Press of Glencoe, New York.

Rogers, L. E., and R. V. Farace (1975). 'Relational communication analysis: new measurement procedures'. *Human Communication Research*, **1**, 222–239.

Rogoff, N. (1953). *Recent Trends in Occupational Mobility*, Free Press, Glencoe, Illinois.

Rosenbaum, S. (1971). 'A report on the use of statistics in social science research' (with discussion). *J. R. Statist. Soc.*, **A134**, 534–610.

Rowe, S. M., W. G. Wagner, and G. B. Weathersby (1970). 'A control theory solution to optimal faculty staffing'. Paper P-11, Research Program in University Administration, University of California.

Rowland, K. M., and M. G. Sovereign (1969). 'Markov chain analysis of internal manpower supply'. *Industrial Relations*, **9**, 88–99.

Rushton, S., and A. J. Mautner (1955). 'The deterministic model of a simple epidemic for more than one community'. *Biometrika*, **42**, 126–132.

Rutherford, R. S. G. (1955). 'Income distributions: a new model'. *Econometrica*, **23**, 277–294.

Ryan, B., and N. C. Gross (1943). 'The diffusion of hybrid seed corn in two Iowa communities'. *Rural Sociology*, **8**, 15–24.

Sabolo, Y. (1971). 'A structural approach to the projection of occupational categories and its application to South Korea and Taiwan'. *Internat. Labour Review*, **103**, 131–155.

Sahin, I. (1978). 'Cumulative constrained sojourn times in semi-Markov processes with an application to pensionable service'. *J. Appl. Prob.*, **15**, 531–542.

Sahin, I., and D. J. Hendrick (1978). 'On strike durations and a measure of termination'. *Appl. Statist.*, **27**, 319–324.

Sales, P. (1971). 'The validity of the Markov chain model for a branch of the Civil Service'. *Statistician*, **20**, 85–110.

Salkin, M. S. (1976). 'A note on the use of Markov chains for population projection'. *J. Reg. Sci.*, **16**, 105–106.

Salkin, M. S., T. P. Lianos, and Q. Paris (1975). 'Population predictions for the western United States: a Markov chain approach'. *J. Reg. Sci.*, **15**, 53–60.

Samuelson P. A. (1965). 'Proof that properly anticipated prices fluctuate randomly'. *Indust. Management Review*, **6**, 41–49.

Savage, L. J., and K. Deutsch (1960). 'A statistical model of the gross analysis of transaction flows'. *Econometrica*, **28**, 551–572.

Schach, E., and S. Schach (1972). 'A continuous time stochastic model for the utilization of health services'. *Socio-econ. Plan. Sci.*, **6**, 263–272.

Schinnar, A. P., and S. Stewman (1978). 'A class of Markov models of social mobility with duration memory patterns'. *J. Math. Sociology*, **6**, 61–86.

Scott, D., and G. Cullingford (1974). 'Transition to a desired manpower structure'. *Omega, the Int. Jl. of. Mngmt. Sci.*, **2**, 793–803.

Seal, H. L. (1945). 'The mathematics of a population composed of *k* stationary strata each recruited from the stratum below and supported at the lowest level by a uniform annual number of entrants'. *Biometrika*, **33**, 226–230.

Seal, H. L. (1970). 'Probability distributions of aggregate sickness durations'. *Skand. Aktuarietidskrift*, Parts 3 and 4, 193–204.

Seneta, E. (1973). *Non-negative Matrices*. George Allen & Unwin, London.

Sernadas, C. S. V. S. (1980). 'Multivariate stochastic models for epidemics'. Ph.D. Thesis, University of London.

Serow, W. J. (1976). 'A note on the use of Markov chain for population projection'. *J. Reg. Sci.*, **16**, 101–104.

Severo, N. C. (1967). 'Generalizations of stochastic epidemic models'. *Bull. Int. Statist. Inst.*, **42**, Book 2, 1064–1066.

Severo, N. C. (1969a). 'Solving non-linear problems in the theory of epidemics'. *Bull. Int. Statist. Inst.*, **43**, Book 2, 226–228.

Severo, N. C. (1969b). 'The probabilities of some epidemic models'. *Biometrika*, **56**, 197–201.

Sewell, W. H., A. O. Haller, and G. W. Ohlendorf (1970). 'The educational and early occupational status attainment process: replication and revision'. *Amer. Sociol. Rev.*, **35**, 1014–1027.

Sheps, M. C. (1971). 'A review of models for population change'. *Rev. Int. Statist. Inst.*, **39**, 185–196.

Sheps, M. C., J. A. Menken, and A. P. Radick (1969). 'Probability models for family building: an analytical review'. *Demography*, **6**, 161–183.

Shorrocks, A. F. (1975). 'On stochastic models of size distributions'. *Rev. Econ. Studies*, **42**, 631–641.

Shorrocks, A. F. (1976). 'Income mobility and the Markov assumption'. *Econ. J.*, **86**, 566–578.

Shorrocks, A. F. (1978). 'The measurement of mobility'. *Econometrica*, **46**, 1013–1024.

Silcock, H. (1954). 'The phenomenon of labour turnover'. *J. R. Statist. Soc.*, **A117**, 429–440.

Simon, H. A. (1955). 'On a class of skew distribution functions'. *Biometrika*, **42**, 425–440.

Simon, H. A. (1957a). 'The compensation of executives'. *Sociometry*, **20**, 32–35.

Simon, H. A. (1957b). *Models of Man, Social and Rational: Mathematical Essays on Rational Human Behaviour in a Social Setting*, Wiley, New York.

Simon, H. A. and C. P. Bonini (1958). 'The size distribution of business firms'. *Amer. Econ. Rev.* **48**, 607–617.

Singer, B., and J. E. Cohen (1980). 'Estimating malaria incidence and recovery rates from panel surveys'. *J. Math. Biosciences*, **49**, 273–305.

Singer, B., and S. Spilerman (1975). 'Identifying structural parameters of social processes using fragmentary data'. *Bull. Int. Statist. Inst.*, **46**, 681–697.

Singer, B., and S. Spilerman (1976a). 'Some methodological issues in the analysis of longitudinal surveys'. *Ann. of Econ. and Social Measurement*, **5**, 447–474.

Singer, B., and S. Spilerman (1976b). 'The representation of social processes by Markov models'. *Amer. J. Sociol.*, **82**, 1–54.

Singer, B., and S. Spilerman (1977a). 'Fitting stochastic models to longitudinal survey data – some examples in the social sciences'. *Bull. Int. Statist. Inst.*, **47**, Book 3, 283–300.

Singer, B., and S. Spilerman (1977b). 'Trace inequalities for mixtures of Markov chains'. *Adv. Appl. Prob.*, **9**, 747–764.

Singer, B., and S. Spilerman (1979). 'Clustering on the main diagonal in mobility matrices'. In *Sociological Methodology* (1979), K. Schuessler (Ed.), Jossey-Bass, San Francisco.

Singh, A. and G. Whittington (1975). 'The size and growth of firms'. *Rev. Econ. Studies*, **42**, 15–26.

Siskind, V. (1965). 'A solution of the general stochastic epidemic'. *Biometrika*, **52**, 613–616.

Slivka, R. T., and F. Cannavale (1973). 'An analytical model of the passage of defendants through a court system'. *J. Res. in Crime and Delinquency*, June, 132–140.

Smith, A. R. (1967). 'Manpower planning in management of the Royal Navy'. *J. Management Studies*, **4**, 127–139.

Smith, A. R. (Ed.) (1971). *Models for Manpower Systems*, English Universities Press, London.

Smith, A. R. (Ed.) (1976). *Manpower Planning in the Civil Service*, Civil Service Studies No. 3, H.M.S.O., London.

Smith, P. E. (1961). 'Markov chains, exchange matrices and regional development'. *J. Reg. Sci.*, **3**, 27–36.

Smith, W. L. (1958). 'Renewal theory and its ramifications'. *J. R. Statist. Soc.*, **B20**, 243–302.

Solomanoff, R. (1952). 'An exact method for the computation of the connectivity of random nets'. *Bull. Math. Biophysics*, **14**, 153–157.

Solomanoff, R., and A. Rapoport (1951). 'Connectivity of random nets'. *Bull. Math. Biophysics*, **13**, 107–117.

Sommers, P. M., and J. Conlisk (1979). 'Eigenvalue immobility measures for Markov chains'. *J. Math. Sociology*, **6**, 169–234.

Sørenson, A. B. (1972). 'The occupational mobility process: an analysis of occupational Careers'. Johns Hopkins University, Ph.D. dissertation.

Sørenson, A. B. (1975a). 'Models of social mobility'. *Social Science Research*, **4**, 65–92.

Sørenson, A. B. (1975b). 'The structure of intragenerational mobility'. *Amer. Sociol. Rev.*, **40**, 456–471.

Sørenson, A. B., and M. T. Hallinan (1977). 'A stochastic model for change in group structure'. In *Mathematical Models of Sociology* (Ed. P. Krishnan), Sociological Review Monograph 24, University of Keele, Keele, Staffs, U.K., 143–166.

Spilerman, S. (1970). 'The causes of racial disturbances: a comparison of alternative explanations'. *Amer. Sociol. Rev.*, **35**, 627–649.

Spilerman, S. (1972a). 'The analysis of mobility processes by the introduction of independent variables into a Markov chain'. *Amer. Sociol. Rev.*, **37**, 277–294.

Spilerman, S. (1972b). 'Extensions of the mover–stayer model'. *Amer. J. Sociol.*, **78**, 599–626.

Staff, P. J., and M. K. Vagholkar (1971). 'Stationary distributions of open Markov processes in discrete time with application to hospital planning'. *J. Appl. Prob.*, **8**, 668–680.

Staff, P. J., and M. K. Vagholkar (1974). 'Multivariate distributions in open Markovian systems'. *Trab Estadist.*, **25**, 119–140.

Stafford, J. (1977). 'Urban growth as an absorbing Markov process'. In *Mathematical Models of Sociology*, (Ed. P. Krishnan). Sociological Review Monograph 24, University of Keele, Keele, Staffs, U.K., 135–142.

Steindhl, J. (1965). *Random Processes and the Growth of Firms*, Griffin, London.

Steindhl, J. (1974). 'Der Arbeitsplatzwechsel als Erneurungsproze β'. *Empirica*, **1**, 33–71.

Stewman, S. (1975a). 'An application of the job vacancy chain model to a Civil Service internal labour market'. *J. Math. Sociology*, **4**, 37–59.

Stewman, S. (1975b). 'Two Markov models of system occupational mobility: underlying conceptualizations and empirical tests'. *Amer. Sociol. Rev.*, **40**, 298–321.

Stewman, S. (1976). 'Markov models of occupational mobility theoretical development and empirical support: Part 1 Careers, Part 2 Continuously operative job systems'. *J. Math. Sociology*, **4**, 201–278.

Stirzaker, D. (1980). 'Inheritance'. *Adv. Appl. Prob.*, **12**, 574–590.

Stoikov, V. (1971). 'The effect of changes in quits and hires on the length-of-service composition of employed workers'. *Brit. J. Indust. Rel.*, **9**, 225–233.

Stone, R. (1966). *Mathematics in the Social Sciences and other Essays*, Chapman and Hall, London.

Stone, R. (1971). 'An integrated system of demographic, manpower and social statistics and its links with the system of National Economic Accounts'. *Sankhya*, **B33**, 1–184.

Stone, R. (1972). 'A Markovian educational model and other examples linking social behaviour to the economy' (with discussion). *J. R. Statist. Soc.*, **A135**, 511–543.

Styan, G. P. H., and H. Smith (1964). 'Markov chains applied to marketing'. *J. Marketing Res.*, **1**, 50–55.

Svalagosta, K. (1959). *Prestige, Class and Mobility*, Heinemann, London.

Sverdrup, E. (1965). 'Estimates and test procedures in connexion with stochastic models of deaths, recoveries and transfers between different states of health'. *Skand. Aktuarietidskrift.*, **46**, 184–211.

Sykes, Z. M. (1969). 'Some stochastic versions of the matrix model for population dynamics'. *J. Amer. Statist. Ass.*, **64**, 111–130.

Taeuber, K. E. (1961). 'Duration of residence analysis of internal migration in the United States'. *Milbank Memorial Fund Quarterly*, **39**, 116–131.

Taeuber, K. E., W. Haenszel, and M. Sirkin (1961). 'Residence histories and exposure residences for the United States population'. *J. Amer. Statist. Ass.*, **56**, 824–834.

Taga, Y. (1963). 'On the limiting distributions in Markov renewal processes with finitely many states'. *Ann. Inst. Statist. Math.*, **15**, 1–10.

Taga, Y., and K. Isii (1959). 'On a stochastic model concerning the pattern of communication-diffusion of news in a social group'. *Ann. Inst. Statist. Math.*, **11**, 25–43.

Taga, Y., and T. Suzuki (1957). 'On the estimation of average length of chains in the communication pattern'. *Ann. Inst. Statist. Math.*, **9**, 149–156.

Takacs, L. (1970). 'On the fluctuations of election returns'. *J. Appl. Prob.*, **7**, 114–123.

Takahashi, Y. (1969). 'Markov chains with random transition matrices'. Kodai Mathematical Seminar Reports.

Takahashi, Y. (1973). 'On the effects of small deviations in the transition matrix of a finite Markov chain'. *J. Operat. Res. of Japan*, **16**, 104–129.

Tarver, J. D., and W. R. Gurley (1965). 'A stochastic analysis of geographic mobility and population projections of the census divisions in the United States'. *Demography*, **2**, 134–139.

Taylor, G. C. (1971a). 'Moments in Markovian systems with lumped states'. *J. Appl. Prob.*, **8**, 599–605.

Taylor, G. C. (1971b). 'Sickness: a stochastic process'. *J. Inst. Actuar.*, **97**, 69–83.

Teece, D. J. (1980). 'The diffusion of an administrative innovation'. *Man. Sci.*, **26**, 464–470.

Theil, H. (1970). 'The cube law revisited'. *J. Amer. Statist. Ass.*, **65**, 1213–1219.

Theil, H., and G. Rey (1966). 'A quadratic programming approach to the estimation of transition probabilities'. *Man. Sci.*, **12**, 714–721.

Theodorescu, R., and I. Vaduva (1967). 'A mathematical demographic model concerning the dynamics of specialists'. *Studii Cereetari Mat.*, **19**, 329–337.

Thompson, M. E. (1980). 'Estimation of mobility parameters from longitudinal data over a fixed time interval using semi-Markov models'. Private communication.

Thonstad, T. (1969). *Education and Manpower: Theoretical Models and Empirical Applications*, Oliver and Boyd, Edinburgh and London.

Tintner, G. (1973). 'Some aspects of stochastic economics'. *Stochastics*, **1**, 71–86.

Tintner, G. (1975). 'Probabilistic economics'. *Bull. Int. Statist. Inst.*, **46**, 698–710.

Tintner, G., and J. K. Sengupta (1972). *Stochastic Economics: Stochastic Processes, Control and Programming*, Academic Press, New York.

Toikka, R. S. (1976) 'A Markovian model of labour market decisions'. *Amer. Econ. Rev.*, **66**, 821–834.

Tsiang, S. C. (1978). 'The diffusion of reserves and the money supply multiplier'. *Econ. J.*, **88**, 269–284.

Tuma, N. B. (1976). 'Rewards, resources and the rate of mobility: a non-stationary multivariate stochastic model'. *Amer. Sociol. Rev.*, **41**, 338–360.

Tuma, N. B., M. T. Hannan and L. P. Groenveld (1979). 'Dynamic analysis of event histories'. *Amer. J. Sociol.*, **84**, 820–854.

Tuma, N. B., and P. K. Robins (1980). 'A dynamic model of employment behaviour an application to the Scattle and Denver income maintenance experiments'. *Econometrica*, **48**, 1031–1052.

Tweedie, M. C. K. (1957). 'Statistical properties of inverse Gaussian distributions'. *Ann. Math. Statist.*, **28**, 362–377 and 696–705.

Upton, G. J. G. (1977). 'A memory model for voting transitions in British elections'. *J. R. Statist. Soc.*, **A140**, 86–94.

Upton, G. J. G. (1978). 'A note on the estimation of voter transition probabilities'. *J. R. Statist. Soc.*, **A141**, 507–512.

Usher, M. B., and M. H. Williamson (1970). 'A deterministic matrix model for handling the birth, death and migration process of spatially distributed populations'. *Biometrics*, **26**, 1–12.

Uyar, K. M. (1972). 'Markov chain forecasts of employee replacement needs'. *Industrial Relations*, **11**, 96–106.

Vajda, S. (1947). 'The stratified semi-stationary population'. *Biometrika*, **34**, 243–254.

Vajda, S. (1948). 'Introduction to a mathematical theory of a graded stationary population'. *Bull. de l'Ass. Actuair. Suisses*, **48**, 251–273.

Vajda, S. (1975). 'Mathematical aspects of manpower planning'. *Operat. Res. Quart.*, **26**, 527–542.

Vajda, S. (1978). *Mathematics of Manpower Planning*. John Wiley, Chichester.

Valliant, R., and G. T. Milkovich (1977). 'Comparison of semi-Markov and Markov models in a personnel forecasting application'. *Decision Sciences*, **8**, 465–477.

Van der Merwe, R., and S. Miller (1971). 'The measurement of labour turnover'. *Human Relations*, **24**, 233–253.

Vandome, P. (1958). 'Aspects of the dynamics of consumer behaviour'. *Bull. Oxford University. Inst. Statist.*, **20**, 65–105.

Van Kampen, N. G. (1973). 'Birth and death processes in large populations'. *Biometrika*, **60**, 419–420.

Van Korff, M. (1979). 'A statistical model of the duration of mental hospitalization: the mixed exponential distribution'. *J. Math. Sociology*, **6**, 169–175.

Vassiliou, P. C. G. (1976). 'A Markov chain model for wastage in manpower systems'. *Operat. Res. Quart.*, **27**, 57–70.

Vroom, V. H., and K. R. MacCrimmon (1968). 'Towards a stochastic model of managerial careers'. *Admin. Sci. Quart.*, **13**, 26–46.

Wang, F. J. S. (1977). 'Gaussian approximation of some closed stochastic epidemic models'. *J. Appl. Prob.*, **14**, 221–231.

Wasserman, S. S. (1978). 'Models for binary directed graphs and their applications'. *Adv. Appl. Prob.*, **10**, 803–818.

Watson, R. K. (1972). 'On an epidemic in a stratified population'. *J. Appl. Prob.*, **9**, 659–666.

Waugh, W. A. O'N. (1971). 'Career prospects in stochastic social models with time-varying rates'. *Fourth Conference on the Mathematics of Population*, East–West Population Institute, Honolulu.

Wederwang, F. (1965). *Development of a Population of Industrial Firms*, Scandinavian University Books, Oslo.

Wegman, E. J., and C. R. Kukuk (1971). 'A time series approach to the life table'. *Ann. Math. Statist.*, **42**, 1491.

Weidlich, W. (1971). 'The statistical description of polarization phenomena in society'. *Brit. J. Statist. Psychol.*, **24**, 251–266.

Weinberg, C. B. (1971). 'Response curves for a leaflet distribution—further analysis of the De Fleur data'. *Operat. Res. Quart.*, **22**, 177–179.

Weiss, G. H. (1965). 'On the spread of epidemics by carriers'. *Biometrics*, **21**, 481–490.

Weiss, G. H., and M. Zelen (1965). 'A semi-Markov model for clinical trials'. *J. Appl. Prob.* **2**, 269–285.

Weiss, H. K. (1963). 'Stochastic models for the duration and magnitude of a "deadly quarrel"'. *Operat. Res.*, **11**, 101–121.

Wessels, J., and J. A. E. E. van Nunen (1976). 'Forecasting and recruitment in manpower systems'. *Statistica Neerlandica*, **30**, 173–193.

White, H. C. (1962). 'Chance models of systems of causal groups'. *Sociometry*, **25**, 153–172.

White, H. C. (1963). 'Cause and effect in social mobility tables'. *Behavioural Science*, **8**, 14–27.

White, H. C. (1969). 'Control and evolution of aggregate personnel: flows of men and jobs'. *Admin. Sci. Quart.*, **14**, 4–11.

White, H. C. (1970a). *Chains of Opportunity*, Harvard University Press, Cambridge, Mass.

White, H. C. (1970b). 'Matching, vacancies and mobility'. *J. Pol. Econ.*, **78**, 97–105.

White, H. C. (1970c). 'Stayers and movers'. *Amer. J. Sociol.*, **76**, 307–324.

White, H. C. (1971). 'Multipliers, vacancy chains and filtering in housing'. *J. Amer. Inst. Planners*, **37**, 88–94.

Whitmore, G. A. (1976). 'Management applications of the inverse Gaussian distribution'. *Omega*, **4**, 215–223.

Whitmore, G. A. (1979). 'An inverse Gaussian model for labour turnover'. *J. R. Statist. Soc.*, **A142**, 468–478.

Whittle, P. (1955). 'The outcome of a stochastic epidemic a note on Bailey's paper'. *Biometrika*, **42**, 116–122.

Whittle, P. (1977). 'Cooperative effects in assemblies of stochastic automata'. *In Proc. Symp. to honour Jerzy Neyman*, Polish Scientific Publishers, Warsaw, 335–343.

Whittle, P. (1980). 'Polymerization processes with intra-polymer bonding: III. several types of unit'. *Adv. Appl. Prob.*, **12**, 135–153.

Widder, D. V. (1946). *The Laplace Transform*, Princeton University Press. Princeton, N.J.

Williams, E. J. (1967). 'The development of biomathematical models'. *Bull. Int. Statist. Inst.*, **42**, Book 1, 131–143.

Williams, T. (1965). 'The simple stochastic epidemic curve for large populations of susceptibles'. *Biometrika*, **52**, 571–579.

Wilson, N. A. B. (Ed.) (1969). *Manpower Research*, English Universities Press, London.

Winick, C. (1961). 'The diffusion of an innovation among physicians in a large city'. *Sociometry*, **24**, 384–396.

Winthrop, H. (1968). 'Experimental results in relation to a mathematical theory of behavioural diffusion'. *J. Soc. Psych.*, **47**, 85–100.

Wise, D. A. (1975). 'Personal attributes, job performance and probability of promotion'. *Econometrica*, **43**, 913–931.

Wold, H. O. A. (1967). 'Non-experimental statistical analysis from the general point of view of scientific method'. *Bull. Int. Statist. Inst.*, **42**, Book 1, 391–427.

Wold, H. O. A., and P. Whittle (1957). 'A model explaining the Pareto distribution of wealth'. *Econometrica*, **25**, 591–595.

Wood, F. (1969). 'An investigation into the introduction of progressive patient care'. M.Sc. Dissertation, University of Essex.

Woodroofe, M., and B. M. Hill (1975). 'On Zipf's Law'. *J. Appl. Prob.*, **12**, 425–434.

Woods, A. J. (1974). 'Epidemic models in non-homogeneous populations'. *Adv. Appl. Prob.*, **6**, 239–240.

Wynn, H. P. (1973). 'Limiting second moments for transient states of Markov chains'. *J. Appl. Prob.*, **10**, 891–894.

Wynn, H. P., and P. Sales (1973a). 'A simple model for projecting means and variances of population grade sizes'. In *Stochastic Analysis of National Manpower Problems*, Research Report, University of Kent.

Wynn, H. P., and P. Sales (1973b). 'The mover–stayer model and the 1963 labour mobility survey'. In *Stochastic Analysis of National Manpower Problems*, Research Report, University of Kent.

Yamaguchi, K. (1980). 'A mathematical model of friendship choice distribution'. *J. Math. Sociology.*, **7**, 261–287.

Yang, G.Lo. (1968). 'Contagion in stochastic models for epidemics'. *Ann. Math. Statist.*, **39**, 1863–1889.

Yang, M. C. K., and C. J. Hursch (1973). 'The use of a semi-Markov model for describing sleep patterns'. *Biometrics*, **29**, 667–676.

Young, A. (1965). 'Models for planning recruitment and promotion of staff'. *Brit. J. Indust. Rel.*, **3**, 301–310.

Young, A. (1971). 'Demographic and ecological models for manpower planning'. In D. J. Bartholomew and B. R. Morris (Eds.) (1971), 75–97.

Young, A., and G. Almond (1961). 'Predicting distributions of staff'. *Comp. J.*, **3**, 246–250.

Young, A., and P. C. G. Vassiliou (1974). 'A non-linear model on the promotion of staff'. *J. R. Statist. Soc.*, **A137**, 584–595.

Zahl, S. (1955). 'A Markov process model for follow-up studies'. *Human Biology*, **27**, 90–120.

Zanakis, S. H., and M. W. Maret (1980). 'A Markov chain application to manpower supply planning'. *J. Operat. Res. Soc.*, **31**, 1095–1102.

Zipf, G. K. (1949). *Human Behaviour and the Principle of Least Effort*. Addison-Wesley Reading, Mass.

Author Index

Subject Index